UNESCO

CBBC

# Biosaline Agriculture and High Salinity Tolerance

Edited by Chedly Abdelly, Münir Öztürk, Muhammad Ashraf
and Claude Grignon

Birkhäuser
Basel • Boston • Berlin

Chedly Abdelly
Institut National de Recherche
Scientifique et Technique
Laboratoire d'Adaptation des Plantes
aux Stress Abiotiques
Hammam-Lif 2050
Tunisia

Münir Öztürk
Ege University
Botany Department
35100 Bornova
Izmir
Turkey

Muhammad Ashraf
Botany Department
University of Agriculture
Faisalabad 38040
Pakistan

Claude Grignon
Institut de Biologie Intégrative des Plantes
SUPAGRO-INRA
Université Montpellier 2
Place Pierre Viala
34060 Montpellier Cedex
France

Library of Congress Control Number: 2007942660

**Bibliographic information published by Die Deutsche Bibliothek**
Die Deutsche Bibliothek lists this publication in the Deutsche Nationalbibliografie;
detailed bibliographic data is available in the internet at http://dnb.ddb.de

ISBN 978-3-7643-8553-8 Birkhäuser Verlag AG, Basel – Boston – Berlin


Basel · Boston · Berlin
P.O. Box 133, CH-4010 Basel, Switzerland
Part of Springer Science+Business Media
Printed on acid-free paper produced from chlorine-free pulp. TCF ∞
Cover illustrations: From the top: *Salsola soda*; *Matthiola tricuspidata; Salicornia europaea* (all M. Öztürk)
Printed in Germany
ISBN 978-3-7643-8553-8 e-ISBN 978-3-7643-8554-5

9 8 7 6 5 4 3 2 1 www.birkhauser.ch

# Contents

List of contributors ........................................................ ix

Foreword ........................................................ vxii

**Section I: Physiology and biochemistry**

*Hans-Werner Koyro, Nicole Geißler, Sayed Hussin and Bernhard Huchzermeyer*
Survival at extreme locations: Life strategies of halophytes – the long way from system ecology, whole plant physiology, cell biochemistry and molecular aspects back to sustainable utilisation at field sites ........ 1

*Muhammad Ashraf, K. Nawaz, Habib-u-Rehman Athar and S.H. Raza*
Growth enhancement in two potential cereal crops, maize and wheat, by exogenous application of glycinebetaine ........................................ 21

*Rosa M. Pérez-Clemente, Almudena Montoliu, Patricia López, María F. López-Climent, Vicent Arbona and Aurelio Gómez-Cadenas*
*In vitro* tissue culture approaches for the study of salt stress in citrus .... 37

*Annalisa Incerti, Riccardo Izzo, Adalgisa Belligno and Flavia Navari-Izzo*
Sea water effects on antioxidant production in berries of three cultivars of tomato (*Lycopersicon esculentum* Mill.) ........................ 43

*Silvia Donnini, Graziano Zocchi, Antonella Castagna, Chedley Abdelly and Annamaria Ranieri*
Identification of morphological, biochemical and physiological parameters for characterizing nutritional stress status in arboreous species differently tolerant to chlorosis ........................................ 53

*Riccardo Izzo, Annalisa Incerti and Claudio Bertolla*
Sea water irrigation: effects on growth and nutrient uptake of sunflower plants ........................................................ 61

*Karim Ben Hamed, Dorsaf Messedi, Annamaria Ranieri and Chedly Abdelly*
Diversity in the response of two potential halophytes (*Batis maritima* and *Crithmum maritimum*) to salt stress.................. 71

*Virginia M. Luna, Analía S. Llanes, Laura R. Sosa, Mariana A. Reginato and Herminda E. Reinoso*
Differential effects of sodium salts on the germination of a native halophytic species from South America: *Prosopis strombulifera* (Lam.) Benth.......................................................... 81

*Haythem Mhadhbi and Mohamed Elarbi Aouani*
Growth and nitrogen-fixing performances of *Medicago truncatula-Sinorhizobium meliloti* symbioses under salt (NaCl) stress: Micro- and macro-symbiont contribution into symbiosis tolerance ....... 91

*Hela Mahmoudi, Houneida Attia, Imen Tarchoun, Zeineb Ouerghi and Mokhtar Lachaâl*
Physiological responses of two *Arabidopsis thaliana* isolates, *N1438* and *Col*, to different salts ............................................ 99

*Hela Ben Ahmed, Dorra Ben Ammar and Ezzeddine Zid*
Physiology of salt tolerance in *Atriplex halimus* L. ......................... 107

*Salma Daoud, Chérif Harrouni, Bernard Huchzermeyer and Hans-Werner Koyro*
Comparison of salinity tolerance of two related subspecies of *Beta vulgaris*: The sea beet (*Beta vulgaris* ssp. *maritima*) and the sugar beet (*Beta vulgaris* ssp. *vulgaris*) ................................................. 115

*Adalgisa Belligno, Marco Russo, Vito Sardo and Ju Ying Wu*
Salinity influence on soil microbial population metabolism and enzymatic activities in lysimeter-grown *Olea europaea* and *Nicotiana glauca* ........................................................ 131

**Section II: Ecology**

*Münir Öztürk, Aykut Güvensen, Serdal Sakçalı and Güven Görk*
Halophyte plant diversity in the Irano-Turanian phytogeographical region of Turkey ............................................................... 141

*M. Ajmal Khan and Raziuddin Ansari*
Potential use of halophytes with emphasis on fodder production in coastal areas of Pakistan ...................................................... 157

*M. Ajmal Khan, Farhat Agha and Bilquees Gul*
Role of seed bank in the dynamics of desert communities ................ 163

*Mohammad Kafi and Majid Jami-al-Ahmadi*
Study of kochia (*Kochia scoparia*) as a forage crop ........................ 177

*Cristina M. Monteverdi, Marco Lauteri and Riccardo Valentini*
Biodiversity of plant species and adaptation to drought and salt conditions. Selection of species for sustainable reforestation activity to combat desertification ...................................................... 197

*Hassan M. El-Shaer and Mohamed H. El-Morsy*
Potentiality of salt marshes in Mediterranean coastal zone of Egypt ........ 207

*Irina N. Safronova*
Studies on the halophyte desert vegetation in the Nothern Caspian Region (Caspian Lowland and Mangyshlak) ..................... 221

*Nariman Shamsutdinov and Zebri Shamsutdinov*
Halophyte utilization for biodiversity and productivity of degraded pasture restoration in arid regions of Central Asia and Russia ........... 233

*Mokded Rabhi, Ons Talbi, Abdallah Atia, Chedly Abdelly and Abderrazak Smaoui*
Selection of a halophyte that could be used in the bioreclamation of salt-affected soils in arid and semi-arid regions .......................... 241

*Kristina N. Toderich, Shoaib Ismail, Ekaterina A. Juylova, Abdullo A. Rabbimov, Batyr B. Bekchanov, Elena V. Shyuskaya, Lilya, G. Gismatullina, Kozan Osamu and Toshpulat B. Radjabov*
New approaches for biosaline agriculture development, management and conservation of sandy desert ecosystems .................................... 247

*Aydar B. Nasrulin, Faruh Sh. Shaazizov and Helmut Lieth*
Computer supported system for the risk assessment and action recommendation for the water objects in Uzbekistan based on the already developed databank .................................................. 265

**Section III: Molecular biology**

*Isacco Beritognolo, Maurizio Sabatti, Mikael Brosché and Giuseppe Scarascia Mugnozza*
Functional genomics to discover genes for salt tolerance in annual and perennial plants ........................................................... 273

*Arianna Latini, Maria Sperandei, Sandeep Sharma, Cristina Cantale, Massimo Iannetta, Marco Dettori, Karim Ammar and Patrizia Galeffi*
Molecular analyses of a dehydration-related gene from the *DREB* family in durum, wheat and triticale ........................................ 287

*Samia Daldoul, Michael U. Höfer, Claudia Linhard, Neila Jallouli, Ahmed Mliki, Götz M.Reustle and Abdelwahed Ghorbel*
Expression analysis of salt stress responsive genes in grapevines ......... 297

*Mohsen Hanana, Olivier Cagnac, Toshio Yamaguchi, Saïd Hamdi, Abdelwahed Ghorbel and Eduardo Blumwald*
Molecular biology and transport properties of grapevine $Na^+/H^+$ antiporter ........................................................................ 305

*Mohamed Ali Ghars, Elodie Parre, Anne-Sophie Leprince, Marianne Bordenave, Delphine Lefebvre-De Vos, Luc Richard, Chedly Abdelly and Arnould Savouré*
Opposite lipid signaling pathways tightly control proline accumulation in *Arabidopsis thaliana* and *Thellungiella halophila* ........................ 317

*Chamseddine Mediouni, Guy Houlné, Marie-Edith Chabouté, Mohamed Habib Ghorbel and Fatma Jemal*
Cadmium and copper genotoxicity in plants ................................ 325

*Falleh Hanen, Riadh Ksouri, Wided Megdiche, Nejla Trabelsi, Mondher Boulaaba and Chedly Abdelly*
Effect of salinity on growth, leaf-phenolic content and antioxidant scavenging activity in *Cynara cardunculus* L. ............................... 335

**Closing remarks**

*Helmut Lieth, Brigitte Herzog and Jens Varnskuehler*
Contributions to the 2006 meeting on sustainable utilization of cash crop halophytes in Tunis ............................................. 345

Index ............................................................................ 363

# List of contributors

Chedly Abdelly, Laboratoire d'Adaptation des Plantes aux Stress Abiotiques, Centre de Biotechnologie du Technopole de Borj-Cédria, Hammam-Lif, 2050 Tunisia; e-mail: chedly.abdelly@cbbc.rnrt.tn

Farhat Agha, Department of Botany, University of Karachi, Karachi-75270, Pakistan; e-mail: goga@super.net.pk

Karim Ammar, CIMMYT Centro Internacional de Mejoramiento de Maíz y Trigo, El Batán, México; e-mail: k.ammar@cgiar.org

Raziuddin Ansari, Institute of Sustainable Halophyte Utilization,University of Karachi, Karachi-75270, Pakistan; e-mail: halophyte_razi@yahoo.com

Mohamed Elarbi Aouani, Laboratoire Interactions Légumineuses Microorganismes (LILM), Centre de Biotechnologie de Borj Cedria (CBBC), BP 901, 2050 Hammam lif, Tunisie

Vicent Arbona, Departament de Ciències Agràries, Universitat Jaume I, Campus de Riu Sec (E.S.T.C.E.), 12071 Castelló de la Plana, Spain

Muhammad Ashraf, Department of Botany, University of Agriculture, Faisalabad 38040, Pakistan; e-mail: ashrafbot@yahoo.com

Abdallah Atia, Laboratoire d'Adaptation de Plantes aux Stress Abiotiques, Centre de Biotechnologie de Borj Cédria, Technopole de Borj Cédria, B. P. 901, 2050 Hammam-lif, Tunisie

Habib-ur-Rehman Athar, Institute of Pure and Applied Biology, Bahauddin Zakariya University, Multan, Pakistan; e-mail: habibathar@yahoo.com

Houneida Attia, Unité de Physiologie et Biochimie de la Tolérance au Sel des Plantes, Département de Biologie, FST, Campus Universitaire, 2092 Tunis, Tunisie

Batyr B. Bekchanov, Uzbek Institute of Karakul Sheep Breeding and Desert Ecology Research, 47, Mirzo Ulugbek street, 703054, Samarkand, Uzbekistan; e-mail: ecokar@rol.uz

Adalgisa Belligno, Dipartimento di Scienze Agronomiche, Agrochimiche e delle Produzioni Animali, Università degli Studi di Catania, Via S. Sofia 98, 95123 Catania, Italia; e-mail: adabelli@unict.it

Hela Ben Ahmed, Unité d'Ecophysiologie et Nutrition des Plantes, Département de Biologie, Faculté des Sciences Mathématiques, Physiques et Naturelles, Université Tunis El Manar, Campus Universitaire, 1060 Tunis, Tunisie; e-mail: benahmed_hela@yahoo.fr

Dorra Ben Ammar, Unité d'Ecophysiologie et Nutrition des Plantes, Département de Biologie, Faculté des Sciences Mathématiques, Physiques et Naturelles, Université Tunis El Manar, Campus Universitaire, 1060 Tunis, Tunisie

Karim Ben Hamed, Laboratory of Plant Adaptation to Abiotic Stresses, Center for Biotechnology in Borj Cedria Ecopark. BP 901, Hammam Lif 2050, Tunisia; e-mail: kbenhamed@yahoo.fr

Isacco Beritognolo, DI.S.A.F.RI – Università della Tuscia, Via S. Camillo de Lellis, snc, 01100 Viterbo, Italy

Claudio Bertolla, Dipartimento di Chimica e Biotecnologie Agrarie, Università di Pisa, Via del Borghetto 80, 56124 Pisa, Italy

Eduardo Blumwald, Plant Reproductive Biology, Extension Drive Center, Room 1201, University of California Davis, CA 95616, USA; email: eblumwald@ucdavis.edu

Marianne Bordenave, Physiologie Cellulaire et Moléculaire des Plantes, UMR 7180 CNRS, Université Pierre et Marie Curie-Paris 6, case 156, 4 place Jussieu, 75252 Paris cedex 05, France

Mondher Boulaaba, Laboratoire d'Adaptation des Plantes aux Stress Abiotiques, Centre de Biotechnologie à la Technopole de Borj Cédria (CBBC), BP 901, 2050 Hammam-lif, Tunisia; e-mail: mondher_82@yahoo.fr

Mikael Brosché, Deptartment of Biological and Environmental Sciences, University of Helsinki, 00014 Helsinki, Finland

Olivier Cagnac, Plant Reproductive Biology, Extension Drive Center, Room 1201, University of California Davis, CA 95616, USA; e-mail: ocagnac@ucdavis.edu

Cristina Cantale, ENEA BIOTEC-GEN, Via Anguillarese 301, 00123 Roma, Italy; e-mail: cantale@casaccia.enea.it

Antonella Castagna, Dipartimento di Chimica e Biotecnologie Agrarie, Pisa University, Via del Borghetto 80, Pisa, 56124, Italy; e-mail: castagna@agr.unipi.it

Marie-Edith Chabouté, Equipe Signalisation de la Réparation de l'ADN, Institut de Biologie Moléculaire des Plantes, 12 rue du Général Zimmer, 67084 Strasbourg Cedex, France; e-mail: marie-edith.chaboute@ibmp-ulp.u-strasbg.fr

Samia Daldoul, Centre de Biotechnologie de Borj Cédria, Laboratoire de Physiologie Moléculaire de la Vigne; B.P.95, 2050 Hammam-Lif, Tunisia; e-mail: samiabiotech@yahoo.fr

Salma Daoud, Faculté des Sciences, Université Ibn Zohr, Dakhla, Agadir, Marocco; e-mail: daoudsalma@yahoo.fr

Marco Dettori, CRAS, (Centro Regionale Agrario Sperimentale) Cagliari, Sardinia, Italy; e-mail: erbacee@cras.sardegna.it

Silvia Donnini, Dipartimento di Produzione Vegetale, Milan University, Via Celoria 2, Milano, 20133 Italy; e-mail: silvia.donnini@unimi.it

Mohamed H. El-Morsy, Desert Research Center, 1 Mathaf El Mataria St., P. O. Box 11753, Mataria, Cairo, Egypt

Hassan M. El-Shaer, Desert Research Center, 1 Mathaf El Mataria St., P. O. Box 11753, Mataria, Cairo, Egypt; e-mail: drc_elshaer@hotmail.com

Patrizia Galeffi, ENEA BIOTEC-GEN, Via Anguillarese 301, 00123 Roma, Via Anguillarese 301, 00123 Rome, Italy; e-mail: galeffi@casaccia.enea.it

Nicole Geißler, Institute of Plant Ecology, Justus Liebig University Giessen, Heinrich-Buff-Ring 26-32, 35392 Giessen, Germany; e-mail: nicole.geissler@bot2.bio.uni-giessen.de

Mohamed Ali Ghars, Physiologie Cellulaire et Moléculaire des Plantes, UMR 7180 CNRS, Université Pierre et Marie Curie-Paris 6, case 156, 4 place Jussieu, 75252 Paris cedex 05, France

Abdelwahed Ghorbel, Centre de Biotechnologie de Borj Cédria, Laboratoire de Physiologie Moléculaire de la Vigne; B.P.95, 2050 Hammam-Lif, Tunisia; e-mail: a.ghorbel@cbbc.rnrt.tn

Mohamed Habib Ghorbel, Unité de Recherche "Nutrition et Métabolisme Azotés et Protéines de Stress" UR 99/09-20, Faculté des Sciences de Tunis, Université Tunis EL-MANAR, 1060 Tunisie; e-mail: ghorbel.mohamed@voila.fr

Lilya G. Gismatullina, Department of Desert Ecology and Water Resources Research, Samarkand Division of the Academy of Sciences of Uzbekistan. 3, Timur Malik str., 703000 Samarkand, Uzbekistan

Aurelio Gómez-Cadenas, Departament de Ciències Agràries, Universitat Jaume I, Campus de Riu Sec (E.S.T.C.E.), 12071 Castelló de la Plana, Spain; e-mail: aurelio.gomez@uji.es

Güven Görk, Mugla University, Faculty of Science & Arts, Biology Department, Kötekli-Mugla, Turkey; e-mail: guven66@yahoo.com; guvengork@mu.edu.tr

Aykut Güvensen, Ege University, Science Faculty, Botany Department, 35100-Bornova-Izmir, Turkey; e-mail: aykutguven@yahoo.com

Bilquees Gul, Department of Botany, University of Karachi, Karachi-75270, Pakistan; e-mail: ishu_gul@yahoo.com

Saïd Hamdi, Laboratoire de Génomique Fonctionnelle et de la Qualité de la Baie de Raisin, Université de Bordeaux1, 71 Av. Edouard Bourleaux, BP 81, Bordeaux 33883, France

Mohsen Hanana, Laboratoire de Physiologie Moléculaire de la Vigne, CBBC, BP 901, Hammam-lif 2050, Tunisia; email: punto80@yahoo.com

Falleh Hanen, Laboratoire d'Adaptation des Plantes aux Stress Abiotiques, Centre de Biotechnologie à la Technopole de Borj Cédria (CBBC), BP 901, 2050 Hammam-lif, Tunisia; e-mail: hanenfalleh@yahoo.fr

Chérif Harrouni, Inst.Agronomique et Vétérinaire Hassan II, Compl. Horticole d'Agadir, Agadir, Morocco; e-mail: c.harrouni@iavcha.ac.ma

Brigitte Herzog, Daimlerstr. 26, 74189 Weinsberg, Germany; e-mail: brigitte.m.herzog@web.de

Michael U. Höfer, RLP-Agroscience GmbH/Alplanta-Institute for Plant research; Breitenweg 71, 67435 Neustadt an der Weinstraße, Germany

Guy Houlné, Equipe Signalisation de la réparation de l'ADN, Institut de Biologie Moléculaire des Plantes, 12 rue du Général Zimmer, 67084 Strasbourg Cedex, France; e-mail: guy.houlne@ibmp-ulp.u-strasbg.fr

Bernhard Huchzermeyer, Institute for Botany, Leibniz University Hannover, Herrenhäuser Str. 2, 30419 Hannover, Germany; e-mail: huchzermeyer@botanik.uni-hannover.de

Sayed Hussin, Institute of Plant Ecology, Justus Liebig University Giessen, Heinrich-Buff-Ring 26-32, 35392 Giessen, Germany; e-mail: sayed.hussin@bot2.bio.uni-giessen.de

Massimo Iannetta, ENEA CR Casaccia, BIOTEC-DES, Rome, Italy; e-mail: miannetta@casaccia.enea.it

Annalisa Incerti, Dipartimento di Scienze Agronomiche, Agrochimiche e delle Produzioni Animali, Università degli Studi di Catania, Via Santa Sofia 98, 95123 Catania, Italy; e-mail: annalisa_inc@yahoo.it

Shoaib Ismail, International Center for Biosaline Agriculture (ICBA), PO Box 14660, Dubai, UAE; e-mail: s.ismail@biosaline.org.ae

Riccardo Izzo, Dipartimento di Chimica e Biotecnologie Agrarie, Università di Pisa, Via del Borghetto 80, 56124 Pisa, Italia; e-mail: ricizzo@agr.unipi.it

Neila Jallouli, Centre de Biotechnologie de Borj cédria, Laboratoire de Physiologie Moléculaire de la Vigne; B.P.95, 2050 Hammam-Lif, Tunisia; e-mail:

Majid Jami-al-Ahmadi, Faculty of Agriculture, Birjand University, Iran; e-mail: mja230@yahoo.com

Fatma Jemal, Unité de Recherche "Nutrition et Métabolisme Azotés et Protéines de Stress" UR 99/09-20, Faculté des Sciences de Tunis, Université Tunis EL-MANAR, 1060 Tunis, Tunisie; e-mail: fatmadaoudi@yahoo.fr

Ekaterina A. Juylova, Department of Desert Ecology and Water Resources Research, Samarkand Division of the Academy of Sciences of Uzbekistan. 3, Timur Malik str., 703000 Samarkand, Uzbekistan

Mohammad Kafi, Center of Excellence for Special Crops, Ferdowsi University of Mashhad, Iran; e-mail: mkafi2003@yahoo.com

M. Ajmal Khan, Institute of Sustainable Halophyte Utilization,University of Karachi, Karachi-75270, Pakistan; e-mail: majmalk@uok.edu.pk

Hans-Werner Koyro, Institute of Plant Ecology, Justus Liebig University Giessen, Heinrich-Buff-Ring 26-32, 35392 Giessen, Germany; e-mail: hans-werner.koyro@bot2.bio.uni-giessen.de

Riadh Ksouri, Laboratoire d'Adaptation des Plantes aux Stress Abiotiques, Centre de Biotechnologie à la Technopole de Borj Cédria (CBBC), BP 901, 2050 Hammam-lif, Tunisia; e-mail: physiologiste@yahoo.com

Mokhtar Lachaâl, Physiologie et Biochimie de la Tolérance au Sel des Plantes, Faculté des Sciences de Tunis, Campus Universitaire, 2092 Tunis El Manar, Tunisia; e-mail: mokhtar.lachaal@fst.rnu.tn

Arianna Latini, ENEA BIOTEC-GEN, Via Anguillarese 301, 00123 Roma, Italy; e-mail: arianna.latini@casaccia.enea.it

Marco Lauteri, CNR, Institute of Agro-Environmental and Forest Biology, Via G. Marconi 2, 05010, Porano (TR), Italy; e-mail: marco.lauteri@ibaf.cnr.it

Delphine Lefebvre-De Vos, Physiologie Cellulaire et Moléculaire des Plantes, UMR 7180 CNRS, Université Pierre et Marie Curie-Paris 6, case 156, 4 place Jussieu, 75252 Paris cedex 05, France

Anne-Sophie Leprince, Physiologie Cellulaire et Moléculaire des Plantes, UMR 7180 CNRS, Université Pierre et Marie Curie-Paris 6, case 156, 4 place Jussieu, 75252 Paris cedex 05, France

Helmut Lieth, The Institute of Environmental Systems Research (USF), ESR and FB Biology University of Osnabrück, Osnabrück, Germany; Wipperfürther Str.147, 51515 Kürten-Dürscheid, Germany; e-mail: helmut.lieth@t-online.de

Claudia Linhard, RLP-Agroscience GmbH/Alplanta-Institute for Plant research; Breitenweg 71, 67435 Neustadt an der Weinstraße, Germany

Analía S. Llanes, Laboratorio de Fisiología Vegetal, Departamento de Ciencias Naturales, Universidad Nacional de Río Cuarto, 5800 Río Cuarto, Argentina; e-mail: allanes@exa.unrc.edu.ar

Patricia López, Departament de Ciències Agràries, Universitat Jaume I, Campus de Riu Sec (E.S.T.C.E.), 12071 Castelló de la Plana, Spain

María F. López-Climent, Departament de Ciències Agràries, Universitat Jaume I, Campus de Riu Sec (E.S.T.C.E.), 12071 Castelló de la Plana, Spain

Virginia M. Luna, Laboratorio de Fisiología Vegetal, Departamento de Ciencias Naturales, Universidad Nacional de Río Cuarto, 5800 Río Cuarto, Argentina; e-mail: vluna@exa.unrc.edu.ar

Hela Mahmoudi, Unité de Physiologie et Biochimie de la Tolérance au Sel des Plantes, Département de Biologie, FST, Campus Universitaire, 2092 Tunis, Tunisie

Chamseddine Mediouni, Unité de Recherche "Nutrition et Métabolisme Azotés et Protéines de Stress" UR 99/09-20, Faculté des Sciences de Tunis, Université Tunis EL-MANAR, 1060 Tunis, Tunisie; e-mail: chamsedinne.mediouni@yahoo.fr

Wided Megdiche, Laboratoire d'Adaptation des Plantes aux Stress Abiotiques, Centre de Biotechnologie à la Technopole de Borj Cédria (CBBC), BP 901, 2050 Hammam-lif, Tunisia; e-mail: widedksouri@yahoo.fr

Dorsaf Messedi, Laboratory of Plant Adaptation to Abiotic Stresses, Center for Biotechnology in Borj Cedria Ecopark. BP 901, Hammam Lif 2050, Tunisia; e-mail: dorsaf.messedi@cbbc.rnrt.tn

Haythem Mhadhbi, Laboratoire Interactions Légumineuses Microorganismes (LILM), Centre de Biotechnologie de Borj Cedria (CBBC), BP 901, 2050 Hammam lif, Tunisie; e-mail: mhadhbihay@yahoo.fr

Ahmed Mliki, Centre de Biotechnologie de Borj cédria, Laboratoire de Physiologie Moléculaire de la Vigne; B.P.95, 2050 Hammam-Lif, Tunisia

Cristina M. Monteverdi, Department of Forest Sciences and Resources, University of Tuscia, Via S. Camillo De Lellis s.n.c., 01100, Viterbo, Italy; e-mail: monteverdicris@libero.it; c.monteverdi@unitus.it

Almudena Montoliu, Departament de Ciències Agràries, Universitat Jaume I, Campus de Riu Sec (E.S.T.C.E.), 12071 Castelló de la Plana, Spain; e-mail:

Aydar B. Nasrulin, The Institute of Water Problems of the Uzbek Academy of Science, 100041, Toshkent sh. F.Khojayev kochasi, 25“A“, Tashkent, Uzbekistan; e-mail: nasrulin@mail.ru

Flavia Navari-Izzo, Dipartimento di Chimica e Biotecnologie Agrarie, Università di Pisa, Via del Borghetto 80, 56124 Pisa, Italia; e-mail: fnavari@agr.unipi.it

Khalid Nawaz, Department of Botany, University of Agriculture, Faisalabad, Pakistan

Kozan Osamu, Disaster Prevention Research Institute, Kyoto University, Japan; e-mail: kozan@flood.dpri.kyoto-u.ac.jp

Zeineb Ouerghi, Unité de Physiologie et Biochimie de la Tolérance au Sel des Plantes, Département de Biologie, FST, Campus Universitaire, 2092 Tunis, Tunisie

Münir Öztürk, Ege University, Science Faculty, Botany Department, 35100-Bornova-Izmir, Turkey; e-mail: munirozturk@gmail.com

Elodie Parre, Physiologie Cellulaire et Moléculaire des Plantes, UMR 7180 CNRS, Université Pierre et Marie Curie-Paris 6, case 156, 4 place Jussieu, 75252 Paris cedex 05, France

Rosa M. Pérez-Clemente, Departament de Ciències Agràries, Universitat Jaume I, Campus de Riu Sec (E.S.T.C.E.), 12071 Castelló de la Plana, Spain

Abdullo A. Rabbimov, Uzbek Institute of Karakul Sheep Breeding and Desert Ecology Research, 47, Mirzo Ulugbek street, 703054, Samarkand, Uzbekistan; e-mail: ecokar@rol.uz

Mokded Rabhi, Laboratoire d’Adaptation de Plantes aux Stress Abiotiques, Centre de Biotechnologie de Borj Cédria, Technopole de Borj Cédria, B. P. 901, 2050 Hammam-lif, Tunisie; e-mail: mokdedrabhi@yahoo.fr

Toshpulat B. Radjabov, Department of Desert Ecology and Water Resources Research, Samarkand Division of the Academy of Sciences of Uzbekistan. 3, Timur Malik str., 703000 Samarkand, Uzbekistan; e-mail: tradjabov@mail.ru

Annamaria Ranieri, Dipartimento di Chimica e Biotecnologie Agrarie, Pisa University, Via del Borghetto 80, Pisa, 56124 Italy; e-mail: aranieri@agr.unipi.it

Syed Hammad Raza, Department of Botany, GC University, Faisalabad, Pakistan

Mariana A. Reginato, Laboratorio de Fisiología Vegetal, Departamento de Ciencias Naturales, Universidad Nacional de Río Cuarto, 5800 Río Cuarto, Argentina; e-mail: mreginato@exa.unrc.edu.ar

Herminda E. Reinoso, Laboratorio de Morfología Vegetal, Departamento de Ciencias Naturales, Universidad Nacional de Río Cuarto, 5800, Río Cuarto, Argentina

Götz M. Reustle, RLP-Agroscience GmbH/Alplanta-Institute for Plant research; Breitenweg 71, 67435 Neustadt an der Weinstraße, Germany

Luc Richard, Physiologie Cellulaire et Moléculaire des Plantes, UMR 7180 CNRS, Université Pierre et Marie Curie-Paris 6, case 156, 4 place Jussieu, 75252 Paris cedex 05, France

Marco Russo, Dipartimento di Scienze Agronomiche, Agrochimiche e delle Produzioni Animali, University of Catania, Via S. Sofia 98, 95123 Catania, Italy; e-mail: marcoanton.russo@tiscali.it

Maurizio Sabatti, DI.S.A.F.RI – Università della Tuscia, Via S. Camillo de Lellis, snc, 01100 Viterbo, Italy

Irina N. Safronova, Komarov Botanical Institute RAS, Professor Popov Street 2, St.-Petersburg, 197376, Russia; e-mail: irasafronova@yandex.ru

Serdal Sakçalı, Fatih University, Faculty of Science & Arts, Biology Department, Istanbul, Turkey; e-mail: ssakcali@hotmail.com; sakcali@fatih.edu.tr

Vito Sardo, Department of Agricultural Engineering, University of Catania, Via S. Sofia 100, 95123 Catania, Italy; e-mail: sardov@unict.it

Arnould Savouré, Physiologie Cellulaire et Moléculaire des Plantes, UMR 7180 CNRS, Université Pierre et Marie Curie-Paris 6, case 156, 4 place Jussieu, 75252 Paris cedex 05, France; e-mail: savoure@ccr.jussieu.fr

Giuseppe Scarascia Mugnozza, DI.S.A.F.RI – Università della Tuscia, Via S.Camillo de Lellis, snc, 01100 Viterbo, Italy; e-mail: gscaras@unitus.it

Faruh Sh. Shaazizov, Institute of Water Problems of the Uzbek Academy of Sciences Uzbekistan, 100041, Toshkent sh. F.Khojayev kochasi, 25“A“ Tashkent, Uzbekistan; e-mail: iwp@mail.ccc.uz

Nariman Shamsutdinov, All-Russian Res. Institute of Hydraulic Eng. & Land Reclamation, Moscow, Russia; e-mail: aridland@mtu-net.ru

Zebri Shamsutdinov, All-Russian Fodder Research Institute, Lobnya, Moscow, Russia; e-mail: darplant@mtu-net.ru

Sandeep Sharma, ENEA BIOTEC-GEN, Via Anguillarese 301, 00123 Roma, Italy; e-mail: sandeep-kaushik2002@rediffmail.com

Elena V. Shyuskaya, K.A. Timiriazeva Plant Physiology Institute of Russian Academy of Sciences, Botanicheskaya st. 35, Moscow, 127276, Russia; e-mail: evshuya@mail.ru

Abderrazak Smaoui, Laboratoire d'Adaptation de Plantes aux Stress Abiotiques, Centre de Biotechnologie de Borj Cédria, Technopole de Borj Cédria, B. P. 901, 2050 Hammam-lif, Tunisie

Laura R. Sosa, Laboratorio de Fisiología Vegetal, Facultad de Química, Bioquímica y Farmacia, Universidad Nacional de San Luis, 5600, San Luis, Argentina; e-mail: lrsosa@unsl.edu.ar

Maria Sperandei, ENEA BIOTEC-GEN, Via Anguillarese 301, 00123 Roma, Italy; e-mail: maria.sperandei@casaccia.enea.it

Ons Talbi, Laboratoire d'Adaptation de Plantes aux Stress Abiotiques, Centre de Biotechnologie de Borj Cédria, Technopole de Borj Cédria, B. P. 901, 2050 Hammam-lif, Tunisie

Imen Tarchoun, Unité de Physiologie et Biochimie de la Tolérance au Sel des Plantes, Département de Biologie, FST, Campus Universitaire, 2092 Tunis, Tunisie

Kristina N. Toderich, Department of Desert Ecology and Water Resources Research, Samarkand Division of the Academy of Sciences of Uzbekistan, 3 Timur Malik street, 703000 Samarkand, Uzbekistan; e-mail: ktoderich@cgiar.org

Nejla Trabelsi, Laboratoire d'Adaptation des Plantes aux Stress Abiotiques, Centre de Biotechnologie à la Technopole de Borj Cédria (CBBC), BP 901, 2050 Hammam-lif, Tunisia; e-mail: nejlatrabe@yahoo.fr

Riccardo Valentini, Department of Forest Sciences and Resources, University of Tuscia, Via S. Camillo De Lellis s.n.c., 01100, Viterbo, Italy; e-mail: rik@unitus.it

Jens Varnskuehler, Plant Science Services GmbH, Heerstr. 49, 49492 Westerkappeln, Germany

Ju Ying Wu, Beijing Research Center for Grass and Environment, Banjing Haidian District, 100097 Beijing, China; e-mail: wujuying1@263.net

Toshio Yamaguchi, Plant Reproductive Biology, Extension Drive Center, Room 1201, University of California Davis, CA 95616, USA

Ezzeddine Zid, Unité d'Ecophysiologie et Nutrition des Plantes, Département de Biologie, Faculté des Sciences Mathématiques, Physiques et Naturelles, Université Tunis El Manar, Campus Universitaire, 1060 Tunis, Tunisie; e-mail: ezzeddine.zid@fst.rnu.tn

Graziano Zocchi, Dipartimento di Produzione Vegetale, Milan University, Via Celoria 2, Milano, 20133 Italy; e-mail: graziano.zocchi@unimi.it

# Foreword

The water crisis is one of the most critical challenges facing the world today. The arid zones are exposed to serious freshwater shortages, and this includes the entire Arab Region.

In the same time, all Arab countries have access to abundant seawater, as well as saline waste-water, and groundwater. Hypersaline soils are widespread. Based on science, and with carefully established good practices, large areas of saline soils can be converted into high productivity man-made agro-ecosystems.

One of the visible success-stories is the replacement of freshwater dependant bermuda grass *Cynodon dactylon* by the seawater tolerant groundcover *Sesuvium portulacastrum*. Most of these halophytes are still being irrigated with fresh-, or low-salinity water, however, there are several plots in the United Arab Emirates, impressively demonstrating saline irrigation with *Sesuvium* and other halophytes.

We hope that this volume will be a meaningful contribution, adding to the availability of correct and adequate scientific knowledge, which is essential, to develop good biosaline management practices.

It is with this in mind, that the UNESCO Office in Doha, and in agreement with the UNESCO Division of Ecological and Earth Sciences, decided to support the "International Conference on Biosaline Agriculture and Salinity Tolerant Plants", Tunisia, in November 2006, and this important publication.

The book has three sections: the first section deals with physiology and biochemistry aspects of salt tolerance. The second section deals with halophyte ecology, and the third section offers papers on molecular biology.

UNESCO congratulates the editors and authors of this book, for producing this valuable scholarly work, which contributes towards improvement of scientific research, education, and environmental management.

November 2007

BENNO BÖER
*Ecological Sciences Advisor*
*– Arab Region*
*UNESCO Doha Office*
*Doha, POBox 3945, State of Qatar*
*b.boer@unesco.org*

MIGUEL CLÜSENER-GODT
*Programme Specialist – Division of*
*Ecological and Earth Sciences*
*UNESCO Headquarters*
*Natural Sciences Sector*
*1 rue Miollis, Paris, France*
*m.clusener-godt@unesco.org*

Biosaline Agriculture and High Salinity Tolerance
Edited by Chedly Abdelly, Münir Öztürk, Muhammad Ashraf and Claude Grignon


# Survival at extreme locations: Life strategies of halophytes – The long way from system ecology, whole plant physiology, cell biochemistry and molecular aspects back to sustainable utilization at field sites

Hans-Werner Koyro[1], Nicole Geißler[1], Sayed Hussin[1] and Bernhard Huchzermeyer[2]

[1] *Institute of Plant Ecology, Justus Liebig University Giessen, Heinrich-Buff-Ring 26–32, 35392 Giessen, Germany*
[2] *Institute for Botany, Leibniz University Hannover, Herrenhäuser Str. 2, 30419 Hannover, Germany*

## Abstract

High concentrations of sodium are toxic to most plant species. Drought and soil salinity are the major abiotic stresses in plant productivity worldwide. Many glycophytic crop species are negatively affected. Physiological and biochemical research – with an accelerating emphasis on molecular biological studies – has shown that salt tolerance in halophytes depends on a range of adaptations. The multifactorial response embraces many aspects such as gas exchange, water relations (osmotic adaptation), selective transport and uptake of ions, ion compartmentalization (homeostasis), osmolyte production, enzyme activities, ion excretion and genetic control. The ability of plant cells to maintain low cytosolic sodium concentrations is an essential process for the ability of plants to grow in salty habitats and depends on several plant-specific interactions. Unfortunately, there are few investigations that combine studies of growth with the individual partially intracellular plant characteristics. Such joint investigations are the basis for the discovery of traits that present the ability to produce cash crops in saline environments. One possible solution could be rapid advances in the genetic transfer of halophyte salt tolerance traits to crop plants. Another is the breeding of cash crop halophytes.

## Introduction

Abiotic stresses, such as drought, salinity, extreme temperatures and oxidative stress are serious threats to agriculture and the natural status of the environment. They are recurring features of nearly all the world‘s climatic regions since various critical environmental threats with global implications have linkages to water crises [1, 2]. These threats are collaterally catalyzed by global warming and population growth.

The latest scientific data confirm that the earth‘s climate is rapidly changing. Global temperatures have increased by about 1 degree Fahrenheit over the course of the last century, and will likely rise even more rapidly in coming decades. Scientists say that temperatures could rise another 3–9 degrees by the end of the century with far-reaching effects. Increased drought and salinization of arable land are expected to have devastating global effects [3]. The area of arable land lost per year could double by the end of the century because of global warming, threatening the survival of millions of people around the world.

Rapid population growth increasingly generates pressure on existing cultivated land and other resources [4]. Population migration to those arid and semi-arid areas increases the problems of water shortage and worsens the situation of land degradation in the destination, and in turn causes severe problems of poverty, social instability, and population health threats [5].

Water scarcity and desertification could critically undermine efforts for sustainable development, introducing new threats to human health, ecosystems and national economies of several countries. Therefore, solutions are required as an integrated approach, seeking the opportunities of economically, socially and environmentally sustainable development [6].

## Cause and effect

Water is onc of the essential resources in arid and semi-arid regions, where one-sixth of the world population lives [7, 8]. The current developments tighten the assessment that the causes for these problems seem to be invincible, at least over the next few centuries. However, time is running fast and if we cannot solve the undoubted global causes, we have at least to try to reduce the adverse effects. In the light of the progressive shortage of fresh water resources and soil salinization, a major aim is to evaluate the potential of local (xero-) halophytic species to be widely and economically used in arid and semi-arid regions. Major research topics are to identify and select plant species tolerant to drought and salt stress, to study their potential in the field of human or animal nutrition and to evaluate the possible use of non-conventional waters, such as seawater, brackish water and pre-treated wastewater.

One possible concept of sustainable agriculture is based on the use of so-called "cash crop halophytes" irrigated with saline waters up to seawater salinity [9, 10]. A major advantage of this concept is that oceans contain most of the water on earth. Major problems are the high salinity itself, imbalance of nutrient ions, and the extreme toxicity of sodium and chloride ions for living organisms at relatively low salinity. Even if we disagree with the use of seawater, the study of salt resistance is essential. Knowledge about plants tolerant to drought and salinity can help to transfer the actual accelerating destructive exertion of saline irrigation waters into sustainable systems.

About 7% of the world's total land area is affected by salt, as is a similar percentage of its arable land [11, 12]. When soils in arid regions of the world are irrigated, solutes from the irrigation water can accumulate and eventually reach levels that have an adverse effect on plant growth. Of the current 230 million ha of irrigated land, 45 million ha are salt affected (19.5%) and of the 1500 million ha under dryland agriculture, 32 million ha are salt affected to varying degrees (2.1%).

Salinity is a multifactorial problem and the use or the breeding of salt-resistant crop varieties will require a clear understanding of the complex mechanisms of salt-stress resistance, which is still lacking despite intensive research during the last decade [13, 14].

## Characterization of halophytes

The first step of this study was to investigate the method of adjustment of the experts – the halophytes. Halophytes grow naturally in very salty soil. In contrast to most glycophytic crops, they still have not lost their resistance mechanisms to salt-stress conditions. This may take the form of salt avoidance or tolerance [15]. Some plants avoid the effects of high salt by fancy tricks, such as by completing the reproductive life cycle during rainy seasons (facultative halophytes). Nevertheless, the bandwidth of exploitation and of resistance mechanisms is larger in obligatory halophytes or xerohalophytes (drought-tolerant halophytes). These are plants tolerating salinities higher than 0.5% NaCl [16].

Halophytes can be grouped into a specific halophyte type according to the most relevant mechanisms of salt resistance. They can be characterized by the dominance of processes regulating salt transport in plant organs, according to the accumulation and secretion of salts. [17]. A simplified definition distinguishes between halophytes accumulating NaCl up to concentrations similar to or higher than the water salinity level (includers) and the NaCl-excluding species (excluders). A small number of plant lineages in numerous, related families have evolved structural, phenological, physiological, and biochemical mechanisms for salt resistance.

## Complexity of salt tolerance

Most crop plants do not fully express their original genetic potential for growth, development, and yield under salt stress, and their economic value declines as salinity levels increase [18, 19]. Numerous attempts have been made to improve the salt tolerance of crops by traditional breeding programs. However, commercial success has been very limited due to the complexity of the trait: salt resistance is genetically and physiologically complex [20]. At present, major efforts are being directed towards the genetic transformation of plants to raise their tolerance [21].

Improving salt resistance of crop plants is of major concern in agricultural research. A potent genetic source for the improvement of salt resistance in crop plants resides among wild populations of halophytes [22, 23]. These can be either domesticated into new, salt-resistant crops, or used as a source of genes to be introduced into crop species by classical breeding or molecular methods.

## State of the art in sustainable utilization with saline irrigation waters

There are already several examples of the utilization of halophytes for industrial, ecological, or agricultural purposes [10]. Because of their diversity, halophytes have been tested as vegetable, forage, and oilseed crops in agronomic field trials [24]. The most productive species yield 10–20 t/ha of biomass on seawater irrigation, equivalent to conventional crops. The oilseed halophyte, *Salicornia bigelovii*, yields 2 t/ha of seeds containing 28% oil and 31% protein, similar to soybean yield [22].

In several countries, specific plant species are used for wastewater treatment. Some halophytes can be used for bio-remediation of salt-contaminated soils and even pharmaceutical values of their plant products are described [10]. Halophyte forage and seed products already replace conventional ingredients in animal feeding systems, with some restrictions on their use due to high salt content and anti-nutritional compounds present in some species.

## Mechanisms of salt resistance

Salt tolerance involves physiological and biochemical adaptations for maintaining protoplasmic viability, with cells compartmentalizing electrolytes. Salt avoidance involves structural and physiological adaptations to minimize salt concentrations of the cells or physiological exclusion by root membranes. In principle, salt tolerance can be achieved by salt exclusion or salt inclusion as mentioned above.

At the whole-plant level, plant resistance may be the process of salt regulation, but at the cellular level it may be the salt tolerance of the cytoplasm [17]. Physiological and biochemical research has shown that salt resistance in halophytes depends on a range of adaptations embracing many aspects of plant physiology, including regulation of $H_2O/CO_2$ gas exchange at the leaves, osmotic adaptation, selective transport and uptake of ions (salt balance), exclusion of NaCl from the symplast (compartmentation) to maintain homeostasis and enhanced synthesis of organic solutes (osmolyte production).

## Morphological advantages of some halophytes

In many cases various mechanisms and special morphological structures are advantageous for halophytes since they help to reduce the salt concentrations [17, 25].

- Excretive halophytes have glandular cells capable of secreting excess salts from plant organs. A simple system with two-celled trichomes has evolved as collecting chambers for salts, e.g., in cordgrass, and a complex type of salt glands is known, e.g., in several common mangroves.
- Not only glands but also bladder hairs can remove salts from salt-sensitive metabolic sites. Some halophytes like *Atriplex* or *Chenopodium* have vesiculated trichomes on the surfaces of the leaves that release the salt back into the environment when they are ruptured.
- Halophytes quite often have reflecting surfaces (by wax or trichomes) preventing ultraviolet light from reaching the leaf tissues and thus minimizing the development of reactive molecules [reactive oxygen species (ROS) as well as nitrogen radicals].
- Succulence is demonstrated in many genera of plants that inhabit saline environments. By depositing ions of salts in vacuoles, the toxicity is partitioned from the cytoplasm and organelles of the cells.
- Many halophytes exhibit a rather rapid turnover of their leaves; salts are removed from the plant when the leaf is shed.

## Course of screening halophytes

Most halophytes do not possess special morphological structures, but still tolerate high seawater salinity. Growth and survival of vascular plants at high salinity depends on adaptation to both low water potentials and high sodium concentrations.

This combination of stressors leads to a situation in which halophytes are on the horns of a dilemma: Salt exclusion minimizes ion toxicity but accelerates water deficit and indirectly diminishes $CO_2$ uptake. Salt absorption (inclusion) facilitates osmotic adjustment but can lead to toxicity and nutritional imbalance.

The situation above is difficult enough. It is also an exception that just a single parameter is of major importance for the ability to survive at high NaCl salinity. Therefore, a comprehensive study in a quick check system (QCS) with the analysis of at least a combination of several parameters is a necessity to get a survey about the combination of mechanisms finally leading to the salinity tolerance of individual species [26]. The specific reaction on the major constraints of plant growth on saline substrates (water deficit, restriction of $CO_2$ uptake, ion toxicity and nutrient imbalance) needs to be captured. To uncover the individual mechanisms for salt resistance or avoidance, it is essential to conduct investigations at the cellular, tissue and whole-plant levels [27].

The QCS enables a detailed record of individual tolerance criteria under reproducible conditions and is the first step on the way to the sustainable use of halophytes (Fig. 1). A further aim of the QCS is comparative research on the existing constitution of mechanisms supporting salt resistance.

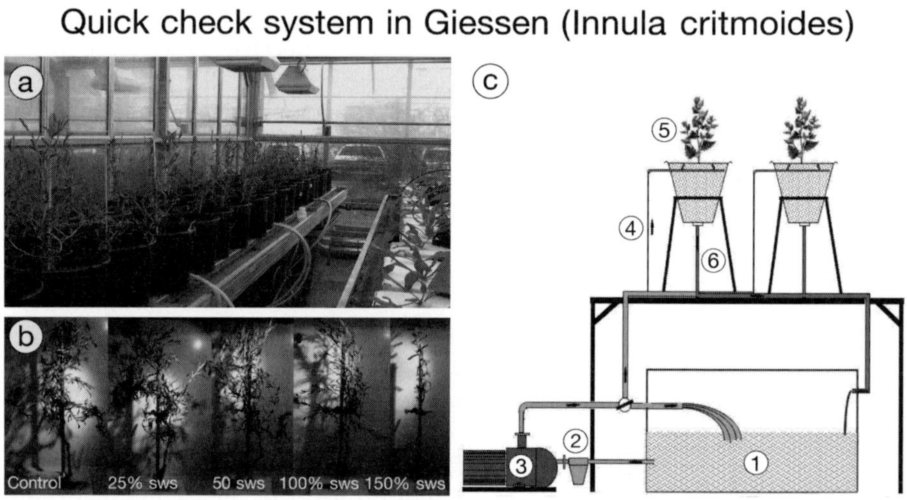


**Figure 1.** A quick-check-system (QCS) of *Innula critmoides* under photoperiodic conditions in a growth cabinet.
(a) Gravel/hydroponic quick check system with automatic drip irrigation. (b) Habitus of *Innula critmoides* plants grown in different salinity levels. Control plants (ctr) are visible on the left side, plants of the seawater salinity treatment (500 mol $m^{-3}$ NaCl) on the right side. (c) Scheme of the quick check system (QCS). (1) Nutrient solution; (2) filter; (3) pump; (4) irrigation tube; (5) *Innula* plant and (6) draining tube.

To facilitate comparisons, the limit of salinity resistance is defined physiologically as the NaCl salinity at which the yield of a crop under saline conditions, relative to its yield under non-saline conditions, reaches less than 50% [28]. The limit of salt resistance can be easily detected by growth development of halophytes at different salinities (Fig. 2).

In the next sections, topics ranging from the reactions of plants growing on saline substrates to the major constraints of plant growth are described. Reduction in yield under salt-stress conditions can be attributed to reduction in photosynthesis as a result of leaf injury, induced nutrient deficiencies and physiological water stress. Salt stress may reduce plant growth by water deficit, ion toxicity, ion imbalance, or a combination of these factors [29].

## Balance of water loss and $CO_2$ uptake

Salt tolerance is not exclusively correlated with adaptations to $Na^+$ toxicity *per se* but also reflects adaptations to secondary effects of salinity, such as water deficit and impaired nutrient acquisition [30]. Terrestrial plants in saline habitats are often surrounded by low water potentials in the soil solution and atmosphere. It is important to prevent the water loss by transpiration from

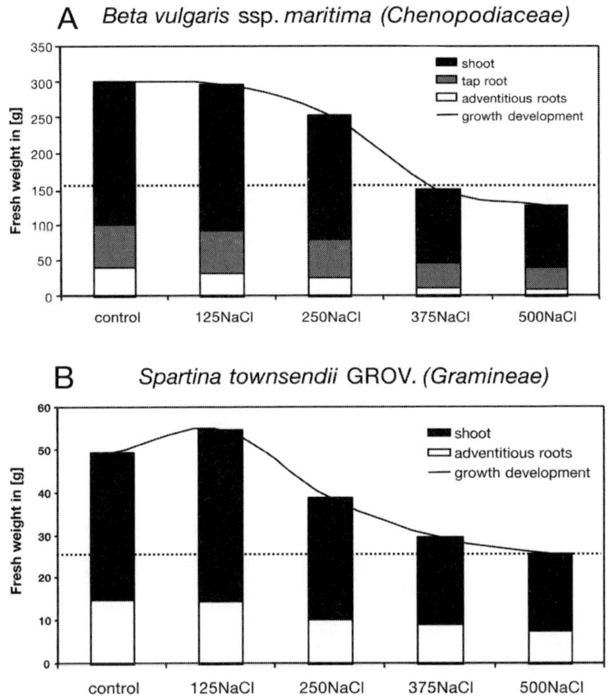


**Figure 2.** Development of the plant fresh weight at treatments with different percentages of seawater salinity.
The crossover of the dotted and the continuous lines reflects the NaCl salinity where the growth depression falls down to 50 % of the control plant (threshold of NaCl salinity according to [28]). (a) *Beta vulgaris* ssp. *maritima*: 75% seawater salinity. (b) *Spartina townsendii*: 100% seawater salinity. 0% seawater salinity = control, 25% = 125NaCl, 50% = 250NaCl, 75% = 375NaCl and 100% = 500NaCl

being higher than the influx rate. This is only possible if the water potential is lower in the plant than in the soil.

Data of leaf water potentials clearly demonstrate that leaf water potential of halophytes does not correlate alone as a single factor with salinity tolerance [31]. They do not correlate at all with the respectively existing salt resistance in species such as *Aster tripolium*, *Beta vulgaris* ssp. *maritima*, *Spartina townsendii* and *Sesuvium portulacastrum*. All of theses species have a sufficiently low water potential even at high salinity, although their salt resistance (definition s. a.) varies between 0.5-fold (*Aster tripolium*) and 1.5-fold seawater salinity (*Sesuvium portulacastrum*). Furthermore, the osmotic potentials of each of these four halophytes (and many other) were sufficiently low to explain the full turgescence of the leaves at all salinity levels.

However, plant water loss has to be minimized at low soil water potentials, since biomass production depends mainly on the ability to keep a high

A

| *Beta vulgaris* ssp. maritima | A ($\mu mol*m^{-2}s^{-1}$) | E ($mmol*m^{-2}s^{-1}$) | WUE (A/E) | rs ($m^2*s*mol^{-1}$) |
|---|---|---|---|---|
| control | 2,79 | 1,82 | 1,53 | 6,44 |
| | +0,22 | +1,30 | +0,69 | +2,90 |
| 125NaCl | 3,13 | 1,87 | 1,68 | 4,53 |
| | +0,20 | +0,64 | +0,56 | +1,24 |
| 375NaCl | 2,90 | 1,17 | 2,47 | 9,67 |
| | +0,10 | +0,11 | +0,17 | +1,23 |
| 500NaCl | 1,58 | 0,44 | 3,62 | 31,45 |
| | +1,01 | +0,27 | +2,44 | +16,73 |

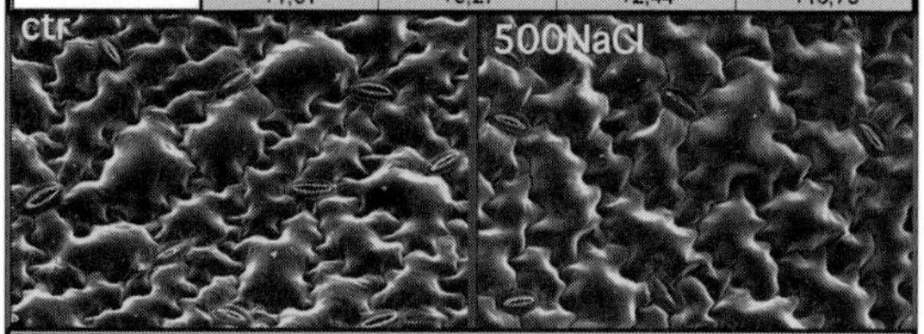


B

| *Spartina townsendii* | A ($\mu mol*m^{-2}s^{-1}$) | E ($mmol*m^{-2}s^{-1}$) | WUE (A/E) | rs ($m^2*s*mol^{-1}$) |
|---|---|---|---|---|
| control | 6,25 | 3,30 | 1,90 | 0,82 |
| | ±1,04 | ±0,20 | ±0,45 | ±0,20 |
| 125NaCl | 8,17 | 3,17 | 2,57 | 1,46 |
| | ±1,01 | ±0,34 | ±2,34 | ±0,59 |
| 375NaCl | 7,25 | 2,29 | 3,17 | 2,49 |
| | ±2,26 | ±0,20 | ±1,31 | ±0,74 |
| 500NaCl | 7,52 | 1,25 | 6,01 | 5,06 |
| | ±1,54 | ±0,26 | ±1,92 | ±1,99 |

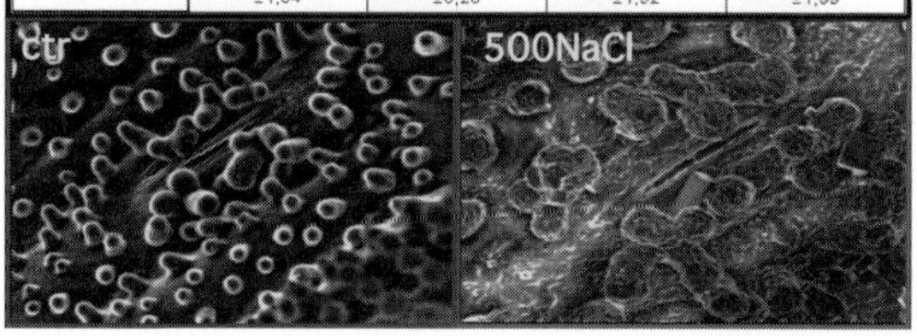


**Figure 3.** Influence of NaCl salinity on the apparent photosynthesis (A), the adaxial transpiration (E), the water-use efficiency (WUE) and the stomatal conductance (rs) of *Beta vulgaris* ssp. *maritima* and *Spartina townsendii*.
The ultrastructures (SEM micrograph) of the stomates on the leaf surfaces (left side controls, right side seawater salinity treatments) are presented next to the responding table.

net photosynthesis at low water loss rates. In this field of tension, biomass production of a plant has always to be seen in connection to the $CO_2/H_2O$ gas exchange, which can be estimated based on water-use efficiency (WUE) of photosynthesis. A critical point for a plant is reached if the $CO_2$ fixation (apparent photosynthesis) falls below the $CO_2$ production (compensation point). Therefore, one crucial aspect of the screening procedure is the study of growth reduction, water consumption and net photosynthesis, especially at the threshold of salinity tolerance (Fig. 3).

Many plants such as *Aster tripolium*, *Beta vulgaris* ssp. *maritima* or *Spartina townsendii* reveal a combination of low (but positive) net photosynthesis, minimum transpiration, high stomatal resistance and minimum internal $CO_2$ concentration at their threshold salinity tolerance [32]. However, there is a big bandwidth between halophytes. In particular, succulent halophytes such as *Sesuvium portulacastrum* and *Avicennia marina* have alternatives if the water balance is still positive (water uptake minus water loss) and not the limiting factor for photosynthesis. In case of *Sesuvium* net photosynthesis and WUE increase but stomatal resistance decreases. These results show that it is quite important to describe the regulation of gas exchange at high salinity in relation to other parameters (such as water relationships).

## Ion excess, deficiency and imbalance

A number of substances including $Na^+$ or $Cl^-$ can be found in irrigation water, which can cause toxic reactions in plants. Ion toxicity and nutrient imbalance are two major constraints of growth in saline habitats and are, therefore, of special importance for the salt tolerance of halophytes. For some ions (such as $K^+$ and $Na^+$), either their excess or deficiency has been found to be toxic to freshwater and marine organisms. Adverse effects can occur in plants on saline substrates when common ions exceed a certain concentration, when the normal composition (ratio) of ions is not correct, or in some cases, when ion concentrations are too low.

Studies have shown that halophytes exhibit very different ways of adjusting to high NaCl salinity. Some salt-tolerant plants differ from salt-sensitive relatives in having a low rate of $Na^+$ and $Cl^-$ transport to leaves [33]. However, some halophytes (includers) even need an excess of salts for maximum growth and for attaining low solute potentials [34, 35]. The accumulation of NaCl consumes much less energy than is needed for the osmotic adjustment with organic substances [15]. Nevertheless, high substrate salinities can lead to toxic effects of salt even inside these includers [36]. The cause of injury is probably the salt load exceeding the ability of cells to compartmentalize salts in the vacuole. Salts would then build up rapidly in the cytoplasm and inhibit enzyme activity. Alternatively, they might build up in the cell walls and dehydrate the cell.

Alternatively, high concentrations can be avoided by filtering out most of the salt (excluders, see above). These halophytes adapt to saline conditions by ion exclusion so that osmotically active solutes have to be synthesized within the plant to meet turgor pressure demands [37]. This adaptive feature can be of importance even in species that have salt glands or bladders.

For the sake of completeness, it has to be said that it is quite important to distinguish between both ions to uncover the individual mechanisms for salt tolerance. The mechanisms of salt exclusion are discussed in the literature mainly as if a common reaction of both ions ($Na^+$ and $Cl^-$) leads to a salt injury. This is not always the case. Some halophytes, such as *Laguncularia racemosa* (with salt glands), are typical Na excluders, but with high $Cl^-$ accumulation in the leaves [38], and *Spartina townsendii* is a typical $Cl^-$ excluder with high $Na^+$ accumulation in the leaves [28, 39].

In $Na^+$- and/or $Cl^-$-excluding species, however, a lack of solutes may result in adverse effects on water balance, so that water deficiency rather than salt toxicity may be the growth-limiting factor [35, 37].

*Beta vulgaris* ssp. *maritima*

| adaxial leafepidermis | vacuole | | | | cytoplasm | | | |
|---|---|---|---|---|---|---|---|---|
| | control | | 480 NaCl | | control | | 480 NaCl | |
| Cl | 22,6 | ±4,7 | 654,3 | ±54,8 | 0,0 | | <5 | |
| P | 22,2 | ±4,6 | 6,4 | ±4,2 | 58,2 | ±8,4 | 28,1 | ±6,7 |
| S | 40,4 | ±7,9 | 2,7 | ±1,9 | 10,5 | ±4,0 | <5 | |
| Na | 12,2 | ±3,1 | 724,9 | ±65,1 | <5 | | <5 | |
| Mg | 24,0 | ±1,2 | 4,3 | ±1,7 | <5 | | <5 | |
| K | 282,5 | ±14,7 | 9,3 | ±6,1 | 88,9 | ±9,5 | 66,7 | ±7,2 |
| Ca | <5 | | <5 | | <5 | | <5 | |

*Spartina townsendii*

| adaxial leafepidermis | vacuole | | | | cytoplasm | | | |
|---|---|---|---|---|---|---|---|---|
| | control | | 480 NaCl | | control | | 480 NaCl | |
| Cl | 21,2 | ±3,2 | 324,3 | ±64,8 | <5 | | <5 | |
| P | 11,1 | ±1,1 | 5,3 | ±2,1 | 81,5 | ±6,8 | 71,6 | ±10,8 |
| S | 24,4 | ±6,9 | 20,8 | ±2,6 | 8,8 | ±3,0 | <5 | |
| Na | 16,4 | ±3,1 | 521,0 | ±54,9 | <5 | | 15,3 | ±3,2 |
| Mg | 29,6 | ±2,1 | 18,8 | ±1,6 | <5 | | <5 | |
| K | 212,5 | ±34,8 | 71,5 | ±6,9 | 92,7 | ±12,7 | 78,4 | ±6,9 |
| Ca | <5 | | <5 | ±2,06 | <5 | | <5 | |

**Figure 4.** Element concentrations (in mol $m^{-3}$) in the vacuoles and the cytoplasm of adaxial epidermis cells of *Beta vulgaris* ssp. *maritima* and *Spartina townsendii* grown at fresh (control) or seawater (480 NaCl = 480 mol $m^{-3}$ NaCl) conditions.

A metabolic response to salt stress, especially in salt excluding species, is the synthesis of compatible osmolytes. These mediate osmotic adjustment and therefore achieve a low water potential and/or a charge balance, protect sub-cellular structures and reduce oxidative damage caused by free radicals that are produced in response to high salinity [40, 41]. The solute potential in these species is decreased by the synthesis of organic solutes such as sugar-alcohols (e.g., mannitol in leaves of *Laguncularia racemosa*), soluble carbohydrates (e.g., sucrose in the taproots of *Beta vulgaris* ssp. *maritima*), organic acids (including amino acids) or by reducing the matric potential (e.g., with soluble proteins in leaves of *Beta vulgaris* ssp. *maritima*; results not shown). However, the synthesis of organic solutes is energy demanding (see above, also includers) and the formation of these solutes decreases the energy status of the plant [15]. Thus, for plant survival, growth depression can be a necessary compromise in $Na^+$- and /or $Cl^-$-excluding species and not a sign of toxicity or nutrient imbalance.

## Protection of metabolism

Mechanisms for tolerance of the salt-specific effects of salinity are of two main types: those minimizing the entry of salt into the plant; and those minimizing the concentration of salt in the cytoplasm (Fig. 4). Root and leaf cytosolic $Na^+$ and $Cl^-$ concentrations are in the order of 10–30 mM [42–44].

The destruction of metabolism by $Na^+$ or $Cl^-$ has to be avoided if plants have to grow on saline habitats. Therefore, the protection of the responsible enzymes is of major importance. The ability of plant cells to maintain low cytosolic sodium concentrations is an essential process for halophytes [21]. Leaves being fed by the transpiration stream receive large quantities of sodium, which must be regulated. Plant cells respond to salt stress by increasing sodium efflux at the plasma membrane and sodium accumulation in the vacuole. Thus, the proteins, and ultimately genes, involved in these processes can be considered as salt-tolerance determinants. The cloning experiments of $Na^+/H^+$ antiporters have demonstrated the role of intracellular sodium [45] compartmentalization in plant salt resistance. Such compartmentalization of sodium and chloride in leaf vacuoles can only be attained by an active transport into the vacuole and low tonoplast permeability to these ions.

The transport of ions across the plasma membrane and tonoplast requires energy, which is provided by vacuolar and plasma membrane ATPase [46–49]. Sodium ions exchanged for hydrogen ions across membrane $Na^+/H^+$ antiporters take advantage of a proton gradient formed by these pumps. Salt stress was shown to increase $Na^+/H^+$ activity in glycophytes and halophytes [13]. The activation of such antiporters is likely to operate to reduce sodium toxicity in salt-tolerant plants under saline conditions.

## Genes promoting salt resistance

The salt resistance mechanisms shown above are a basis for improving salt tolerance in glycophytic crops [36]. Molecular control mechanisms for abiotic stress tolerance are based on the activation and regulation of specific stress-related genes. This section summarizes the recent advances in elucidating stress-response mechanisms and their biotechnological applications.

It is known that large apoplastic concentrations of $Na^+$ and $Cl^-$ produce hyperosmotic stress and ionic imbalance [29]. Thus, it is vital for the plant to re-establish cellular ionic homeostasis for metabolic functioning and growth to respond to the saline environment. In this context, a common molecular approach used to define salt tolerance mechanisms in plants is to identify cellular processes and genes whose activity or expression is affected by salt stress [50, 51]. Additionally, this leads to a better understanding of the complexity of salt tolerance in higher plants.

Genes that could increase salt tolerance fall into three main functional groups [36]:

- Genes that control salt uptake and transport
- Genes with an osmotic or unknown protective function
- Genes that control cell and tissue growth rates

In connection with the exclusion of salts and the preservation of homeostasis in the cytoplasm (avoidance of a salt load leading to an inhibition of enzyme activities) three mechanisms of halophytes are in front line of the current research:

- Sodium extrusion and the salt-overly sensitive (SOS) pathway enabling ion homeostasis in the cytoplasm
- Vacuolar compartmentalization
- Sodium uptake and recirculation

Since Na toxicity is the principal stress component in saline soils, much research has focused on the identification of ion transporters and regulatory mechanisms that mediate $Na^+$ homeostasis and maintenance of a high cytoplasmic $K^+/Na^+$ ratio (Fig. 5). It was shown that a salt-sensitive rice cultivar that expresses the vacuolar-type $Na^+/H^+$ antiporter gene from the halophytic plant *Atriplex gmelini* (Ag NHX1 = vacuolar $Na^+/H^+$ antiporter) was much more salt resistant than the wild-type rice [45], indicating that overexpression of the $Na^+/H^+$ antiporter significantly enhances transgenic rice salt tolerance. Extending such research could improve crop plant salt tolerance in the near future if successful genetic transfer can occur on a large scalc. The SOS signaling pathway, composed of the SOS1, 2 and 3 proteins, has emerged as a key factor in the detection of tolerance to salt stress. Evidence suggests that a protein kinase complex of SOS3 and SOS2 is activated by a salt stress-elicited calcium signal [52]. This protein kinase complex phosphorylates and activates various ion transporters, such as the plasma membrane $Na^+/H^+$ antiporter SOS1 responsible for excreting $Na^+$ into the apoplast.

The tonoplast $Na^+$ transport may also be regulated through SOS3-like calcium binding proteins. Vacuolar compartmentalization of $Na^+$ ions is, beside the excretion across the plasmalemma (see SOS1 above), an efficient mechanism to maintain a lower concentration of $Na^+$ in the cytosol. Transport of $Na^+$ into the vacuole is mediated by a $Na^+/H^+$ antiporter. The proton motive force for this transport is provided by both an ATPase and an $H^+$-pyrophosphatase. Overexpression of a vacuolar $H^+$-pyrophosphatase (AVP1) and of the tonoplast $Na^+/H^+$ antiporter AtNHX1 enhanced sequestration of $Na^+$ into vacuoles and led to increased salt tolerance [53].

A gene locus has been identified that corresponds to the HKT1 gene, which is a sodium influx transporter. The HKT1 expression is restricted to the phloem tissue in all organs, strongly reducing sodium concentration in the phloem sap. HKT1 is probably involved in the recirculation of sodium from the shoots to the roots by mediating sodium loading into the phloem sap in shoots and unloading in the roots. The high-affinity carrier HKT1, a $Na^+$-$K^+$ symporter, is also required for continued $K^+$ uptake during salt stress [54]. High external $Na^+$ concentrations inhibited HKT1-mediated $K^+$ influx [55] and transgenic wheat with reduced HKT1 expression exhibited significantly less $Na^+$ uptake and enhanced tolerance to salinity [56]. This suggests that

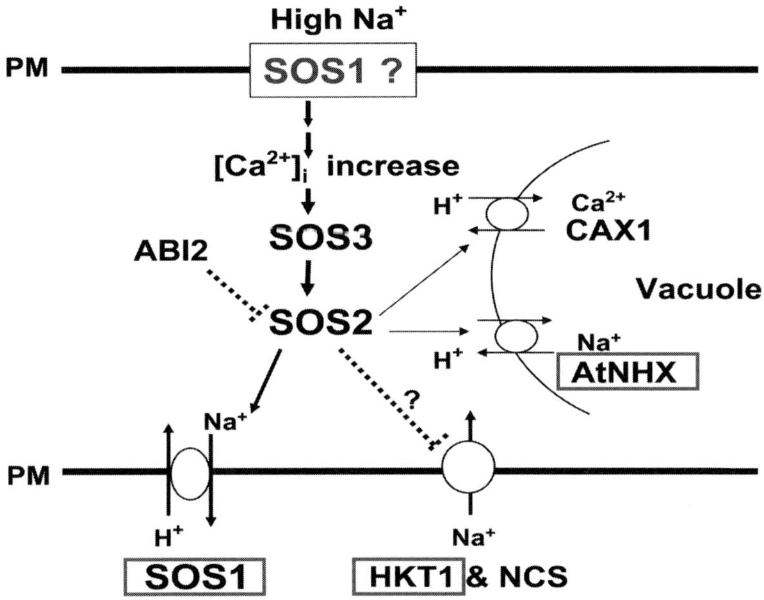


**Figure 5.** Scheme of the SOS signaling pathway for the regulation of $Na^+$ homeostasis and salt tolerance in *Arabidopsis* modified after [71].
SOS3: Ca-binding protein; SOS2: sucrose kinase; SOS1: plasma membrane $Na^+/H^+$-antiporter and effector of the $Na^+$-homeostasis; HKT1: high affinity $Na^+$-coupled $K^+$ transporter; AtNHX1: vacuolar $Na^+/H^+$ antiporter overexpression confers salt tolerance.

HKT1 is an important determinant of salt tolerance [57]. However, attempts to improve the salt tolerance of crops through conventional breeding programs have met with very limited success. This has been due to the complexity of the salt tolerance trait both genetically and physiologically.

Application of the knowledge about candidate genes is currently hampered by our lack of understanding of their function at the cell, tissue and whole-plant levels [36]. It is obvious that only both together – the quick check of salt resistance mechanisms and molecular studies – have a joint potential to improve the performance of crop plants under saline conditions.

## Energetics of salt-stress response

As outlined above, active transmembrane ion export out of the cytosol, either into the apoplast or into the vacuoles, is among the major strategies of halophytes thriving in salty environments. Sequestration of salt to the vacuoles is the strategy of choice for succulent plants, and *Salicornia* is a typical example, well characterized in this respect. Succulent plants can survive if they manage to produce new vacuolar space fast enough to keep vacuolar salt concentra-

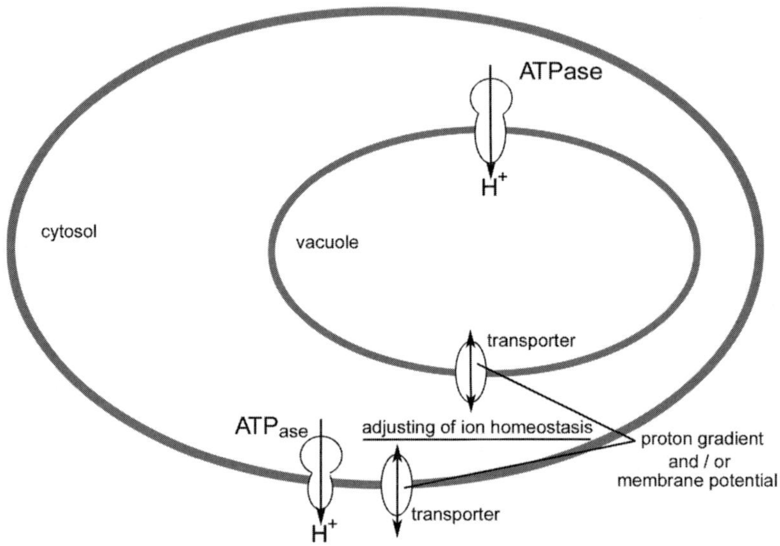


**Figure 6.** An overview of the situation in the cytosol.
ATP-driven export of surplus ions may occur at both, the plasma membrane and the tonoplast.

tion below a threshold value. This value can be easily calculated based on energetics of membrane transport.

Salt is transported at the expense of ATP hydrolysis, catalyzed by the V-type ATPase of the tonoplast. Based on Mitchell's calculations [58], ATP hydrolysis builds up a proton motive transmembrane gradient. The energy of this gradient consists of the osmotic pressure of a proton concentration gradient and the energy of a transmembrane electric potential difference (initially based on the positive charges of protons). If the pH and cytosolic cation concentration are known, the hydrolysis energy of an ATP molecule can be calculated. This energy will equal the energy consumed for the translocation of a proton or any other ion transported across the membrane at the threshold concentration gradient. Thus, a maximal concentration difference between the cytosolic and „external" compartments for any ion can be calculated (Fig. 6).

Based on these considerations it is obvious that mere overexpression of ATPase genes, $Na^+/H^+$ transporters, etc., does not at all solve the problem of plants growing under salt stress. The transgenic plants would be capable of higher transport rates than the wild type, but the capacity of salt sequestration in the external compartment will be unchanged (unless the external compartment is increased by a concomitant effect). Such a double response can occur, for instance, if a halophyte can increase growth rate (thus building extra vacuolar space) under stress. Such observations have been made with *Cakile maritima*, a plant that tends to develop succulence under salt stress [59].

## Metabolite flow and primary events in photosynthesis

It is well documented that plant performance under stress can be controlled by monitoring gas exchange. These data correlate with parameters like pigment bleaching, sugar export rates from source to sink tissues and intensity of chlorophyll fluorescence. This observation is highly significant if the experimental plant is a non-succulent $C_3$ plant. This situation can be explained in terms of strictly coupled metabolic pathways in plants, as described in the literature [60], and comes along with extremely high protein concentrations (more than 150 mg/ml) found inside chloroplasts.

Extremely close neighborhoods of reaction partners are essential for functioning of photosynthesis to allow high turnover rates of substrates. This is essential because the low number of coenzymes (NADP and ADP) would be consumed within seconds without recycling from NADPH and ATP by Calvin cycle events. Intruding salt interferes with structuring of enzyme complexes, thus resulting in inhibition of metabolite flow and export of sugar. This adverse salt effect can be reduced by interactions of proteins with compatible solutes or chaperons. The respective genes respond in all plants tested so far under salt stress (see above).

If plant stress response is not sufficient to guarantee high turnover rates of metabolites, feed back inhibition will eventually result in an overreduction of the photosynthetic electron transport chain. This situation can be documented by measuring chlorophyll fluorescence (for references see [61]). If no oxidized redox partners are available, light-activated chlorophyll has various options to return to its ground state again: (a) heat production, (b) emission of light quanta (red fluorescence), or (c) transferring an electron to oxygen (production of an oxygen radical), i.e., induction of a reaction sequence producing reactive oxygen and nitrogen species [62], which might lead to critical situations like pigment bleaching inside the chloroplast (Fig. 7).

All three reactions are permanently competing with the transfer of electrons to the photosynthetic redox chain. However, under optimal growth conditions, photosynthetic electron transport is dominant over the other options. Moreover, depending on their natural habitat, plants have evolved ROS scavengers like ascorbic acid, glutathione and other antioxidant systems. These systems are well known from plant species thriving in sunny and arid areas. Halophytes are well equipped with such systems, too, and these compounds interfere with some laboratory experiments like sugar and protein concentration measurements, causing unexpected problems.

Structuring of protein-to-protein interactions to bring about high turnover of metabolic pathways as well as biosynthesis of antioxidants and ROS scavengers is under the control of many genes. No successful molecular genetic approach leading to improved stress tolerance of crops has been published so far. However, the target of researchers has been the overproduction of compatible solutes (trehalose and betaines, for instance) that help to stabilize interaction among enzymes of a pathway [63].

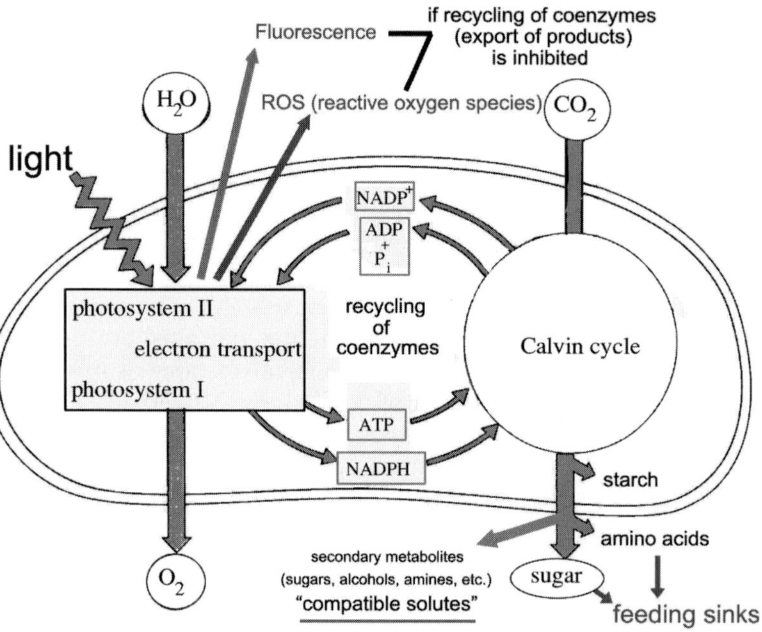


**Figure 7.** Under salt stress toxic ion concentrations inside the chloroplasts have not been observed. However, salt and drought stress-induced inhibition of photosynthate export may result in an over-reduction of the chloroplast. Such a situation will favor the transfer of electrons to molecular oxygen as well as chlorophyll fluorescence. Over-production of compatible solutes in such a situation may function as an energy valve in addition to the protective effects of these compounds.

## Conclusion

Although tolerance of saline conditions by plants is variable, crop species are generally intolerant when the concentration of salts is 30% of that found in seawater. This is the gap for the utilization of cash crop halophytes. Although early conclusions showed that halophytes differ from glycophytes in quantity rather than in quality [64], their halophyte compartmentalization, water-use efficiency and ion selectivity are still of superior importance and are still not well understood.

It is conceivable that approaches to identify specific genes that are up- or down-regulated either through the analysis of RNA [65] or proteins [66] might provide more information. More than 700 patents have already been granted that invoke salt tolerance as part of the invention, most involving specific genes [67]. However, the successful use of these genes to produce a salt-tolerant cultivar is very slow.

Up to now, in spite of the complexity of the trait, the transfer of a single gene or a few genes, such as genes involved in the control of $Na^+$ transport [53,

68, 69], has led to claims of little improvement in salt tolerance. However, the nature of the genetically complex mechanisms of abiotic stress tolerance, and potential detrimental side effects, make this task extremely difficult [3, 20].

The road to engineering such tolerance into sensitive species will be long, and the lack of public acceptance of genetic engineering means that other approaches raising tolerance to salt need to be considered at present [70]. Useful knowledge about genes would be increased if studies on plant transformation included measurements of biochemical or physiological processes in which the transgene is acting, and targeted particular tissues and salinity-specific responses [36].

Until the time when such breeding of salt tolerant crops is successful and accepted, we need to reduce the pressure on fresh water resources by acquiring and testing candidate halophyte species, screening germplasm under highly saline conditions, and developing management techniques for productive use of halophytes. Further development of halophytes is the only available way for a sustainable utilization, and represents an efficient resource for the reduction of the water crisis.

Time is running out fast and the last decade has witnessed, especially in the arid and semi-arid regions, a sharp increase in losses of arable land due to salinization. As shown above, salinity is an ever-present threat to crop yield, especially in countries where irrigation is an essential aid to agriculture [20]. Irrigation farming is expanding fast and many fields have reached a soil salinity level that prevents farmers from raising common crops. If we do not react soon, there will be no further necessity to study salt-resistance genes because soil salinity worldwide will reach levels where no halophyte can grow at all.

## References

1 Gleick P (1994) Water, war, and peace in the Middle East. *Environment* 36: 7–41

2 Gleick PH (2000) *The world's water 2000–2001. The Biennial Report on Freshwater Resources*, Island Press, Washington, D.C., 309

3 Wang XW, Vinocur B, Altman A (2003) Plant responses to drought, salinity and extreme temperatures: Towards genetic engineering for stress tolerance. *Planta* 218: 1–14

4 Ericson J, Freudenberger M, Boege E (1999) Population dynamics, migration, and the future of the Calakmul Biosphere Reserve. *Occasional Paper No. 1, Program on Population and Sustainable Development (PSD).* American Association for the Advancement of Science (AAAS), Washington, D.C., 1–40

5 Moench M (2002) Water and the potential for social instability: Livelihoods, migration and the building of society. *Nat Resour Forum* 26: 195–204

6 Duda A, El Ashry MTE (2000) Addressing the global water and environment crises through integrated approaches to the management of land, water and ecological resources. *Water Int* 25: 115–126

7 World Bank (1999) *World Development Report.* Washington

8 United Nations Development Program (UNDP) (1999) *Human development Report.* www.undp/hdro/population.htm

9 Lieth H (1999) Development of crops and other useful plants from halophytes. In: H Lieth, M Moschenko, M Lohmann, H-W Koyro, A Hamdy (eds): *Halophyte uses in different climates. I: Ecological and ecophysiological studies*. Prog. Biometeorol. Vol. 13, Backhuys Publ., Leiden, 1–18

10 Lieth H, Moschenko M, Lohmann M, Koyro HW, Hamdy A (eds) (1999) *Halophyte uses in different climates. I. Ecological and ecophysiological studies*. Prog. Biometeorol. Vol. 13, Backhuys Publ., Leiden

11 Ghassemi F, Jakeman AJ, Nix HA (1995) *Salinisation of land and water resources: Human causes, extent, management and case studies*. USNW Press, Sydney

12 Szabolcs I (1994) Soils and salinisation. In: M Pessarakli (ed): *Handbook of Plant and Crop Stress*. Marcel Dekker, New York., 3–11

13 Apse MP, Blumwald E (2002) Engineering salt tolerance in plants. *Curr Opin Biotech* 13: 146–150

14 Sosa L, Llanese A, Reinoso H, Reginato M, Luna V (2005) Osmotic and specific ion effects on the germination of *Prosopis strombulifera*. *Ann Bot* 96: 261–267

15 Yeo AR (1983) Salinity resistance: Physiologies and prices. *Physiol Plant* 58: 1399–3054

16 Koyro H-W, Lieth L (1998) *Salinity conversion table,* 2nd enlarged edn., H Lieth (ed) Osnabrück. ISSN 09336–3114

17 Breckle S-W (2002) Salinity, halophytes and salt affected natural ecosystems. In: A Läuchli, U Lüttge (eds): *Salinity: Environment – Plants – Molecules*. Kluwer, Dordrecht, 53–77

18 Läuchli A, Epstein E (1990) Plant responses to saline and sodic conditions. In: Tanji KK (ed): *Agricultural salinity assessment and management*. ASCE manual no. 71, New York, 113–137

19 Maas EV (1990) Crop salt tolerance. In: Tanji KK (ed): *Agricultural salinity assessment and management.* ASCE manual no. 71, New York, 262–304

20 Flowers TJ (2004) Improving crop salt tolerance. *J Exp Bot* 55: 307–319

21 Borsani O, Valpuesta V, Botella MA (2003) Developing salt-tolerant plants in a new century: A molecular biology approach. *Plant Cell Tissue Organ Culture* 73: 101–115

22 Glenn E, Brown JJ, Blumwald E (1999) Salt tolerance and crop potential of halophytes. *Crit Rev Plant Sci* 18: 277–255

23 Serrano R, Mulet JM, Rios G, Marquez JA, de Larrinoa IF, Leube MP, Mendizabal I, Pascual-Ahuir A, Proft M, Ros R, Montesinos C (1999) A glimpse of the mechanisms of ion homeostasis during salt stress. *J Exp Bot* 50: 1023–1036

24 Koyro H-W, Geissler N, Hussin S, Huchzermeyer B (2006) Mechanisms of cash crop halophytes to maintain yield and reclaim soils in arid areas. *Task Veg Sci* 40: 345–366

25 Marschner H (1995) *Mineral nutrition of higher plants.* Academic Press, London, 889

26 Koyro H-W (2003) Study of potential cash crop halophytes in a quick check system. *Task Veg Sci* 38: 5–17

27 Epstein E (1980) Responses of plants to saline environments. In: DW Rains, RC Valentine, A Hollaender (eds): *Genetic engineering of osmoregulation*. Plenum Press, New York, 7–21

28 Kinzel H (1982) *Pflanzenökologie und Mineralstoffwechsel*. Eugen Ulmer, Stuttgart, 534

29 Cramer RC, Läuchli A, Epstein E (1986) Effects of NaCl and $CaCl_2$ on ion activities in complex nutrient solutions and root growth of cotton. *Plant Physiol* 81: 792–797

30 Maathuis FJM, Amtmann A (1999) $K^+$ nutrition and $Na^+$ toxicity: The basis of cellular $K^+$/$Na^+$ ratios. *Ann Bot* 84: 123–133

31 Koyro H-W (2006) Effect of salinity on growth, photosynthesis, water relations and solute composition of the potential cash crop halophyte *Plantago coronopus* (L.). *Environ Exp Bot* 56: 136–146

32 Koyro H-W, Huchzermeyer B (2004) Ecophysiological needs of the potential biomass crop *Spartina townsendii* GROV. *Trop Ecol* 45: 123–139

33 Munns R (2002) Comparative physiology of salt and water stress. *Plant Cell Environ* 25: 239–250
34 Flowers TJ, Troke PF, Yeo AR (1977) The mechanisms of salt tolerance in halophytes. *Annu Rev Plant Physiol* 28: 89–121
35 Greenway H, Munns R (1980) Mechanisms of salt tolerance in nonhalophytes. *Annu Rev Plant Physiol* 31: 149–190
36 Munns R (2005) Tansley review: Genes and salt tolerance: Bringing them together. *New Phytol* 167: 645–669
37 Mengel K, Kirkby EA (2001) *Principles of plant nutrition*. Kluwer, Dordrecht, 849
38 Koyro H-W, Wegmann L, Lehmann H, Lieth H (1997) Physiological mechanisms and morphological adaptation of *Laguncularia racemosa* to high salinity. In: H Lieth, A Hamdy, H-W Koyro (eds): *Water management, salinity and pollution control towards sustainable irrigation in the Mediterranean region: Salinity problems and halophyte use*. Tecnomack, Bari, 51–78
39 Koyro H-W, Huchzermeyer B (1999) Influence of high NaCl-salinity on growth, water and osmotic relations of the halophyte *Beta vulgaris* ssp. maritima. Development of a quick check. *Prog Biometeorol* 13: 87–101
40 Hong B, Barg R, Ho TH (1992) Developmental and organ specific expression of an ABA- and stress-induced protein in barley. *Plant Mol Biol* 18: 663–674
41 Hare PD, Cress WA, Van Staden J (1998) Dissecting the roles of osmolyte accumulation during stress. *Plant Cell Environ* 21: 535–554
42 Tester M, Davenport R (2003) $Na^+$ tolerance and $Na^+$ transport in higher plants. *Ann Bot* 91: 503–527
43 Wyn Jones RG, Gorham J (2002) Intra- and inter-cellular compartmentation of ions. In: A Läuchli, U Lüttge (eds): *Salinity: Environment – Plants – Molecules*. Kluwer, Dordrecht, 159–180
44 Koyro H-W, Huchzermeyer B (2004) Ecophysiological mechanisms leading to salinity tolerance – Screening of cash crop halophytes. *Recent Res Dev Plant Sci* 1: 187–207
45 Ohta H (2002) Introduction of a $Na^+/H^+$ antiporter gene from *Atriplex gmelini* confers salt tolerance to rice. *FEBS Lett* 532: 279–282
46 Gordon-Weeks R, Koren'kov VD, Steele SH, Leigh RA (1997) Tris is a competitive inhibitor of $K^+$ activation of the vacuolar $H^+$-pumping pyrophosphatase. *Plant Physiol* 114: 901–905
47 Koyro H.W, Huchzermeyer B (1997) The physiological response of *Beta vulgaris* ssp. *maritima* to sea water irrigation. In: H Lieth, A Hamdy and H-W Koyro (eds): *Water management, salinity and pollution control towards sustainable irrigation in the Mediterranean region. Salinity problems and halophyte use.* Tecnomack, Bari, 29–50
48 Leigh RA (1997) The solute composition of the vacuoles. *Adv Bot Res* 25: 253–295
49 Leigh RA, Sanders D (1997) The plant vacuole. *Adv Bot Res* 25: 463
50 Hasegawa PM, Bressan RA, Pardo JM (2000) The dawn of plant salt tolerance genetics. *Trends Plant Sci* 5: 31–319
51 Hasegawa PM, Bressan RA, Zhu JK, Bohnert HJ (2000) Plant cellular and molecular responses to high salinity. *Annu Rev Plant Physiol* 51: 493–499
52 Zhu JK (2003) Regulation of ion homeostasis under salt stress. *Curr Opin Plant Biol* 6: 441–445
53 Gaxiola RA, Li J, Undurraga S, Dang V, Allen GJ, Alper SL, Fink GR (2001) *Proc Natl Acad Sci USA* 98: 11444–11449
54 Uozumi N, Kim EJ, Rubio F, Yamaguchi T, Muto S, Tsuboi A, Bakker EP, Nakamura T, Schroeder JI (2000) The *Arabidopsis* HKT1 gene homolog mediates inward $Na^+$ currents in *Xenopus laevis* oocytes and $Na^+$ uptake in *Saccharomyces cerevisiae*. *Plant Physiol* 122: 1249–1260

55 Rubio F, Gassmann W, Schroeder JI (1995) Sodium driven potassium uptake by the plant potassium transporter HKT1 and mutations conferring salt tolerance. *Science* 270: 1660–1663

56 Laurie S, Feeney KA, Maathuis FJM, Heard PJ, Brown SJ, Leigh RA (2002) A role for HKT1 in sodium uptake by wheat roots. *Plant J* 32: 139–149

57 Katiyar-Agarwal S, Verslues P, Zhu J-K (2005) Mechanisms of salt tolerance in plants. In: CJ Li, FS Zhang, A Dobermann, P Hinsinger, H Lambers, XL Li, P Marschner, L Maene, S McGrath, O Oenema et al (eds): *Plant nutrition for food security, human health and environmental protection.* Tsinghua University Press, Bejing, 44–45

58 Mitchell P (1974) A chemiosmotic molecular mechanism for proton-translocating adenosine triphosphatases. *FEBS Lett* 78: 189–194

59 Debez A, Saadaoui D, Balasubramian R, Ouerghi Z, Koyro H-W, Huchzermeyer B Abdelly C (2006) Leaf ATPase activity and photosynthetic capacity of *Cakile maritima* in response to increasing salinity. *Environ Exp Bot* 57: 285–295

60 Süss KH, Arkona, C, Manteuffel R, Adler K (1993) Calvin cycle multienzyme complexes are bound to chloroplast thylakoid membranes of higher plants *in situ*. *Proc Natl Acad Sci USA* 90: 5514–5518

61 Strasser RJ (1988) A concept for stress and its application in remote sensing. In: HK Lichtenthaler (ed): *Applications of chlorophyll fluorescence*. Kluwer, Dordrecht, 333–337

62 Mehler AH (1951) Studies on reactions of illuminated chloroplasts. *Arch Biochem Biophys* 33: 65–77

63 Garg AK, Kim J-K, Owens TG, Ranwala AP, Choi YD, Kochian LV, Wu RJ (2002) Trehalose accumulation in rice plants confers high tolerance levels to different abiotic stresses. *Proc Natl Acad Sci USA* 99: 15898–15903

64 Waisel Y (1972) *Biology of halophytes*. Academic Press, New York

65 Kawasaki S, Borchert C, Deyholos M, Wang H, Brazille S, Kawai K, Galbraith D, Bohnert HJ (2001) Gene expression profiles during the initial phase of salt stress in rice. *Plant Cell* 13: 889–905

66 Salekdeh GH, Siopongco J, Wade LJ, Ghareyazie B, Bennett J (2002) A proteomic approach to analyzing drought- and salt responsiveness in rice. *Field Crops Res* 76: 199–219

67 BIOS (2005) Biological Innovation for an Open Society. Canberra, ACT, Australia: CAMBIA (Center for the Application of Molecular Biology to Agriculture). http://www.bios.net

68 Rus AM, Estan MT, Gisbert C, Garcia-Sogo B, Serrano R, Caro M, Moreno V, Bolarin MC (2001) Expressing the yeast HAL1 gene in tomato increases fruit yield and enhances $K^+/Na^+$ selectivity under salt stress. *Plant Cell Environ* 24: 875–880

69 Zhang HX, Blumwald E (2001) Transgenic salt-tolerant tomato plants accumulate salt in foliage but not in fruit. *Nat Biotechnol* 19: 765–768

70 Munns R, Husain S, Rivelli AR, James RA, Condon AG, Lindsay MP, Lagudah ES, Schachtman DP, Hare RA (2002) Avenues for increasing salt tolerance of crops, and the role of physiologically based selection traits. *Plant Soil* 247: 93–105

71 Zhang JZ, Creelman RA, Zhu JK (2004) From laboratory to field. Using information from *Arabidopsis* to engineer salt, cold, and drought tolerance in crops. *Plant Physiol* 135: 615–621

Biosaline Agriculture and High Salinity Tolerance
Edited by Chedly Abdelly, Münir Öztürk, Muhammad Ashraf and Claude Grignon


# Growth enhancement in two potential cereal crops, maize and wheat, by exogenous application of glycinebetaine

Muhammad Ashraf[1], Khalid Nawaz[1], Habib-ur-Rehman Athar[2] and Syed Hammad Raza[3]

[1] *Department of Botany, University of Agriculture, Faisalabad, Pakistan*
[2] *Institute of Pure and Applied Biology, Bahauddin Zakariya University, Multan, Pakistan*
[3] *Department of Botany, GC University, Faisalabad, Pakistan*

## Abstract

Ameliorative effect of exogenously applied glycinebetaine (GB) on growth, photosynthetic and antioxidant capacities of two potential cereals wheat (cv. S-24) and maize (cv. Golden) grown under salt stress was assessed in two different independent experiments. Plants of maize were grown at 0 or 10 dS/m NaCl, while those of wheat were subjected to 2.17 or 14.67 dS/m NaCl salinity. Different levels of GB, i.e., 0 (unsprayed), 50 and 100 mM (in 0.10% Tween-20 solution) were applied as a foliar spray to both wheat and maize plants at the vegetative growth stage. Salt stress reduced the growth and yield of both maize and wheat plants. However, salt-induced reduction in growth and yield of both maize and wheat was ameliorated by exogenous application of GB, but this enhancement effect was more in wheat than that in maize. Furthermore, this GB-induced growth and yield enhancement was positively associated with increased endogenous GB, photosynthetic capacity, and superoxide dismutase (SOD) activity. Although exogenous application of GB improved photosynthetic capacity of both maize and wheat by increasing stomatal conductance, and thus favoring higher $CO_2$ fixation rate, this effect seemed to be partial in maize. In addition, the GB-induced reduction in transpiration rate in wheat compared with that in maize was found to be an additional factor that might have contributed to a better growth and yield of wheat under salt stress. The activity of only SOD was enhanced by GB application in both maize and wheat under saline conditions. Thus, it is likely that both applied GB and intrinsic SOD scavenged reactive oxygen species in these potential cereals under saline conditions. In view of all these findings, it can be concluded that the adverse effects of salt stress on cereals such as maize and wheat can be alleviated by the exogenous application of GB, which in turn enhances photosynthetic capacity and modulates activities of antioxidant enzymes. Furthermore, effectiveness of GB application on regulation of photosynthetic and antioxidant capacities was found to be species specific.

## Introduction

Soil salinity is one of the major environmental stresses causing substantial crop losses worldwide. According to an estimate, salinity reduced the average yield of major crops by more than 50% [1]. In view of rapidly increasing world population, crop production must increase substantially if food security is to be ensured, which is already considerably low for meeting world demands (http://www.unfpa.org/swp/2007/english/introduction.html). It is estimated that there is a need to increase productivity by 20% in the developed countries and by 60% in the developing countries [2]. Thus, reduction in soil salinization and increase in the salt tolerance of crops, particularly that of cereals, the demand for which is growing at 2% per year with growing urban population [2–5 ].

Hexaploid wheat is one of the world's most important cereal crop and it is grown under wide range of climatic conditions, particularly in Pakistan, India, China, the United Kingdom, the United States, Turkey, Australia, Russia, Germany and France (http://www.fao.org/statistics/yearbook/vol_1_1/site_en.asp?page=production). Due to rapid urbanization (http://www.unfpa.org/swp/2007/english/introduction.html) and economic growth, dramatic changes occur in dietary patterns, which have resulted in an increase demand for wheat production. It is calculated that wheat demand worldwide will increase by 40% from 552 metric ton in 1993 to 775 metric ton in 2020 [6]. Similarly, demand for maize is increasing day by day. According to projections, the demand of maize will increase from 526 metric tons to 784 metric tons from 1993 to 2020, particularly in developing countries [7]. In view of this alarming situation, different effective measures need to be adopted to improve crop productivity, particularly in salt-affected soils where crop productivity is reduced by more than 50%.

A plethora of information is available in the literature on salinity tolerance of potential agricultural crops at cellular level as well as on whole plant level, on the basis of which a number of strategies have been devised to overcome the salinity problem [4, 5, 8–10]. These include screening and conventional breeding, wide crossing, and, more recently, marker-assisted selection and the use of transgenic plants. A number of researchers still emphasize the use of conventional selection and breeding, with the help of advanced molecular biology techniques, to improve crop salt tolerance [4, 5, 8–11]. However, complex interactions between stress factors and various molecular, biochemical and physiological phenomena make it difficult to achieve the desired degree of salt tolerance [5, 8]. Thus, some alternative approaches need to be adopted to overcome the problem.

Recently, some rapid and economically feasible shotgun approaches have been proposed to alleviate the adverse effects of salt stress [10, 12]. A number of studies have emphasized that exogenous application of osmoprotectants is a useful approach in inducing salt tolerance in crops [12]. Of the different compatible solutes known, glycinebetaine (GB) is relatively more important as it is capable of promoting plant growth and yields under normal or stress conditions due to its osmoprotective influence on photosynthetic machinery [13–15], and regulation of antioxidant capacity and ion homeostasis [16, 17]. However, the detailed physiological basis of how GB regulates these phenomena is not clearly understood. It is therefore necessary to confirm and further elucidate the mode of action of GB in plants under stress conditions. In the present study, GB was exogenously applied as a foliar spray to assess up to what extent exogenously applied GB could mitigate the adverse effects of salt stress on plant growth and yields of the commonly grown cereal crops, wheat and maize.

## Materials and methods

In our work, the influence of exogenously applied GB on maize and wheat grown under saline conditions was assessed in two independent experiments. Field grown 9-day-old wheat plants were subjected to 2.17 or 14.67 dS/m NaCl salinity for 47 days after which GB treatments (0, i.e., no spray; 50 and 100 mM GB in 0.1% Tween-20 solution) were applied as a foliar spray. At 40 days after the exogenous application of GB (when plants were 96 days old or at the initiation of the boot stage), nine plants from each treatment (three plants/replicate) were uprooted and washed with distilled water. After drying with filter paper, roots were carefully removed and all plant parts were dried at 65°C until constant dry weight. Before harvest, photosynthetic capacity, GB, proline, and activities of superoxide dismutase (SOD), catalase (CAT), and peroxidase (POD) were measured in the second leaf from the top of each plant following standard laboratory protocols.

The experiment with maize was conducted in a similar manner except with different cultural and growth conditions. Fourteen-day-old maize plants grown in plastic pots filled with sands were subjected to 0 or 10 dS/m NaCl salinity. When plants were 3 weeks old, GB treatments (0, 50 and 100 mM GB in 0.1% Tween-20 solution) were applied as a foliar spray. At 2 weeks after the exogenous application of GB, two plants from each replicate were harvested and their fresh weights recorded. After oven-drying at 65°C dry weights recorded. Before harvest, photosynthetic capacity, GB, proline, and activities of SOD, CAT, and POD were measured in the second leaf from top of each plant following standard laboratory protocols.

## Influence of exogenously applied GB on wheat

The growth and grain yield of wheat plants were improved due to exogenous application of GB under saline conditions (Fig. 1a–d). This growth and yield enhancement in the salt-stressed plants of wheat was positively associated with enhanced endogenous level of GB, resulting from GB application (Fig. 2a, b). Thus, higher endogenous levels of GB in wheat due to exogenous application of GB can be related to enhanced salt tolerance. These findings are similar to some earlier studies in which it has been observed that exogenous application of GB alleviated the adverse effects of salt stress on the growth and/or yield of different crops, e.g., wheat [18, 19], tomato [20], and rape [14].

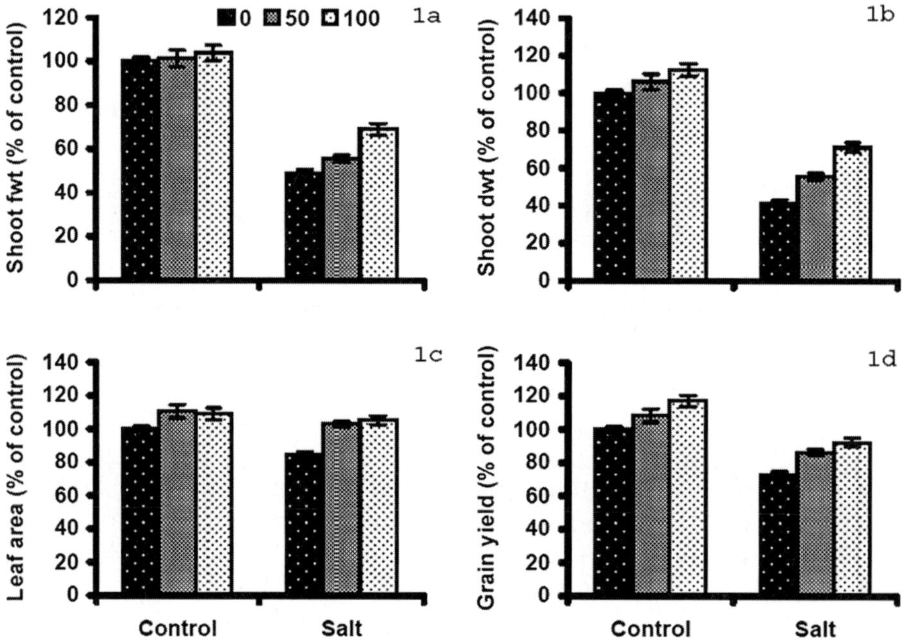


**Figure 1.** Fresh and dry weight of shoot, leaf area and grain yield of wheat when different amounts of glycinebetaine (GB) were exogenously applied to salt-stressed or non-stressed plants (0, no spray; 50, 50 m*M* GB; and 100, 100 m*M* GB foliar spray).

From previous reports (e.g., [21]), it is evident that GB-induced increase in salt tolerance is associated with improved photosynthetic capacity of most crops under saline conditions (Fig. 3a ). The same was demonstrated in wheat in our present study. The GB-induced increase in photosynthetic capacity in wheat may have been due to stomatal or non-stomatal limitations (Fig. 3a–e), which are major controlling factors of photosynthetic rate [22, 23]. Furthermore, increase in photosynthesis was found to be primarily due to an increase in stomatal conductance, which caused higher $CO_2$ diffusion inside the leaf, thereby favoring higher photosynthetic rate [23]. These results are similar to those of Mäkela et al. [14, 20] in which increase in salt tolerance of field-grown tomatos due to GB application was linked with increased net CO2 assimilation rate and stomatal conductance under salt or water stress. However, the mechanism by which GB application reversed to some extent salt-induced injurious effects on photosynthesis through stomatal conductance was not clear [14, 20]. More importantly, rate of transpiration with GB application decreased in wheat under saline conditions, resulting in improved water-use efficiency (WUE). Another possibility for the GB-induced enhancement in photosynthetic rate could be a protective effect of GB on photosynthetic pigments of wheat. However, in the present study, only chlorophyll 'b' content

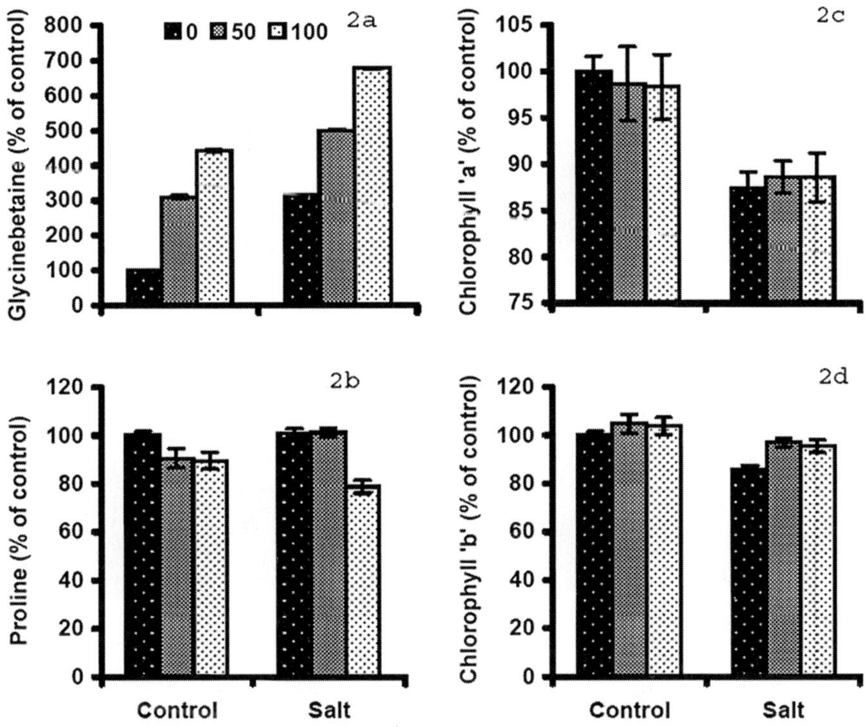


**Figure 2.** Leaf GB, proline, chlorophyll 'a' and 'b' of wheat when different amounts of GB were exogenously applied to salt-stressed or non-stressed plants (0, no spray; 50, 50 m*M* GB; and 100, 100 m*M* GB foliar spray).

was improved under saline conditions as a result of GB application (Fig. 2c, d). As chlorophyll 'b' is mainly associated with PS-II antenna, GB-induced improvement in Chl 'b' concentration might have been due to structural/conformational changes in the PS-II antennae, as suggested by Kocheva et al. [24]. This could be one of the additional factors causing an increase in photosynthetic capacity due to exogenously applied GB.

It is now evident that under saline conditions plants up-regulate antioxidant enzymes to detoxify salt-induced reactive oxygen species (ROS). A better antioxidant system can protect the plants from the adverse effects of salt stress [16, 25–27]. However, in the present study, the activity of only SOD was enhanced by GB application under saline conditions (Fig. 4a–c). From these results, it is possible that both SOD and GB scavenged ROS in wheat. Similarly, Demiral and Turkan [16], in a study on rice, found that 15 mM GB applied to the roots enhanced the SOD activity under saline conditions. These results are also in agreement with those of Shalata and Tal [28], who reported constitutively higher SOD and ascorbic acid peroxidase (APX) activities in

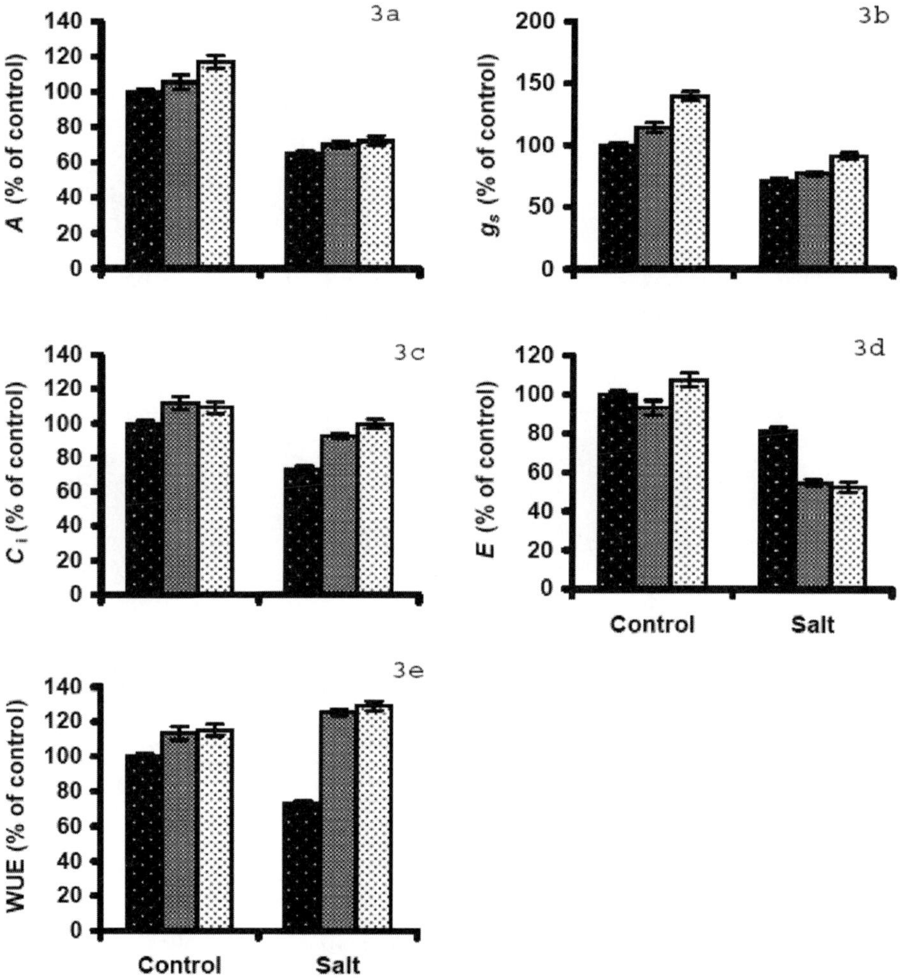


**Figure 3.** Net $CO_2$ assimilation rate (*A*), transpiration rate (*E*), sub-stomatal $CO_2$ ($C_i$), stomatal conductance ($g_s$), water use efficiency (WUE measured as *A*/*E*) of wheat when different levels of GB were exogenously applied to salt-stressed or non-stressed plants (0, no spray; 50, 50 m*M* GB; and 100, 100 m*M* GB foliar spray).

wild salt-tolerant tomato plants. If we draw parallels between the endogenous level of GB and antioxidant activities of each enzyme, it is evident that endogenous level of GB may have a significant protective effects on membranes by decreasing the levels of ROS in wheat plants, which thereby results in lower activities of other antioxidant enzymes (CAT and POD) under salt stress, because GB is known to scavenge hydroxyl radicals [29].

The results presented here, as well as those from some previous studies, clearly show that foliar application of GB improved the growth and yield

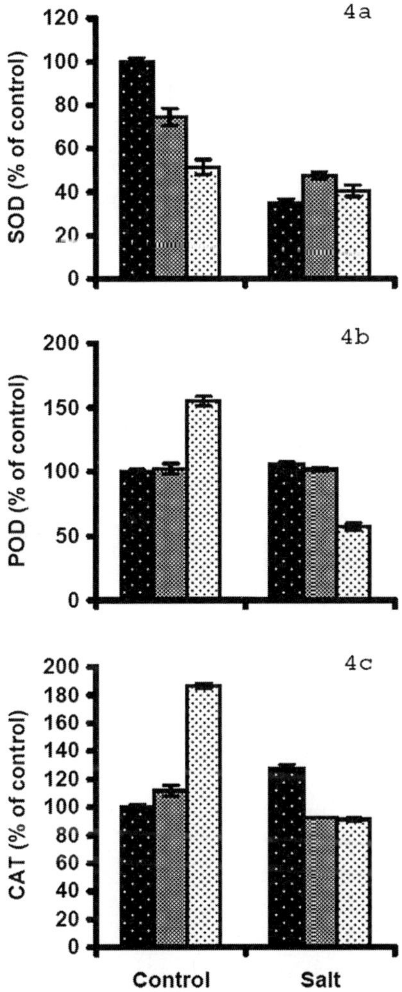


**Figure 4.** Activities of superoxide dismutase (SOD), catalase (CAT) and peroxidase (POD) of wheat when different amounts of GB were exogenously applied to salt-stressed or non-stressed plants (0, no spray; 50, 50 m*M* GB; and 100, 100 m*M* GB foliar spray).

of wheat plants by improving photosynthetic capacity and alleviating the adverse effects of salt-induced oxidative stress.

## Influence of exogenously applied GB on maize

Undoubtedly, in the present study, salt stress caused a reduction in growth and yield of maize; however, this reduction in growth and yield was com-

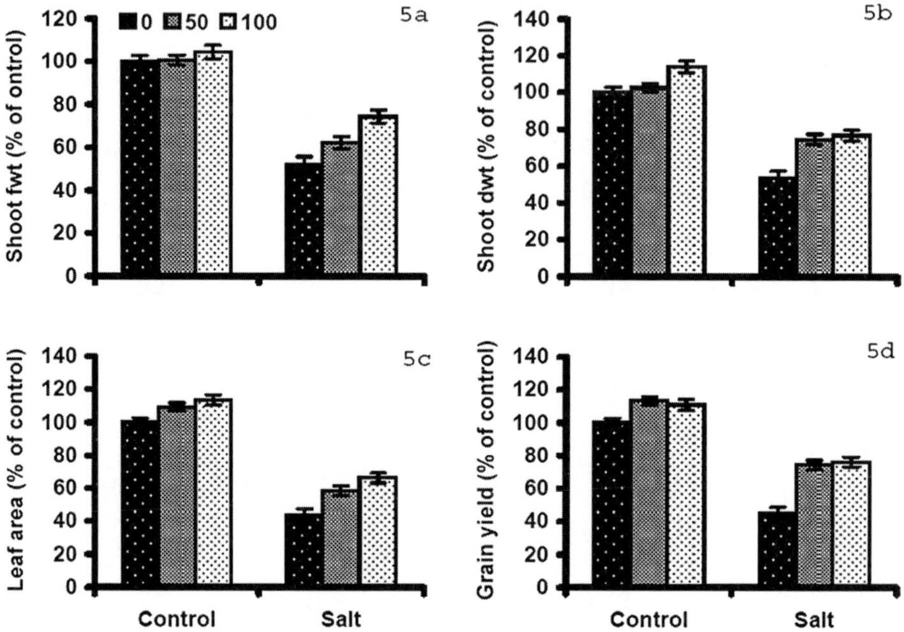


**Figure 5.** Fresh and dry weight of shoots, leaf area and grain yield of maize when different levels of GB were exogenously applied to salt-stressed or non-stressed plants (0, no spray; 50, 50 m*M* GB; and 100, 100 m*M* GB foliar spray).

pensated by foliar application of GB (Fig. 5a–d). These results are similar to some earlier findings in which foliar application of GB resulted in a significant improvement in salt tolerance of maize plants [21, 30]. Similarly, in rice, a marked improvement in salt tolerance was observed on exogenous application of GB [31, 32]. However, these changes in the growth of maize due to imposition of salt stress or foliarly applied GB were found to be associated with leaf growth, as biomass production is closely related to leaf area index (LAI) in different agricultural crops [33] and other vegetation types [34]. Exogenous application of GB also increased the endogenous level of GB of both salt-stressed and non-stressed plants of maize plants. It is widely accepted that GB protects higher plants against salt/osmotic stresses by stabilizing many functional units, like the oxygen-evolving PS-II complex [31], membranes [35], quaternary structures of complex proteins [13], and enzymes such as rubisco [36]. Furthermore, increased salt tolerance of maize may to be due to enhanced endogenous GB level caused by exogenously applied GB (Fig. 6a, b). Such a relationship between salt tolerance and endogenous GB level has been reported previously in different crops such as tomato and rape [20], kidney beans [37], and rice [16].

Photosynthesis in maize was reduced due to salt stress; however, foliarly applied GB ameliorated this inhibitory effect (Fig. 7a). It has already been

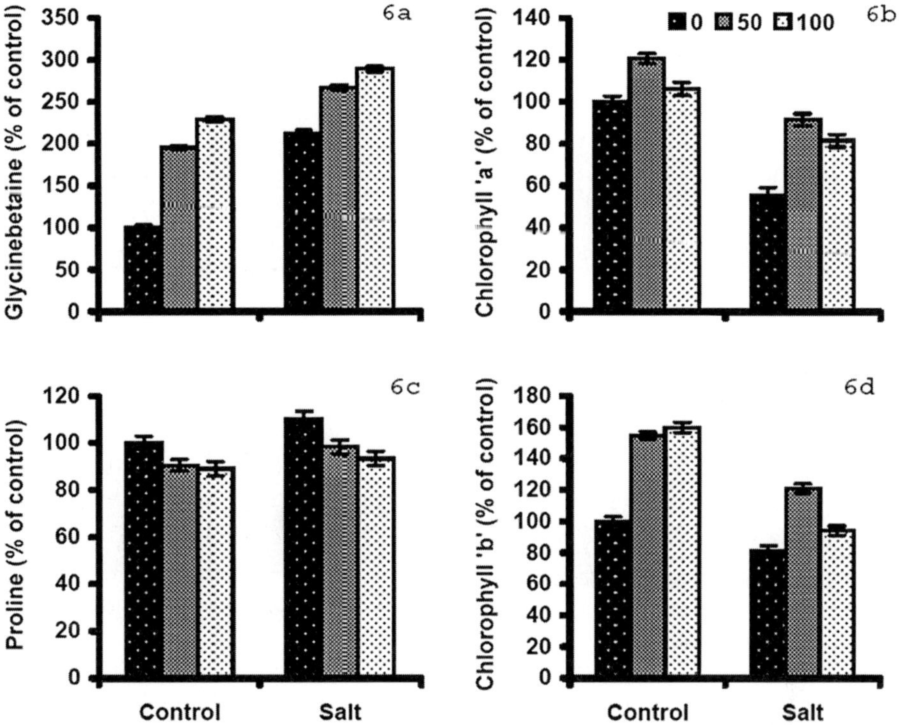


**Figure 6.** Leaf glycinebetaine, proline, chlorophyll 'a' and 'b' of maize when different levels of GB were exogenously applied to salt-stressed or non-stressed plants (0, No spray; 50, 50 mM GB; and 100, 100 m*M* GB foliar spray).

reported that the enhancement in net $CO_2$ assimilation rate due to foliarly applied GB is correlated with the degree of salt tolerance of different crops, e.g., in tomato [14] and wheat [15]. In the present study, foliar application of GB significantly increased the stomatal conductance in maize plants. Furthermore, net $CO_2$ assimilation rate ($A$) and stomatal conductance ($g_s$) showed a significant positive relationship. Similarly, $A$ and $g_s$ were also positively correlated with sub-stomatal $CO_2$ ($C_i$) (Fig. 7a–d). Foliar application of GB also improved WUE in salt-stressed plants (Fig. 7e). Furthermore, enhanced WUE in the salt-stressed maize plants due to GB application under saline conditions showed a positive relationship with growth and yield. These results can be related to those of Ashraf and Bashir [38] that demonstrated a positive relationship between WUE and grain yield of wheat.

Although the relationship between $A$, $g_s$ and $C_i$ was significant, the pattern of increase in $A$, along with $g_s$ and $C_i$ was not consistent with increasing level of GB applied (Fig. 7a–d), indicating that this effect was partial. These findings are similar to those of Mäkela et al. [14, 36], who reported that exogenous application of GB increased photosynthetic capacity of tomato through

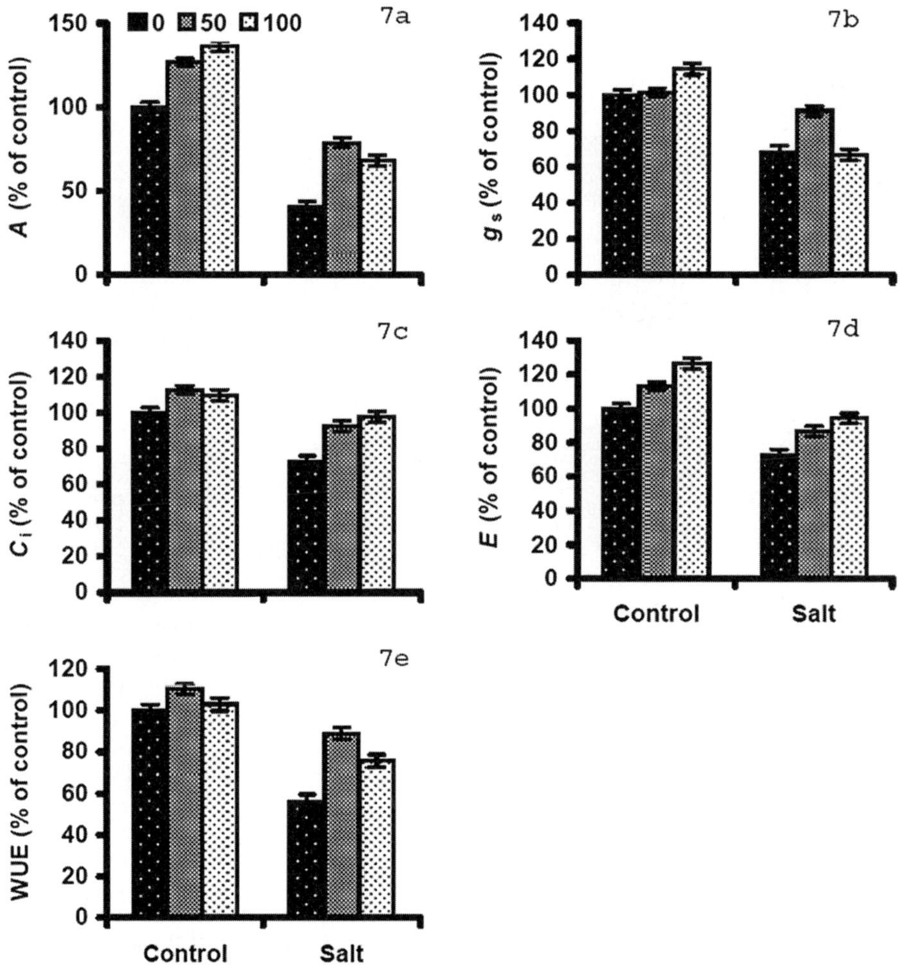


**Figure 7.** Net $CO_2$ assimilation rate ($A$), transpiration rate ($E$), sub-stomatal $CO_2$ ($C_i$), stomatal conductance ($g_s$), water use efficiency (WUE measured as $A/E$) of maize when different levels of GB were exogenously applied to salt-stressed or non-stressed plants (0, no spray; 50, 50 m*M* GB; and 100, 100 m*M* GB foliar spray).

stomatal limitations as well as by improving activity of rubisco. Likewise, Yang and Lu [21] reported that 10 mM GB applied to the roots enhanced salt tolerance of wheat plants by improving photosynthesis through stomatal conductance. In another study with maize, Yang and Lu [30] reported that GB improved photosystem-II (PS-II) efficiency in maize plants. This view was further supported by the beneficial effect that foliar application of GB had on photosynthetic pigments (chlorophyll '*a*' and '*b*') in the salt-stressed plants of maize (Fig. 6c, d). In addition, leaf chlorophyll 'a' of both cultivars

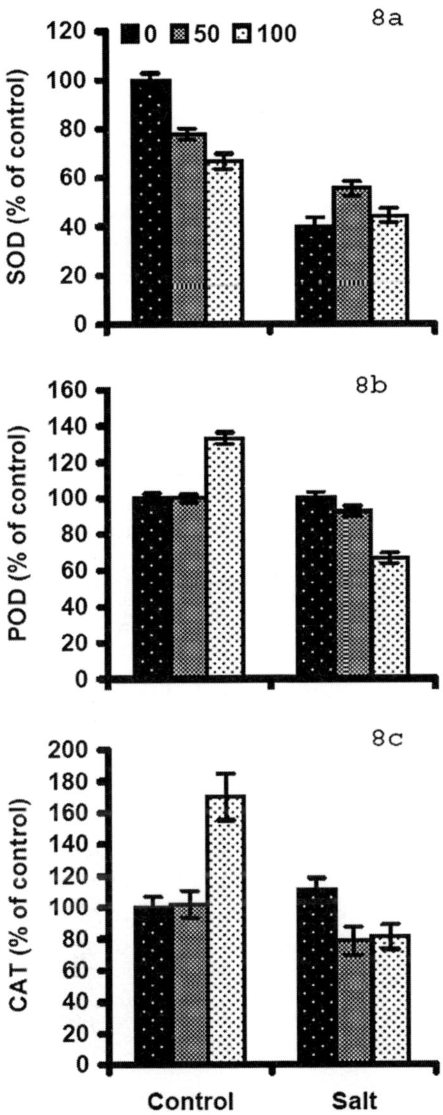


**Figure 8.** Activities of SOD, CAT and POD of maize when different levels of GB were exogenously applied to salt-stressed or non-stressed plants (0, no spray; 50, 50 m*M* GB; and 100, 100 m*M* GB foliar spray).

was positively correlated with *A*. A similar positive relationship between *A* and chlorophyll '*a*' has already been observed in sunflower [39], and wheat [15]. Thus, GB-induced enhancement of photosynthetic capacity was due to both stomatal and non-stomatal limitations.

Environmental stresses such as drought and salinity are known to increase the production of ROS, such as $H_2O_2$ (hydrogen peroxide), $O^{\bullet}{}_2^-$ (superoxide), $^1O_2$ (singlet oxygen) and $OH^{\bullet}$ (hydroxyl), by enhanced leakage of electrons from electron transport chains in the chloroplasts and mitochondria to molecular oxygen [40]. It is known that cytotoxic ROS can destroy normal metabolism through oxidative damage of lipids, proteins and nucleic acids [41]. Membrane injury induced by salt stress is related to an enhanced production of highly toxic ROS [42]. To scavenge these ROS, plants either synthesize different antioxidant compounds or activate antioxidant enzymes [42]. SOD is an important antioxidant enzyme [43, 44] because it is present in different cellular compartments such as chloroplasts, mitochondria, microsomes, glyoxysomes, oxysomes, apoplasts, and cytosol [44]. In the present study, salt stress caused the reduction in SOD activity of maize plants, but the exogenous application of GB enhanced its activity (Fig. 8a). In contrast, the activities of POD and CAT were inconsistently increased or decreased due to both salt stress and foliar application of GB (Fig. 8c, d). These results suggest that GB-induced enhancement in SOD activity may have protected photosynthetic machinery from salt-induced oxidative damage. These results can be related to those of Ma et al. [45] who found that GB-treated wheat plants that exhibited increased SOD and APX activities showed higher photosynthetic activity and water stress tolerance. In view of these findings, it is suggested that higher SOD activity in the salt-stressed plants of maize due to GB applied at the vegetative stage might be one of the additional factors in improving salt tolerance in maize cultivars as recently proposed by Cuin and Shabala [46].

From these results, it is clear that foliar application of GB was effective in ameliorating the adverse effects of salinity on photosynthesis and yield of maize plants. GB-induced enhancement in photosynthetic capacity of maize was found to be associated with both stomatal and non-stomatal limiting factors. Furthermore, exogenous application of GB also scavenged free radicals generated by salt stress directly and by enhancing SOD activity.

## Conclusions

Although exogenous application of GB induced salt tolerance in potential cereals (wheat and maize) by enhancing photosynthetic and antioxidant capacities, its effect seems to be greater in wheat plants assessed as percent increase in growth and yield under saline conditions. However, GB-induced improvement in photosynthetic capacity in wheat was found to be mainly through stomatal limitations, while that in maize it was through both stomatal and non-stomatal factors, suggesting that the mechanism by which GB-induced salt tolerance is species specific. Furthermore, because naturally produced GB does not normally break down in plants [1], it can easily be extracted from high-producing plants such as sugar beets. According to an estimate, exogenous application of GB, with a cost of US$ 20–25 $kg^{-1}$ $ha^{-1}$, the net benefit appears to be as high as US$ 580 $ha^{-1}$ [47]. Thus, the easy extraction and its exogenous application make the use of GB an economically feasible approach to counteract adverse effects of environmental stresses on crop productivity.

Overall, exogenous application of GB is a promising means to improve growth and crop yield under salt stress.

*Acknowledgements*

This research was partially supported by a grant no. 403 from the Pakistan Academy of Sciences, and by the Pakistan Science Foundation and the University of Agriculture, Faisalabad, Pakistan.

## References

1 Bray EA, Bailey-Serres J, Weretilnyk E (2000) Responses to abiotic stress. In: B Buchanan, W Gruissem, R Jones (eds): *Biochemistry and Molecular Biology of Plants*. American Society of *Plant Physiol*ogy, Rockville, 1158–1203

2 Owen S (2001) Salt of earth. Genetic engineering may help to reclaim agricultural land lost due to salinization. *EMBO Rep* 2: 877–879

3 Skovmand B, Reynolds MP, DeLacy IH (2001) Searching genetic resources for physiological traits with potential for increasing yield. In: MP Reynolds, I Ortiz-Monasterio, A. McNab (eds): *Application of physiology in wheat breeding*, CIMMYT, DF, Mexico, 17–28

4 Flowers TJ (2004) Improving crop salt tolerance. *J Exp Bot* 55: 307–319

5 Munns R (2007) Utilising genetic resources to enhance productivity of salt-prone land. *CAB Reviews: Perspectives in Agriculture, Veterinary Science, Nutrition and Natural Resources,* 2, No. 009

6 Hussain SS, Revandi A (2007) Molecular breeding for drought tolerance in plants: Wheat perspective. *Proc Pak Acad Sci* 44: 35–62

7 Rosegrant MW, Leach N, Gerpacio RV (1999) Alternative futures for world cereal and meat consumption. Summer meeting of the Nutrition Society. Guildford, UK, 29 June–2 July 1998. *Proc Nutr Soc* 58: 1–16

8 Ashraf M (2004) Some important physiological selection criteria for salt tolerance in plants. *Flora* 199: 361–376

9 Munns R (2005) Genes and salt tolerance: Bringing them together. *New Phytol* 167: 645–663

10 Cuartero J, Bolarin MC, Asinns MJ, Moreno V (2006) Increasing salt tolerance in the tomato. *J Exp Bot* 57: 1045–1058

11 Flowers TJ, Flowers SA (2005) Why does salinity pose such a difficult problem for plant breeders? *Agric Water Manage* 78: 15–24

12 Ashraf M, Foolad MR (2007) Improving plant abiotic-stress resistance by exogenous application of osmoprotectants glycine betaine and proline. *Environ Exp Bot* 59: 206–216

13 Murata N, Mohanty PS, Hayashi H, Papageorgiou GC (1992) Glycinebetaine stabilizes the association of extrinsic proteins with the photosynthetic oxygen-evolving complex. *FEBS Lett* 296: 187–189

14 Mäkela P, Kontturi M, Pehu E, Somersalo S (1999) Photosynthetic response of drought and salt stressed tomato and turnip rape plants to foliarly applied glycinebetaine. *Physiol Plant* 105: 45–50
15 Raza SH, Athar HR, Ashraf M (2006) Influence of exogenously applied glycinebetaine on the photosynthetic capacity of two differently adapted wheat cultivars under salt stress. *Pak J Bot* 38: 341–251
16 Demiral T, Türkan I (2004) Does exogenous glycinebetaine affect antioxidant system of rice seedlings under NaCl treatment? *J Plant Physiol* 161: 1089–1100
17 Raza SH, Athar HR, Ashraf M, Hameed A (2007) Glycinebetaine-induced modulation of antioxidant enzymes activities and ion accumulation in two wheat cultivars differing in salt tolerance. *Environ Exp Bot* 60: 368–378
18 Bergmann H, Eckert H (1984) Einfluss von Glycinbetain auf die Wasserausnutzung von Winter-Weizen *Triticum aestivum* L. *Biol Plant* 26: 384–387
19 Diaz-Zorita M, Fernández-Canigia MV, Grosso GA (2001) Applications of foliar fertilizers containing glycinebetaine improve wheat yields. *J Agron Crop Sci* 186: 209–215
20 Mäkela P, Jokinen K, Kontturi M, Peltonen-Sainio P, Pehu E, Somersalo S (1998) Foliar application of glycine betaine – a novel product from sugarbeet – as an approach to increase tomato yield. *Ind Crops Prod* 7: 139–148
21 Yang X, Lu C (2005) Photosynthesis is improved by exogenous glycinebetaine in salt stressed maize plants. *Physiol Plant* 124: 343–352
22 Baker NR, Harbinson J, Kramer DM (2007) Determining the limitations and regulation of photosynthetic energy transduction in leaves. *Plant Cell Environ* 30: 1107–1125
23 Sharkey TD, Bernacchi CJ, Farquhar GD, Singsaas EL (2007) Fitting photosynthetic carbon dioxide response curves for C3 leaves. *Plant Cell Environ* 30: 1035–1040
24 Kocheva K, Lambrev P, Georgiev G, Goltsev V, Karabaliev M (2004) Evaluation of chlorophyll fluorescence and membrane injury in the leaves of barley cultivars under osmotic stress. *Bioelectrochemistry* 63: 121–124
25 Gossett DR, Millhollon EP, Lucas MC (1994) Antioxidant response to NaCl stress salt-tolerant and salt-sensitive cultivars of cotton. *Crop Sci* 34: 706–714
26 Gossett DR, Banks SW, Millhollon EP, Lucas MC (1996) Antioxidant response to NaCl stress in a control and a NaCl-tolerant cotton line grown in the presence of paraquat, buthionine sulfoxime and exogenous glutathione. *Plant Physiol* 112: 803–809
27 Bor M, Özdemir F, Türkan I (2003) The effect of salt stress on lipid peroxidation and antioxidants in leaves of sugar beet (*Beta vulgaris* L.) and wild beet (*Beta maritime* L.). *Plant Sci* 164: 77–84
28 Shalata A, Tal M (1998) The effect of salt stress on lipid peroxidation and antioxidants in the leaf of the cultivated tomato and its wild salt-tolerant relative *Lycopersicon penellii*. *Physiol Plant* 104: 169–174
29 Smirnoff N, Cumbes OJ (1989) Hydroxyl radical scavenging activity of compatible solutes. *Phytochemistry* 28: 1057–1060
30 Yang X, Lu C (2006) Effects of exogenous glycinebetaine on growth, $CO_2$ assimilation, and photosystem II photochemistry of maize plants. *Physiol Plant* 127: 593–602
31 Harinasut P, Tsutsui K, Takabe T, Nomura M, Kishitani S (1996) Exogenous glycine betaine accumulation and increased salt tolerance in rice seedlings. *Biosci Biotechnol Biochem* 60: 366–368
32 Lutts S (2000) Exogenous glycine betaine reduces sodium accumulation in salt-stressed rice plants. *Int Rice Res Notes* 25: 39–40
33 Monteith JL (1977) Climate and the efficiency of crop production in Britain. *Philos Trans R Soc Lond* 281: 277–294

34 Hirose T, Ackerly DD, Traw MB, Ramseier D, Bazzaz FA (1997) $CO_2$ elevation, canopy photosynthesis, and optimal leaf area index. *Ecology* 78: 2339–2350

35 Mansour MMF (1998) Protection of plasma membrane of onion epidermal cells by glycine betaine and proline against NaCl stress. *Plant Physiol Biochem* 36: 767–772

36 Mäkela P, Kärkkäinen J, Somersalo S (2000) Effect of glycinebetaine on chloroplast ultrastructure, chlorophyll and protein content, and Rubisco activities in tomato grown under drought or salinity. *Biol Plant* 43: 471–475

37 Lopez CML, Takahashi H, Yamazaki S (2002) Plant water relations of kidney bean plants treated with NaCl and foliarly applied glycinebetaine. *J Agron Crop Sci* 188: 73–80

38 Ashraf M Bashir A (2003) Salt stress induced changes in some organic metabolites and ionic relations in nodules and other plant parts of two legumes differing in salt tolerance. *Flora* 198: 486–498

39 Ashraf M, Sultana R. (2000) Combine effect of NaCl salinity and N-form on mineral composition of sunflower plants. *Biol Plant* 43: 615–619

40 Hernandez JA, Ferrer MA, Jimenez A, Ros Barcelo A, Sevilla F (2001) Antioxidant systems and $O_2^{\bullet-}/H_2O_2$ production in the apoplast of pea leaves. Its relation with salt-induced necrotic lesions in minor veins. *Plant Physiol* 127: 817–831

41 McCord JM (2000) The evolution of free radicals and oxidative stress. *Am J Med* 108: 652–659

42 Mittler R (2002) Oxidative stress, antioxidants and stress tolerance. *Trends Plant Sci* 7: 405–410

43 Elstner EF (1991) *Mechanism of oxygen activation in different compartments of plant cell.* American Society of Plant Physiology, Rockville, 13–25

44 Alscher RG, Erturk N, Heath LS (2002) Role of superoxide dismutases (SODs) in controlling oxidative stress in plants. *J Exp Bot* 53: 1331–1341

45 Ma QQ, Wang W, Li YH, Li DQ, Zou Q (2006) Alleviation of photoinhibition in drought-stressed wheat (*Triticum aestivum*) by foliar-applied glycinebetaine. *J Plant Physiol* 163: 165–175

46 Cuin TA, Shabala S (2007) Compatible solutes reduce ROS-induced potassium efflux in *Arabidopsis* roots. *Plant Cell Environ* 30: 875–885

47 Naidu BP, Cameron DF, Konduri SV (1998) Improving stress tolerance and productivity of plants by a biochemical approach in agronomy and plant breeding. In: *Proceedings of the IX Australian Agronomy Conference*, Wagga Wagga, Australia, 355–358

Biosaline Agriculture and High Salinity Tolerance
Edited by Chedly Abdelly, Münir Öztürk, Muhammad Ashraf and Claude Grignon


# *In vitro* tissue culture approaches for the study of salt stress in citrus

Rosa M. Pérez-Clemente, Almudena Montoliu, Patricia López, María F. López-Climent, Vicent Arbona and Aurelio Gómez-Cadenas

*Departament de Ciències Agràries, Universitat Jaume I, Campus de Riu Sec (E.S.T.C.E.), 12071 Castelló de la Plana, Spain*

## Abstract

On the Mediterranean coast of Spain, where most of the citrus are cultivated, water restrictions together with water overexploitation lead to the increase of salt concentrations in the irrigation water. Therefore, a significant number of citrus trees grow under salt-stress conditions. In this work, an *in vitro* experimental system has been developed to study the toxic effect of NaCl on citrus rootstocks, avoiding the ion filter that represents the root system. For this, internodal stem segments of Carrizo citrange, Cleopatra mandarin and Swingle Citrumelo CPB4475 were disinfected and the explants cultivated in an enriched MS medium. Once the newly emerged shoots reached 1.5 cm, they were transferred to control MB medium or to a medium containing MB salts plus 60 mM NaCl. Using this experimental system, the percentage of plants affected by high salinity was very similar among the three genotypes studied, despite their different tolerance under field conditions, indicating that on eliminating the root system, most of the citrus genotypes will have the same behavior under salt stress. Overall, this *in vitro* culture system is a good tool to study biochemical processes involved in the response of citrus to salt stress.

## Introduction

In citrus, a major crop throughout the world, growth and yield are seriously affected by salinity. Although differences in salt tolerance within cultivars have been described, the citrus and closely related genus can be classified as salt-sensitive plants [1]. Different approaches, including agronomical, physiological and molecular methods have been used to address this problem.

In citrus, the salt-induced toxic effect seems to be caused by chloride ions, in contrast to many other species [2]. The progressive accumulation of these ions in plant tissues leads to well-documented injurious effects [3]. A general effect induced by high salinity in leaf cells is the accumulation of activated oxygen species leading to lipid radical formation and, subsequently, cell membrane breakdown and other severe damage [4]. To prevent this oxidative stress, citrus plants have developed a defense system including detoxification enzymes and antioxidant molecules [5].

## Physiological basis of citrus tolerance to salinity

It is generally accepted that the capability of citrus plants to tolerate high salinity is particularly dependent upon the rootstock [1], and, so far, the differences found among citrus rootstocks regarding salt tolerance have been related to their ability to exclude chloride [2]. On the other hand, salt tolerance in citrus appears to be dependent on several factors. It is known, for example, that growth habit and plant morphology play a pivotal role in tolerating of high concentration of salt [6, 7]. Relationships among vigor, water usage and salt absorption are also apparent from previous research. More recent work provided evidence that, in both sensitive and tolerant rootstocks, chloride uptake and accumulation, and therefore salt tolerance, are strongly related to water usage [7]. Based on previous data, the higher chloride amounts found in the sensitive genotypes, such as Carrizo citrange, in comparison with the tolerant types (e.g., Cleopatra mandarine), are partially due to its higher ability for water uptake and transport, and greater rates of leaf transpiration in both normal and saline conditions [6, 7].

If it is assumed that the amounts of chloride absorbed are linked to the water uptake, rootstocks that use more water may also permit higher chloride influxes. Furthermore, transpiration under saline stress decreased faster in tolerant rootstocks such as Cleopatra [6, 7], indicating that several tolerance responses to salt stress (i.e., readjustments to progressive reductions of chloride uptake) take place more rapidly in the tolerant rootstocks.

Confirming this assumption, we recently demonstrated that exogenous abscisic acid (ABA) regularly added to the watering solution reduces the negative effects caused by high NaCl concentration in citrus plants [8]. The salt-induced defoliation was delayed by ABA treatment, probably through a mechanism that slowed down (probably as a result of an initial reduction of stomatal aperture and transpiration) chloride uptake and, therefore, chloride accumulation in leaves. Interestingly, the decline in photosynthesis associated to salt stress was also lower in ABA-treated plants.

Moreover, it was shown [6] that the root system clearly plays a key role in controlling water and chloride uptake. The exchange of the root system between tolerant and sensitive citrus rootstocks, as a part of a reciprocal grafting experiment, provided the sensitive rootstock with a more tolerant character, making it able to cope with salt stress for a longer time.

However, since differences are not only restricted to the aerial part or the root system, it is very difficult to study, under field conditions, other putative biochemical differences between sensitive and tolerant citrus rootstocks that could be important for salt tolerance. Under a certain stress pressure, the latter will always accumulate less leaf chloride than the former [6, 7]. To overcome this problem, we developed an *in vitro* tissue culture-based system. The hypothesis was that eliminating the root system, chloride uptake should be similar among different citrus genotypes despite their level of salt tolerance. In this way, the importance of salt-induced oxidative stress and other parameters on citrus tolerance to high salinity could be studied.

## Plant tissue culture: The technology

Tissue culture can be defined as the culture and maintenance of plant cells or organs in sterile, nutritionally and environmentally supportive conditions in test tubes, Petri dishes or other containers. Applications of *in vitro* plant tissue culture include:

- Micropropagation using meristem and shoot culture to produce large numbers of identical individuals
- Large-scale growth of plant cells in liquid culture as a source of secondary products
- Crossing distantly related species by protoplast fusion and regeneration of hybrids
- Production of dihaploid plants from haploid cultures to achieve homozygous lines more rapidly in breeding programs
- Removal of viruses by meristematic tissue culture or *in vitro* shoot tip grafting, etc.

In commercial settings, tissue culture is often referred to as micropropagation, which is really only one form of a set of techniques that offer numerous significant benefits over traditional propagation methods [9]: (1) Large number of true-to-type plants can be obtained from a small sample of tissue, (2) establishment and maintenance of "virus-free" mother plant stock, and (3) reductions in the required space for maintenance and multiplication of plants, etc.

Plant tissue culture, a fundamental technique of plant biotechnology, is also an invaluable laboratory technique to study basic aspects of plant growth and development, and to manipulate these processes, as mentioned earlier, it is possible to have a large number of plants in a small space; there is no interference with pathogens, and the growing conditions are uniform year round, etc.

## Plant tissue culture in citrus

Citrus subjected to many biotic and abiotic stresses do not produce good quality fruit, but markets continuously require fruits of higher quality. This poses significant problems in most citrus growing areas, which can only be solved with the establishment of citrus improvement programs to develop new and

**Table 1.** Percentage of plants affected by high salinity (showing necrosis in at least 50% of their leaves) in different citrus rootstocks cultured *in vitro*, in control (CT) medium or in medium supplemented with 60 mM NaCl.

| | | % affected plants | |
|---|---|---|---|
| **Genotypes** | **Days after treatment** | **CT** | **60 mM NaCl** |
| Citrumelo | 20 | 0 | 55.0 |
| | 30 | 0 | 97.2 |
| Carrizo citrange | 20 | 0 | 72.9 |
| | 30 | 0 | 97.0 |
| Cleopatra mandarin | 20 | 0 | 61.0 |
| | 30 | 0 | 98.7 |

healthy genotypes to be used as rootstocks and varieties. The success of traditional citrus breeding strategies is limited by the peculiar genetic and reproductive characteristics of citrus. *In vitro* plant tissue culture techniques offer different approaches that can overcome many of these limitations [10]: (1) Shoot tip grafting *in vitro*, to recover plants free of all known citrus pathogens; (2) embryo rescue, to recover triploid seedless hybrid varieties; (3) protoplast fusion, to produce allotetraploid hybrids; and (4) genetic transformation.

In our work, greenhouse-grown plants of three citrus rootstocks: citrange Carrizo [*Citrus sinensis* (L.) Osb. × *Poncirus trifoliata* (L.) Raf.], Cleopatra mandarin (*Citrus reshni* Hort. Ex Tan.) and citrumelo [*Citrus paradisi* Macf. × *Poncirus trifoliata* (L.) Raf.] were used as sources of tissue. Stem pieces (10 cm long) were stripped of their leaves, disinfected by immersion for 10 min in a 2% (v/v) sodium hypochlorite solution containing 0.1% (v/v) Tween wetting agent, and rinsed three times with sterile water. Node stem segments (1 cm long) were cultured in Petri dishes with basal medium (BM), containing the inorganic salts of Murashige and Skoog (MS) [11]: 100 mg/L *i*-inositol, 1 mg/L pyridoxine-HCl, 0.2 mg/L thiamine-HCl, 1 mg/L nicotinic acid, 30 g/L sucrose and the pH was set at 5.7 ± 0.1 with 0.1 N NaOH. The medium was solidified by the addition of agar (9 g/L) before autoclaving.

Shoots recovered from nodal stem segments were excised from the explant and cultured into 150 × 20 mm tubes on multiplication medium (MM) to promote the development of axillary buds. MM consisted of BM medium supplemented with 2 mg/L 6-benzylaminopurine. During the growth of plants *in vitro*, new shoots are formed from buds located at leaf axils. These shoots were excised and used as plant material in successive experiments. As control, shoots grown on BM were used. Based on previous works carried out in our laboratory, 60 mM NaCl was selected as salt stress treatment. Therefore, salinized shoots were cultured in BM medium supplemented with that concentration of NaCl.

Cultures were maintained under standard environmental conditions in a culture room at 26±1°C, 60% relative humidity and a 16-h/8-h light/dark photoperiod with a daily photosynthetic radiation of 40 μmol $m^{-2}$ $s^{-1}$.

**Table 2.** Chloride content in shoot fresh tissue of different citrus rootstocks cultured *in vitro* in control (CT) medium or in one supplemented with 60 mM NaCl.

| | Chloride concentration(mg/g fresh tissue) | | |
|---|---|---|---|
| **Genotypes** | **Days after treatment** | **CT** | **60 mM NaCl** |
| Citrumelo | 20 | 7.90±0.13 | 38.0±0.34 |
| | 30 | 9.09±0.19 | 48.34±0.44 |
| Carrizo citrange | 20 | 6.06±0.21 | 30.25±0.40 |
| | 30 | 13.20±0.17 | 33.73±0.54 |
| Cleopatra mandarin | 20 | 8.57±0.82 | 43.14±2.12 |
| | 30 | 6.55±0.37 | 39.08±1.13 |

**Table 3.** Malondialdehyde (MDA) content in different citrus rootstocks cultured *in vitro* in control (CT) medium or in one supplemented with 60 mM NaCl.

| | MDA (nmol/g fresh tissue) | | |
|---|---|---|---|
| **Genotypes** | **Days after treatment** | **CT** | **60 mM NaCl** |
| Citrumelo | 20 | 22.04±0.50 | 20.65±0.51 |
| | 30 | 23.75±0.84 | 20.45±1.55 |
| Carrizo citrange | 20 | 16.90±0.85 | 18.42±1.33 |
| | 30 | 17.31±2.92 | 16.93±0.70 |
| Cleopatra mandarin | 20 | 21.34±0.96 | 21.02±0.76 |
| | 30 | 24.83±0.47 | 25.21±0.74 |

When using this experimental system, the percentage of plants affected by high salinity (60 mM NaCl) was very similar among the three genotypes studied despite their different tolerance under field conditions (see Tab. 1). These data indicate that the initial hypothesis was true (on eliminating the root system, most of the citrus genotypes will have the same behavior under salt stress). When chloride concentration was determined in these plants (Tab. 2), no differences were found among the three rootstocks, which confirms the idea that the root system acts as a filter for ion uptake and constitutes an important aspect that contributes to the relative tolerance or sensitivity to high salinity. Finally, the *in vitro* culture system turned out to be a good tool for studying biochemical processes involved in the response of citrus to salt stress. Table 3 shows, as an example, the content of malondialdehyde found in the citrus plant under control or stress conditions. Strikingly, high salinity did not induce an increase of this indirect marker of oxidative stress, as reported before in our group studying Carrizo citrange plants grown in the greenhouse [5]. This observation can be extrapolated to the rest of studied genotypes and indicates that, under the same salt conditions and at the same level of leaf chloride intoxication, no biochemical differences are found among tolerant and sensitive genotypes. This might indicate that the roots are an important part in the system that triggers oxidative stress in shoots.

Current efforts in our laboratory are oriented to studying the hormonal profiling to determine whether stress signaling is also common among different genotypes.

*Acknowledgements*

This work was supported by the Spanish Ministerio de Ciencia y Tecnología (grant no. AGL2007-65437-C04-03/AGR) and Fundació Baucaixa/Universitat Jaume I through grants P1-1B2006-02.

## References

1 Storey R, Walker RR (1999) Citrus and salinity. *Sci Hort* 78: 39–81

2 Bañuls J, Serna MD, Legaz M, Primo-Millo E (1997) Growth and gas exchange parameters of Citrus plants stressed with different salts. *J Plant Physiol* 150: 194–199

3 Gómez-Cadenas A, Tadeo FR, Primo-Millo E, Talón M (1998) Involvement of abscisic acid and ethylene in the response of citrus seedlings to salt shock. *Physiol Plant* 103: 475–484

4 Halliwel B (1987) Oxidative damage, lipid peroxidation and antioxidant protection in chloroplasts. *Chem Phys Lipids* 44: 327 340

5 Arbona V, Flors V, Jacas J, García-Agustín P, Gómez-Cadenas A (2003) Enzymatic and non-enzymatic antioxidant responses of Carrizo citrange, a salt-sensitive citrus rootstock, to different levels of salinity. *Plant Cell Physiol* 44: 388–394

6 Moya JL, Tadeo FR, Gómez-Cadenas A, Primo-Millo E, Talón M (2002) Transmissible salt tolerance traits identified through reciprocal grafts between sensitive Carrizo and tolerant Cleopatra citrus genotypes. *J Plant Physiol* 159: 991–998

7 Moya JL, Gómez-Cadenas A, Primo-Millo E, Talón M (2003) Chloride absorption in salt-sensitive Carrizo citrange and salt-tolerant Cleopatra mandarin citrus rootstocks is linked to water use. *J Exp Bot* 54: 825–833

8 Gómez-Cadenas A, Arbona V, Jacas J, Primo-Millo E, Talon M (2003) Abscisic acid reduces leaf abscission and increases salt tolerance in citrus plants. *J Plant Growth Regul* 21: 234–240

9 George E (1993) *Plant propagation by tissue culture,* 2nd edn, ISBN 0-9509325-4-X. Exegetics Ltd. Edington, UK

10 Navarro L, Olivares-Fuster O, Juarez J, Aleza P, Pina JA, Ballester-Olmos JF, Cervera M, Fagoaga C, Duran-Vila N, Peña L (2004) Applications of biotechnology to citrus improvement in Spain. *Acta Hort* 632: 221–234

11 Murashige T, Skoog F (1962) A revised medium for rapid growth and bioassays with tobacco tissue cultures. *Physiol Plant* 15: 473–497

Biosaline Agriculture and High Salinity Tolerance
Edited by Chedly Abdelly, Münir Öztürk, Muhammad Ashraf and Claude Grignon


# Seawater effects on antioxidant production in berries of three cultivars of tomato (*Lycopersicon esculentum* Mill.)

Annalisa Incerti[1], Riccardo Izzo[2], Adalgisa Belligno[1] and Flavia Navari-Izzo[2]

[1] *Dipartimento di Scienze Agronomiche, Agrochimiche e delle Produzioni Animali, Università degli Studi di Catania, Via S. Sofia 98, 95123 Catania, Italia*
[2] *Dipartimento di Chimica e Biotecnologie Agrarie, Università di Pisa, Via del Borghetto 80, 56124 Pisa, Italia*

## Abstract

Tomato is moderately tolerant to salt. However, under stress conditions, antioxidative defense mechanisms in tomato are activated. The effects of diluted seawater on the antioxidant capacity, namely ascorbate, tocopherols and cellular redox status, have been evaluated in three tomato cultivars. Two salad tomato cultivars, Jama and Gimar and a cherry tomato cv. Naomi were used. Our results indicate that the three cultivars had different salt tolerance. Naomi showed the best adaptive response due to its increased antioxidant pool after salinization.

## Introduction

The increasing demand for water resources in the world, especially in the arid and semi-arid regions, has forced farmers to use low quality water for irrigation, such as agricultural drainage water and marginal quality water. The irrigation with these waters does not always produce high yields. Mixing low quality water with good quality water to keep the salinity of the irrigation water below the threshold of the target crop is an acceptable practice and is used by many researchers [1]. Moreover, some researchers used the good quality water during the sensitive stages of plant growth and the poor quality water during the non-sensitive stages. Recently, seawater utilization has been explored to assess the possibility of obtaining a reasonable yield and quality of products from crops [2]. Seawater salinity delayed germination of tomato seed and reduced germination percentage especially with increasing salinity level [3]. Selection of an appropriate irrigation method and the use of cultivars suitable for growth and providing economic yield even under saline conditions could improve water-use efficiency, thereby reducing the demand for fresh water and constituting a solution to minimize the effects of salinity.

Tomato plants are moderately tolerant to salt, and under saline conditions they produce smaller berries, but with better pigmentation and higher levels of sugars and organic acids [4]. A customary vegetable like tomato, which is one of the most important vegetables worldwide, can fully fit the requirements for a balanced diet. It contains a series of beneficial health compounds and can be easily integrated in daily nutrition. The tomato's importance as a nutraceutical, phytochemical, and chemo-preventive vegetable is based on its different health-promoting ingredients [5]. Tomato could be considered a functional food according to the working definition by the EU: "A food can be regarded as functional if it is satisfactorily demonstrated to affect beneficially one or more target functions in the body" [6]. Regular consumption of tomatoes and tomato products has been associated with a lower incidence of various forms of cancer and cardiovascular diseases [7–9]. The idea of increasing health benefits in daily consumed vegetables is interesting, because the search for newer natural antioxidants is increasing. Some studies show that salinity may improve the antioxidant activity of tomatoes [10–12]. Many of the health compounds belong to the plant defense system, so the well-directed application of a stressful situation to plants could probably increase the concentration of desirable components.

This work focused on the effects of conveniently diluted seawater irrigation on the antioxidant capacity of tomato berries of three cultivars, Jama, Gimar and Naomi. We have utilized a saline water similar to seawater because along the coasts of Italy the water used for irrigation is often polluted by seawater.

## Materials and methods

### *Plant growth*

Plants of three tomato cultivars, Jama and Gimar (salad tomatoes) and Naomi ("cherry" tomato), were grown in a glasshouse in a closed cycle hydroponic culture. The treatments consisted of two different salinity levels (i.e., electric conductivity or EC), 3 mS/cm as a control and 8 mS/cm (10% sea water) for salad tomatoes and 4 mS/cm as a control and 10 mS/cm (12% sea water) for cherry tomatoes, which requires higher salinity to produce high-quality berries.

Crop density was roughly 3 plants/m$^2$. Day/night temperature inside the greenhouse was 28°/16°C, relative humidity 80–90%, global daily radiation 9.2 MJ/m$^2$ and photon flux density 500–700 mmol/(m$^2$ s). The ionic composition of the culture solutions is reported in Table 1; all the solutions contained Hoagland's two concentrations of macronutrients, added with micronutrients. The solutions were supplied for 2 min up to 12 times per day, depending on growth stage and environmental conditions. Salinization was initiated 3 weeks after sowing by adding concentrated saline solution to the recirculating nutri-

**Table 1**. Different nutrients concentration (mmol/L) in the solutions used for rockwool culture for growing tomato plants.

| Treatment | N | P | K | Ca | Mg | Na | Cl | Br | Fe |
|---|---|---|---|---|---|---|---|---|---|
| 3 (mS/cm) | 13.0 | 1.5 | 7.5 | 4.0 | 1.0 | 10.0 | 12.0 | – | 0.02 |
| 8 (mS/cm) | 13.0 | 1.5 | 7.5 | 4.0 | 6.0 | 60.0 | 69.0 | 0.1 | 0.02 |
| 4 (mS/cm) | 13.0 | 1.5 | 7.5 | 4.0 | 2 | 20.0 | 24.0 | – | 0.02 |
| 10 (mS/cm) | 13.0 | 1.5 | 7.5 | 4.0 | 7.0 | 70.0 | 79.0 | 0.1 | 0.02 |

ent solution. To avoid osmotic shock, salinization was stepped up in roughly 1.5 mS/cm daily increments until the final ECs were reached. Berries were collected randomly from the second and third truss of separate plants, at the same time in the morning to avoid variation in the quantities of analyzed compounds.

### *Antioxidant determinations*

The content of ascorbate [reduced, AsA; total, $AsA^{+}$dehydroascorbate (DHA)], tocopherols (α- and γ-) and cellular redox state ($NADPH/NADP^{+}$) have been evaluated in the mature berries of cultivars Jama, Gimar and Naomi harvested in June 2004. Ascorbate (reduced and oxidized forms) was analyzed according to [11]. Tocopherols were determined in the lipid extract [13] and analyzed by HPLC according to [14] using an electrochemical detector equipped with a graphite electrode. NADPH and $NADP^{+}$ were analyzed following [15].

### *Statistical analysis*

For each analysis, three replications of three different berries of the same treatment were used. Data are means ± SD of three replicates. One-way analysis of variance (ANOVA) was applied to the data.

## Result and discussion

### *Ascorbate (AsA and $AsA^{+}DHA$)*

The precise antioxidant response of tomato berries is not known because only very few studies discriminate between reduced and oxidized forms [11]. Indeed, the total ascorbate pool is not the index of cell redox status as only its reduced form represents an antioxidant reserve. However, besides assuming great importance in the adaptive response to stress conditions [16], it may be relevant for human health and can delay berry senescence during shelf-life.

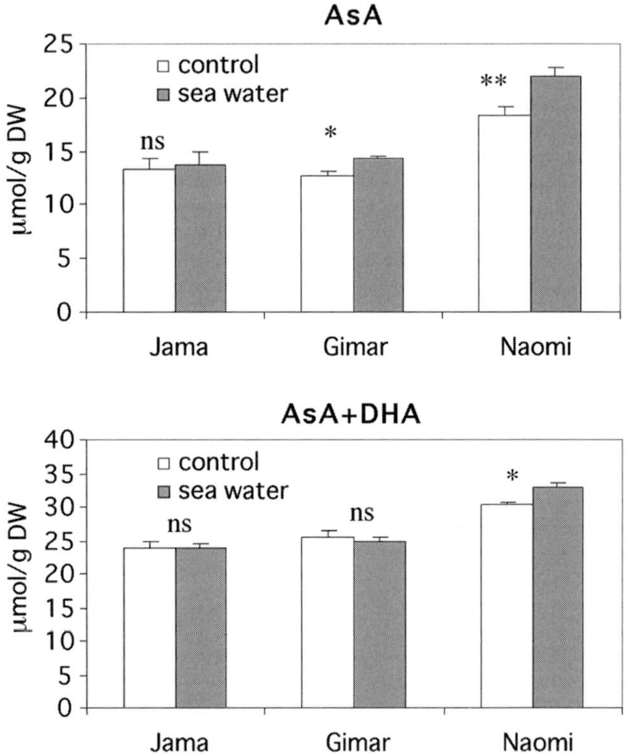


**Figure 1.** Effect of seawater on AsA and AsA+DHA content in tomato berries of cvs Jama, Gimar and Naomi. Significance is shown by: ns, non significant; *, ** and *** significant at $p \leq 0.05$, 0.01 and 0.001, respectively.

Control berries of Naomi showed higher AsA and AsA+DHA contents in comparison to the two salad tomatoes (Fig. 1). In addition, a positive response of the ascorbate pool was observed in Naomi berries, where the total ascorbate and AsA increased with salinity by 9% and 19%, respectively. No significant variations were observed in AsA and AsA+DHA for Jama, but in Gimar an increase in AsA of 14% occurred. In particular, the increase observed in all the cultivars in AsA/DHA ratio with salinity (Tab. 2), testifies to an improvement of those tomato properties that are beneficial, besides for the berry defense system, also for human nutrition [11]. It is the increase in the ratio between the two forms of ascorbate, rather than the increase in total ascorbate, that suggests a positive reaction of tomato berries to sea-water irrigation. The increase in AsA constitutes a potential reserve, which favors a better shelf-life of the berries. This assumes a particular relevance for human nutrition since AsA is one of the antioxidants able to protect animal cells against various diseases [17]. Other authors [10, 12] have also found an

**Table 2.** Effect of seawater on AsA/DHA and NADPH/$NADP^+$ ratios in tomato berries of cvs Jama, Gimar and Naomi.

| | | AsA/DHA | NADPH/$NADP^+$ |
|---|---|---|---|
| Jama | Control | 1.31 | 0.93 |
| | Seawater | 1.40 | 0.86 |
| Gimar | Control | 1.01 | 0.63 |
| | Seawater | 1.42 | 0.36 |
| Naomı | Control | 1.55 | 0.72 |
| | Seawater | 2.00 | 0.94 |

increase in ascorbate following NaCl treatments, although they refer only to total ascorbate content.

## *Cellular redox status*

An increasing body of evidence has suggested that both ripening and salt stress are associated with an increase in reactive oxygen species (ROS) production [18]. ROS are highly reactive and, in the absence of any protective mechanism, can seriously compromise normal metabolism through oxidative damage towards pigments, lipids, proteins and nucleic acids [19].

$NADP^+$ and NADPH were monitored to show the possible association of the oxidative processes accompanying the salt treatment with the ability of tomato berry to regulate ROS. The importance of NADPH, besides in the photosynthetic process, lies in its function as a reducing agent in the regeneration of AsA and reduced glutathione (GSH) through the AsA/GSH cycle [15]. There is only scarce literature on the cell redox status in terms of NADPH/$NADP^+$ ratio and no information concerning tomato berries and salt-stress effects. Salt stress is an oxidative phenomenon, which requires a turnover of ROS such as $H_2O_2$ and superoxide anion [16]. For this, there must be a balance between the production of ROS and their removal by antioxidant systems, and in such circumstances tomato berries seem to use the reducing power of NADPH to produce AsA. In all the berries analyzed the content of $NADP^+$ was higher than that of NADPH, suggesting that NADPH (Fig. 2) is largely used for the regeneration of oxidized antioxidants and, therefore, it is present mainly in its oxidized form. In Jama berries, no significant variations were observed in NADPH and $NADP^+$ with salinity (Fig. 2), while an opposite trend was observed in Gimar and Naomi. In Gimar, $NADP^+$ decreased by 46% and NADPH by 68%, and in Naomi both forms increased by 16% and 52%, respectively, leading to a decrease of the NADPH/$NADP^+$ ratio in Gimar and an increase in Naomi with treatment (Tab. 2).The fact that in Jama the oxidation of NADPH with salt treatment did not occur, whereas Gimar showed a decrease in NADPH, indicates a different tolerance of the two cultivars to salt stress.

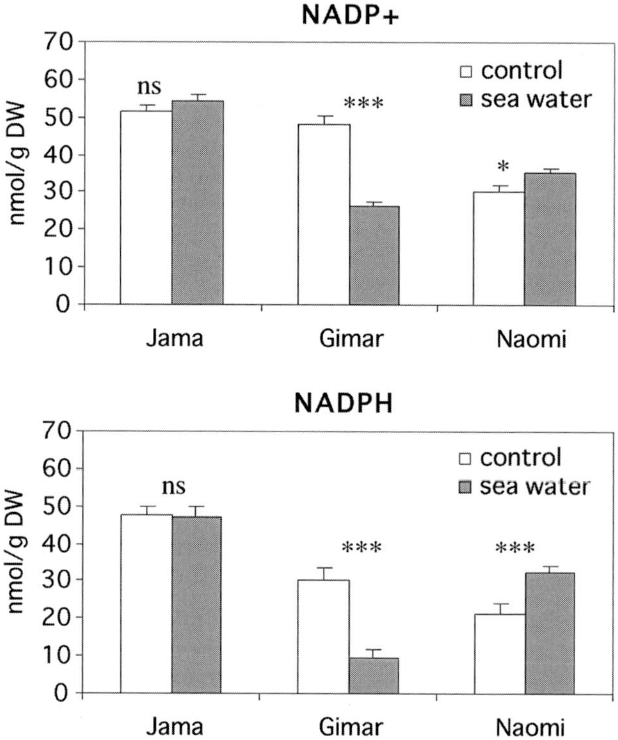


**Figure 2.** Effect of seawater on $NADP^+$ and NADPH content in tomato berries of cvs Jama, Gimar and Naomi. Significance is shown by: ns, non significant; *, ** and *** significant at $p \le 0.05$, 0.01 and 0.001, respectively.

## *Tocopherols*

The term vitamin E refers to a group of tocopherols (α-, β-, γ- and δ-), which differ in structure and bio-potency: tocopherols are ubiquitous in higher plants and essential nutrients for animals. Vitamin E functions as a chain-breaking antioxidant that prevents the propagation of free radical reactions [20] and plants are the only organisms that produce tocopherols. The major tocopherols determined in this study were α- and γ-tocopherols; δ-tocopherol was found only in traces and no β-tocopherol was found. The amount and the variations of different forms of tocopherols in the salt-stressed plants differed among the cultivars, confirming previous findings by Abushita et al. [21] of a dependence of tocopherol concentration on the cultivars. The strong dominance of α-tocopherol is particularly important because it is considered as the most active form of vitamin E. γ-Tocopherol has less antioxidant ability than α-tocopherol, but is considered the promising "other" vitamin E for

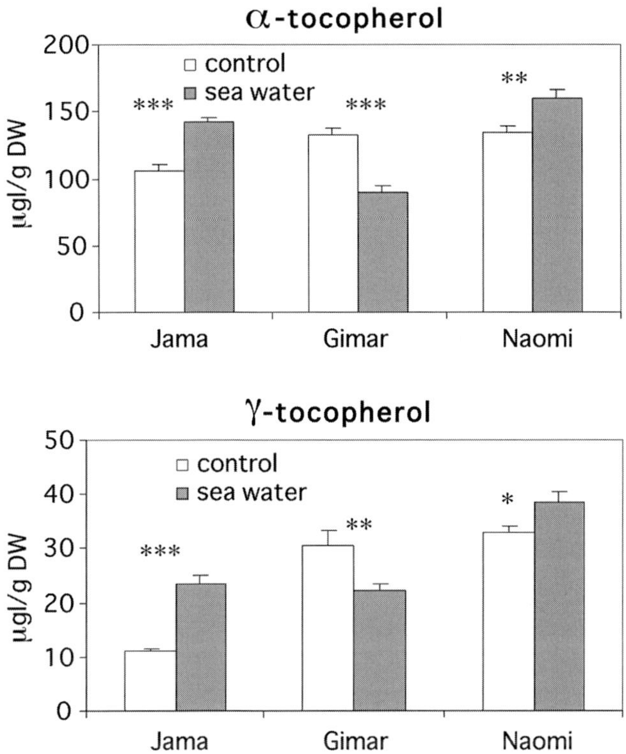


**Figure 3.** Effect of seawater on α- and γ-tocopherols content in tomato berries of cvs Jama, Gimar and Naomi. Significance is shown by: ns, non significant; *, ** and *** significant at $p \leq 0.05$, 0.01 and 0.001, respectively.

human health because its capacity to trap peroxynitrite formed in excess during inflammation [22], so that changes in the relative proportion of α- and γ-tocopherol change the nutritional value of the different cultivars. Total tocopherols, taken as sum of α- and γ-tocopherol, increased with salinity in Jama by 43% and Naomi by 18%, but decreased in Gimar by 31% (Fig. 3).

Both identified forms of tocopherol increased significantly in salted Naomi, paralleling the increase of the $NADPH/NADP^+$ ratio, indicating a better tolerance to salt. Gimar was the cultivar most strongly affected by salt conditions, as shown by the significant decrease in tocopherols together with a dramatic decrease in the $NADPH/NADP^+$ ratio. In this cultivar ascorbate is the predominant antioxidant ensuring the defense against salt stress. In Jama, no significant changes in ascorbate content were induced by salinity, while the content of tocopherols increased and the cellular redox state became more oxidized, confirming that this cultivar tolerates better high salt concentration.

Changes in the content of tocopherols were correlated with changes in the NADPH/NADP$^+$ ratio (Fig. 3, Tab. 2), indicating a possible role of tocopherols as a marker of resistance to oxidative stress [23].

In conclusion, our results indicate that the three cultivars have different salt tolerance. In fact, plants with a high level of constitutive antioxidants have been reported to have greater resistance to oxidative damage caused by environmental stress conditions. In this regard, in Gimar, a decrease in antioxidative properties of the berries occurred, probably because part of the antioxidant power was utilized to oppose the effects of seawater irrigation, so that antioxidants did not accumulate. The very best adaptive response was found in Naomi, which is naturally more tolerant to higher salinity levels than the other cultivars.

### *Acknowledgement*

This paper was supported by the Italian Ministry of University, Scientific and Technological Research and by Pisa University (PRIN 2003).

## References

1 Gawad GA, Arslan A, Gaihbe A., Kadouri F (2005) The effects of saline irrigation water management and salt tolerant tomato varieties on sustainable production of tomato in Syria (1999–2002). *Agric Water Manag* 78: 39–53

2 D'Amico ML, Navari-Izzo F, Sgherri C, Izzo R (2004) The role of lipoic acid in the regulation of the redox status of wheat irrigated with 20% seawater. *Plant Physiol Biochem* 42: 329–334

3 Hajer AS, Malibari AA, Al-Zahrani HS, Almaghrabi OA (2006) Responses of three tomato cultivars to seawater salinity. Effect of salinity on the seedling growth. *Afr J Biotechnol* 5: 855–861

4 Cuartero J, Fernandez-Munoz R (1999) Tomato and salinity. *Sci Hort* 78: 83–125

5 Ferrari CKB (2004) Functional foods, herbs and nutraceuticals: Towards biochemical mechanisms of healthy aging. *Biogerontology* 5: 275–289

6 Diplock AT, Agget PJ, Ashwell M, Bornet F, Fern EB, Roberfroid MB (1999) Scientific concepts of functional foods in Europe: Consensus Document. *Br J Nutr* 81: 1–27

7 Hercberg S, Galan P, Preziosi P, Alfarez MJ, Vazques C (1998) The potential role of antioxidant vitamins in preventing cardiovascular diseases and cancer. *Nutrition* 14: 513–520

8 Blum A, Monir M, Wirsansky I, Ben-Arzi S (2005) The beneficial effects of tomatoes. *Eur J Intern Med* 16: 402–404

9 Willcox JK, Catignani GL, Lazarus S (2003) Tomatoes and cardiovascular health. *Crit Rev Food Sci Nutr* 43: 1–18

10 De Pascale S, Maggio A, Fogliano V, Ambrosino P, Ritieni A (2001) Irrigation with saline water improves carotenoids content and antioxidant activity of tomato. *J Hort Sci Biotechnol* 76: 447–453

11 D'Amico ML, Izzo R, Tognoni F, Pardossi A, Navari-Izzo F (2003) Application of diluted sea water to soil-less culture of tomato (*Lycopersicon esculentum* Mill.): Effects on plant growth, yield, fruit quality and antioxidant capacity. *Food Agric Environ* 1: 112–116

12 Krauss S, Schnitzler WH, Grassmann J, Woitke M (2006) The influence of different electrical conductivity values in a simplified recirculating soilless system on inner and outer fruit quality characteristics of tomato. *J Agric Food Chem* 54: 441–448

13 Quartacci MF, Cosi E, Navari-Izzo F (2001) Lipids and NADPH-dependent superoxide production in plasma membrane vesicles from roots of wheat grown under copper deficiency or excess. *J Exp Bot* 52: 77–84

14 Galatro A, Simontacchi M, Puntaruolo S (2001) Free radical generation and antioxidant content in chloroplasts from soybean leaves exposed to ultraviolet-B. *Physiol Plant* 113: 564–570

15 Sgherri C, Quartacci MF, Izzo R, Navari-Izzo F (2002) Relation between lipoic acid and cell redox status in wheat grown in excess copper. *Plant Physiol Biochem* 40: 591–597

16 Meneguzzo S, Navari-Izzo F, Izzo R (1999) Antioxidative responses of shoots and roots of wheat to increasing NaCl concentrations. *J Plant Physiol* 155: 274–280

17 Olsson ME, Gustavsson KE, Andersson S, Nilsson A, Duan RD (2004) Inhibition of cancer cell proliferation in vitro by fruit and berry extracts and correlations with antioxidant levels. *J Agric Food Chem* 52: 7264–7271

18 Hernandez JA, Jimenez A, Mullineaux P, Sevilla F (2000) Tolerance of pea (*Pisum sativum* L.) to long-term salt is associated with induction of antioxidant defences. *Plant Cell Environ* 23: 853–862

19 Navari-Izzo F, Rascio N (1999) Plant response to water-deficit conditions. In: M. Pessarakli (ed): *Handbook of plant and crop stress.* Marcel Dekker, New York, 231–270

20 Brigelius-Flohè R, Traber MG (1999) Vitamin E: function and metabolism. *FASEB J* 13: 1145–1155

21 Abushita AA, Daood HG, Biacs PA (2000) Change in carotenoids and antioxidant vitamins in tomato as a function of varietal and technological factors. *J Agric Food Chem* 48: 2075–2078

22 Chericoni L (2005) Markers per l'identificazione precoce di stress ossidativo in piante di grano cresciute in eccesso di rame. Degree thesis, University of Pisa, Italy

23 Sgherri C, Kadlecovà Z, Chericoni L, Izzo R, Navari-Izzo F (2004) Ruolo dei tocoferoli in grano e pomodoro: Effetti di stress diversi. *Atti del XXII Convegno SICA*, Perugia, 263–267

Biosaline Agriculture and High Salinity Tolerance
Edited by Chedly Abdelly, Münir Öztürk, Muhammad Ashraf and Claude Grignon


# Identification of morphological, biochemical and physiological parameters for characterizing nutritional stress status in arboreous species differently tolerant to chlorosis

Silvia Donnini[1], Graziano Zocchi[1], Antonella Castagna[2], Chedley Abdelly[3] and Annamaria Ranieri[2]

[1] *Dipartimento di Produzione Vegetale, Milan University, Milano, 20133 Italy*
[2] *Dipartimento di Chimica e Biotecnologie Agrarie, Pisa University, Pisa, 56124 Italy*
[3] *Laboratoire d'Adaptation des Plantes aux Stress Abiotiques, Centre de Biotechnologie du Technopole de Borj-Cédria, Hammam-Lif, 2050 Tunisia*

## Abstract

Lime-induced chlorosis is one of the major abiotic stresses affecting fruit tree crops in the Mediterranean area. However, fruit tree species have been the object of only few studies and the results obtained are insufficient to supply parameters for breeding. Here we report the results of a study carried out to identify morphological and biochemical modifications induced by low iron availability and a high level of bicarbonate in the medium in pear (cv Conference; tolerant genotype) and quince rootstocks (MA and BA29; susceptible genotypes) cultured by *in vitro* and hydroponic culture. Morphological parameters of *in vitro* plantlets were differently influenced by the two stress conditions depending on plant genotype and parameter analyzed, and suggested that the pear cv carried out an adaptive strategy to warrant sufficient iron supply, whereas the two quince rootstocks failed to adapt to conditions typical of calcareous soil. The strong and generalized reduction in chlorophyll and carotenoid content observed only in quince plantlets suggests a down-regulation of the whole chloroplast machinery in iron-deficient quince. Measurement of Fe(III)-chelate reductase activity (FCR) of rooted cuttings from *in vitro* culture grown in hydroponic solution suggests the probable involvement of enhanced FCR activity in the major tolerance of cv Conference to iron chlorosis. Cv Conference was also less sensitive to bicarbonate supply than quince rootstocks in terms of reduction of leaf pigment content and activation of the photoprotective xanthophyll cycle. In conclusion, this study shows that the mechanisms of differential Fe efficiency are associated to differences in leaf pigment content and photoprotective process and that *in vitro* culture could be a valid technique to test rootstock susceptibility to iron chlorosis.

## Iron deficiency is a widespread problem

Iron deficiency occurs in many plant species grown in calcareous soils. In Europe and especially in the Mediterranean area a large amount of fruit tree cultivation is located on these soils [1], resulting in reduced crop yield and fruit quality. Fe chlorosis represents a widespread nutritional disorder in some pear cultivars (*Pyrus communis* L.) when grafted on quince (*Cydonia oblonga* Mill.), which has been largely adopted as it is more attractive from an agronomic viewpoint. Under Fe deficiency commercial yield depends on supplementary iron chelate application, which, besides the high management costs, may represent a hazard for soil and water pollution [2]. Thus, the use of chlorosis-tolerant rootstocks and intercropping association may represent a valid alternative to the employment of fertilizers, due to both low costs and lower ecological impact [3].

## Biological effects of iron deficiency

Woody plant species differ as to their susceptibility to Fe chlorosis and many studies [4, 5] have reported that, similarly to herbaceous dicots (Strategy I plants), some tolerant woody genotypes are able to improve Fe acquisition through enzymatic Fe(III) reduction, associated to the increase of a Fe-chelate reductase (FCR) activity. Since most of the leaf iron (about 80%) is localized in the chloroplast, mainly in the thylakoid membranes that contain about 60% of the total leaf iron [6], iron deficiency primarily affects structure and functioning of chloroplasts [7]. The most evident effect of Fe deficiency is a marked decrease in the amount of chlorophyll [8] and a significant, although less intense, decrease in the carotenoid content. Conversely, an increase in the content of the xanthophyll cycle pigments under iron starvation, together with a shift towards the de-epoxidated forms, has been reported [9], indicating an imbalance between energy absorption and utilization.

## Selection of biomarkers to characterize nutritional stress status

Different techniques and approaches have been developed, from the most conventional such as hydroponic culture to more recent approaches such as *in vitro* culture, to study lime-induced chlorosis from several points of view, as well as to detect parameters that could be used as reliable biomarkers to screen differently chlorosis-susceptible genotypes. Here we report the results of a study carried out to identify morphological and biochemical modifications induced by low iron availability and a high level of bicarbonate in the medium in pear (cv Conference; tolerant genotype) and quince rootstocks (MA and BA29; susceptible genotypes) cultured adopting two different methodological approaches: *in vitro* and hydroponic culture.

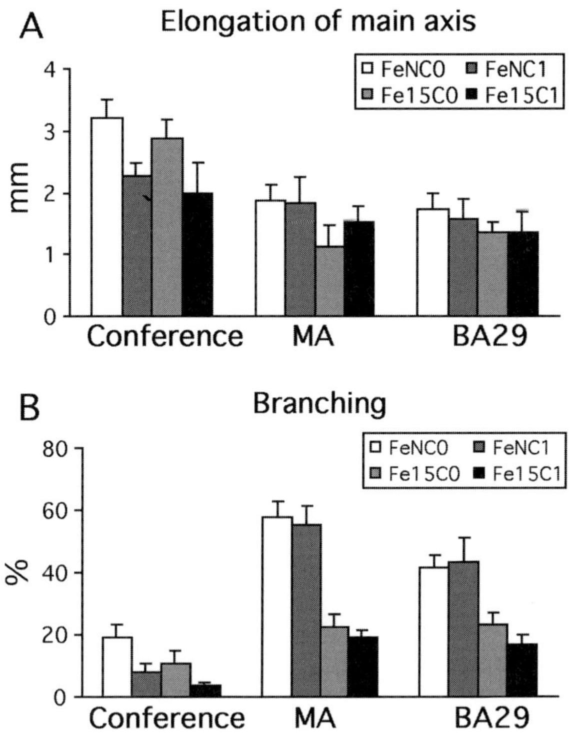


**Figure 1.** Stem elongation (A) and percentage of branching (B) of *in vitro* plantlets of quince (*Cydonia oblonga* Mill.) rootstocks MA and BA29 and pear (*Pyrus communis* L.) cv Conference cultured onto Murashige and Skoog medium [12] supplemented with 15% (Fe15) or 100% (FeN) of iron (0.10 mM FeNaEDTA), and with (C1) or without (C0) 1.0 mM $KHCO_3$.

## *In vitro* culture

Morphological and biochemical alterations were induced by both low iron availability and a high level of bicarbonate in the medium. The stem elongation of *in vitro* plantlets of pear (cv Conference; tolerant genotype) and quince rootstocks (MA and BA29; susceptible genotypes) was found to be differently influenced by the stress conditions and a statistically significant difference was found for the interaction of genotypes and bicarbonate factors. In fact, iron deficiency did not affect the elongation of the main axis of all plantlet genotypes and the only significant difference was found among Conference plantlets due to the presence of bicarbonate (Fig. 1A). Plantlets of BA29 and MA did not show any variations in internode extension under any of the treatments, while Conference plantlets showed a reduction under the two treatments with bicarbonate, indicating a strong susceptibility for this phenological trait to bicarbonate (data not shown).

**Table 1.** Leaf pigment content of *in vitro* plantlets. Chlorophyll and carotenoid content of *in vitro* plantlets of quince (*Cydonia oblonga* Mill.) rootstocks MA and BA29 and pear (*Pyrus communis* L.) cv Conference cultured onto Murashige and Skoog medium supplemented with 15% (Fe15) or 100% (FeN) of iron (0.10 mM FeNaEDTA), and with (C1) or without (C0) 1.0 mM $KHCO_3$. Pigment analysis was performed by HPLC according to Castagna et al. [14]. fw, Fresh weight; V+ A+ Z, sum of the three xanthophylls violaxanthin, antheraxanthin and zeaxanthin. For each pigment different letters indicate significantly different values at $p \leq 0.05$ according to three factorial ANOVA followed by Tukey's multiple range test.

| | | Chlorophyll *a* nmol g$^{-1}$ fw | Chlorophyll *b* nmol g$^{-1}$ fw | β-carotene nmol g$^{-1}$ fw | Neoxanthin nmol g$^{-1}$ fw | Lutein nmol g$^{-1}$ fw | V+A+Z nmol g$^{-1}$ fw |
|---|---|---|---|---|---|---|---|
| cv Conference | FeNC0 | 1876 c | 630 abc | 153 ab | 111 ab | 346 ab | 26 ab |
| | FeNC1 | 1960 bc | 684 ab | 152 ab | 144 a | 377 ab | 30 ab |
| | Fe15C0 | 1735 cd | 620 bc | 158 ab | 115 ab | 360 ab | 28 ab |
| | Fe15C1 | 1696 cd | 630 bc | 139 ab | 114 ab | 321 abc | 25 ab |
| BA29 | FeNC0 | 2242 a | 823 a | 235 a | 151 a | 498 a | 46 a |
| | FeNC1 | 2026 ab | 823 a | 208 a | 127 ab | 393 a | 45 a |
| | Fe15C0 | 803 e | 309 d | 88 bc | 54 c | 198 cd | 17 bc |
| | Fe15C1 | 439 f | 175 d | 49 c | 21 d | 116 d | 6 c |
| MA | FeNC0 | 2000 bc | 702 a | 234 a | 120 ab | 455 a | 38 a |
| | FeNC1 | 1923 c | 714 a | 192 a | 98 b | 399 a | 41 a |
| | Fe15C0 | 1090 d | 496 bcd | 89 bc | 53 c | 215 bcd | 18 bc |
| | Fe15C1 | 964 e | 390 cd | 98 bc | 69 c | 219 bcd | 21 bc |

Data on the branching, expressed as number of new outbreak lateral buds on the total formed nodes of the main axis (Fig. 1B), show that the presence of bicarbonate did not affect the apical dominance. Conversely, in the two quince rootstocks, this parameter was strongly affected by the reduction of iron supply.

As far as the pigment content was concerned, chlorophyll levels of cv Conference plantlets were unaffected in any way by the treatments (Tab. 1). Quince plantlets grown with 0.10 mM FeNaEDTA plus 1.0 mM $KHCO_3$ (FeNC1) were also not affected. β-carotene, lutein and neoxantin content of the different genotypes of pear and quince were differently influenced by the treatments. A similar trend was detected in the two quince rootstocks, which showed a statistically significant reduction in the content of these pigments in the two treatments with the lowest amount of iron in the medium. On the contrary, plantlets of cv Conference did not exhibit any change in carotenoid amount following either iron deprivation or bicarbonate addition. The sum of the three xanthophylls involved in the photoprotective xanthophyll cycle (violaxantin, antheraxanthin, zeaxanthin) remained unchanged in cv Conference, while it significantly decreased in plantlets of the quince genotypes BA29 and MA under reduced FeNaEDTA supply (Fe15 treatment). The analyzed chloroplast pigments are the target compounds for photosystem structures of the *in vitro*-grown plantlets and also play an important role in protecting the chloroplast components against oxidative stress. The strong and generalized reduction observed not only for chlorophyll but also for carotenoid content indicates that, in iron-deficient quince, the whole chloroplast machinery is down-regulated, probably owing to a negative effect on genes involved in pigment synthesis and/or a secondary oxidative stress induced by iron deficiency.

Our results show that in quince rootstocks iron chlorosis induced by reduction of iron supply is associated with morphological modifications, as indicated by reduced branching, while the main axis elongation and internode extension were not affected. In contrast, in the cv Conference the effects on the morphological parameters, but not on the pigment content, were induced only following bicarbonate supply to the medium. The morphological changes occurring in pear (decrease in the main axis elongation and internode extension) under this condition could represent an adaptive strategy carried out to maintain a sufficient iron supply to leaf tissues, whereas the two rootstocks MA and BA29 showed the lack of adaptive strategy to conditions typical of calcareous soil.

## Hydroponic culture

To better investigate the root mechanisms involved in the iron-deficiency responses, rooted cuttings from *in vitro* culture were grown in hydroponic solution and FCR activity, the main biochemical mechanism induced by Fe deficiency in Strategy I species [10], was investigated. The apex roots of cv Conference exhibited higher FCR activity following both iron deprivation and, albeit to a lesser extent, bicarbonate treatment, whereas no increase in FCR activity was shown by the two quince genotypes (Fig. 2). These data suggest the probable involvement of enhanced FCR activity in the major mechanism of tolerance of cv Conference to iron chlorosis.

At the leaf level, the chlorophyll content of both pear and quince genotypes underwent a decrease following the two kind of stress, although in cv Conference this parameter was less affected by bicarbonate supply in comparison with the quince genotypes (Tab. 2). A similar behavior was shown by β-carotene (Tab. 2) and the other carotenoids (data not shown). Many kinds of environmental stress, such as iron deficiency, predispose the photosynthetic apparatus to photoinhibition, which can ultimately lead to degradation of the photosystem core complexes [7]. The activation of the xanthophyll cycle, which involves the de-epoxidation of violaxantin to antheraxanthin and zeaxanthin, represents one of the main processes involved in dissipation of the harmful excess energy in an attempt to prevent photodamage [11]. This is represented by the increase of de-epoxidation index in the two quince rootstocks under both stress treatments and in the cv Conference only under iron-deprived conditions (Tab. 2).

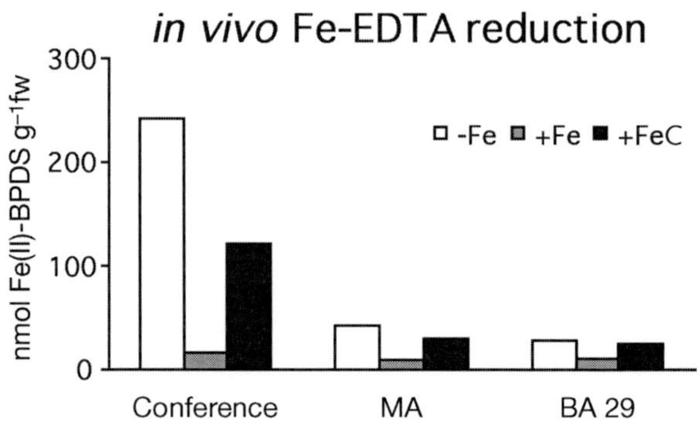


**Figure 2.** Fe(III)-chelate reductase activity of root apex of cuttings of quince (*C. oblonga* Mill.) rootstocks MA and BA29 and pear (*P. communis* L.) cv Conference grown in hydroponic Hoagland solution containing 0.080 mM Fe-EDTA (+Fe), 0 mM Fe-EDTA (–Fe), or 0.80 mM Fe-EDTA + 10 mM $KHCO_3$ (+FeC). The activity was determined using the bathophenanthrolin-edisulfonate (BPDS) reagent [13].

**Table 2.** Leaf pigment content of hydroponically grown cuttings. Chlorophyll and carotenoid content of cuttings of quince (*C. oblonga* Mill.) rootstocks MA and BA29 and pear (*P. communis* L.) cv Conference grown in hydroponic Hoagland solution containing 0.080 mM Fe-EDTA (+Fe), 0 mM Fe.EDTA (-Fe), or 0.80 mM Fe-EDTA + 10 mM $KHCO_3$ (+FeC). Pigment analysis was performed by HPLC according to Castagna et al. [14]. For each genotype, the percentage of variation in comparison to the respective control pigment is reported in parentheses. For each pigment different letters indicate significantly different values at $p \leq 0.05$ according to two-way ANOVA followed by Student's *t*-test.

| | | Chlorophyll *a* nmol $g^{-1}$ fw | Chlorophyll *b* nmol $g^{-1}$ fw | β-carotene nmol $g^{-1}$ fw | V+A+Z nmol $g^{-1}$ fw | DEPS index % (A/2+Z)/ (V +A+Z) |
|---|---|---|---|---|---|---|
| **cv Conference** | **+Fe** | 1658 a | 492 a | 271 ab | 169 b | 3.7 e |
| | **+FeC** | 1128 c (–32%) | 414 ab (–16%) | 152 cd (–44%) | 184 b (+9%) | 3.8 e (+3%) |
| | **–Fe** | 501 d (–70%) | 204 c (–59%) | 44 ef (–84%) | 38 e (–78%) | 11.0 d (+197%) |
| **BA29** | **+Fe** | 1505 b | 503 a | 311 a | 221 a | 2.0 e |
| | **+FeC** | 425 d (–72%) | 230 c (–54%) | 105 de (–66%) | 63 de (–71%) | 39.0 b (+1850%) |
| | **–Fe** | 216 e (–86%) | 75 d (–85%) | 24 f (–92%) | 28 e (–87%) | 22.0 c (+1000%) |
| **MA** | **+Fe** | 1574 ab | 347 b | 219 bc | 132 c | 3.3 e |
| | **+FeC** | 432 d (–72%) | 181 c (–48%) | 113 de (–48%) | 86 d (–35%) | 57.0 a (+1627%) |
| | **–Fe** | 153 e (–90%) | 69 d (–80%) | 17 f (–92%) | 54 de (–59%) | 37.0 b (+1021%) |

## Concluding remarks

In this work we demonstrate that biochemical root responses in quince and pear are differently affected by iron starvation and bicarbonate supply, and that the mechanisms of differential Fe efficiency are also associated to differences in leaf pigment content and photoprotective process. Moreover, this study shows that *in vitro* culture could be a valid technique to test rootstock susceptibility to iron chlorosis and to study the morphological aspects of this syndrome.

### *Acknowledgements*

We are indebted to Dr. Fabrizio Cinelli (Dipartimento di coltivazione e difesa delle specie legnose "G. Scaramuzzi", University of Pisa) and Prof. Rosario Muleo (Dipartimento di Produzione Vegetale, Università degli Studi della Tuscia, Viterbo) for the "*in vitro*" plantlets and advice on morphological analysis. This research was supported by a grant (National Project PRIN) from MIUR Rome, Italy, as well as that from the University of Pisa.

## References

1 Tagliavini M, Rombolà AD (2001) Iron deficiency and chlorosis in orchard and vineyard ecosystems – A review. *Eur J Agron* 15: 71–92
2 Wallace A (1984) Effect of liming of soil on additive, synergistic and protective interactions of some multiple trace element combinations and on iron and calcium interactions for dicot and monocot plant species. *J Plant Nutr* 7: 595–601
3 Socias i Company R, Gomez Aparisi J, Felipe AJ (1995) A genetical approach to iron chlorosis in deciduous fruit trees. In: J Abadía (ed): *Iron nutrition in soil and plants*. Kluwer, Dordrecht, 167–174
4 Vizzotto G, Pinton R, Bomben C, Cesco S, Varanini Z, Costa G (1999) Iron reduction in iron stressed plants of *Actinidia deliciosa* genotypes: Involvement of PM Fe(III)-chelate reductase and $H^+$-ATPase activity. *J Plant Nutr* 22: 479–488
5 Brancadoro L, Rabotti G, Scienza A, Zocchi G (1995) Mechanisms of Fe-efficiency in roots of vitis spp. in response to iron deficiency. *Plant Soil* 171: 229–234
6 Terry N, Abadia J (1986) Function of iron chloroplasts. *J Plant Nutr* 9: 609–646
7 Ferraro F, Castagna A, Soldatini GF, Ranieri A (2003) Tomato (*Lycopersicon esculentum* M.) T3238*FER* and T3238*fer* genotypes. Influence of different iron concentrations on thylakoid pigment and protein composition. *Plant Sci* 164: 783–792
8 Abadia J (1992) Leaf responses to Fe deficiency: A review. *J Plant Nutr* 15: 1669–1713
9 Morales F, Abadia A, Abadia J (1990) Characterization of the xanthophylls cycle and other photosynthetic pigment changes induced by iron deficiency in sugar beet (*Beta vulgaris* L.). *Plant Physiol* 94: 607–613
10 Römheld V, Marschner H (1986) Mobilization of iron in the rhizosphere of different plant species. *Adv Plant Nutr* 2: 155–204

11 Demmig-Adams B, Adams WW (1996) The role of xanthophylls cycle carotenoids in the protection of photosynthesis. *Trend Plant Sci* 1: 21–26
12 Murashige T, Skoog F (1962) A revised medium for rapid growth and bioassays with tobacco tissue cultures. *Physiol Plant* 15: 437–497
13 Chaney RL, Brown JC, Tiffin O (1972) Obligatory reduction of ferric chelates in iron uptake by soybeans. *Plant Physiol* 81: 208–213
14 Castagna A, Nali C, Ciompi S, Lorenzini G, Soldatini GF, Ranieri A (2001) Ozone exposure affects photosynthesis of pumpkin (*Cucurbita pepo*) plants. *New Phytol* 152: 223–229

Biosaline Agriculture and High Salinity Tolerance
Edited by Chedly Abdelly, Münir Öztürk, Muhammad Ashraf and Claude Grignon


# Seawater irrigation: Effects on growth and nutrient uptake of sunflower plants

Riccardo Izzo[1], Annalisa Incerti[2] and Claudio Bertolla[1]

[1] *Dipartimento di Chimica e Biotecnologie Agrarie, Università di Pisa, Via del Borghetto 80, 56124 Pisa, Italy*
[2] *Dipartimento di Scienze Agronomiche, Agrochimiche e delle Produzioni Animali, Università degli Studi di Catania, Via Santa Sofia 98, 95123 Catania, Italy*

## Abstract

The aim of the present research was the evaluation of the effects of irrigation with diluted seawater on main morphological characteristics of sunflower plants (cv. Katharina, Piacenza ecotype). Plants, irrigated with fresh water or with 20% or 30% seawater during the whole biological cycle, were harvested at four growth stages. At each stage, the main growth parameters were measured and the principal nutrients were quantified. In particular, $Cl^-$ and $Na^+$ increased significantly, especially in the plants irrigated with 30% seawater. Both seawater concentrations reduced N content but did not affect P content. $K^+$ and $Ca^{2+}$ decreased during the growth.

## Introduction

Soil salinization processes, occurring in regions with arid and semi-arid climates, are augmented by reduced availability of fresh irrigation water and represent a further menace for production systems [1]. The controlled use of seawater could be a valid tool to save fresh water. Therefore, it is important to understand the defense mechanisms of plants against salt stress to select crops suitable for saline environment.

In tomato, seawater at increasing concentration up to 20% significantly reduced crop water consumption, plant growth and berry yield, but improved berry quality [2]. The demonstrated increase in reduced forms of glutathione and ascorbate further confirmed the positive response of the berry to the increasing salinity. In some cases, 20% seawater led to the appearance of blossom-end rot [2], which was the result of Ca deficiency due to saline water [3]. Lipoic acid, present in the reduced form, played a pivotal role in alleviating seawater effects on tomato berries [2]. In wheat, seawater application reduced plant elongation, affecting shoots more than roots [4]. As concerns nutrients, salinity induced a decrease in $Cl^-$ levels in shoots and an increase in roots,

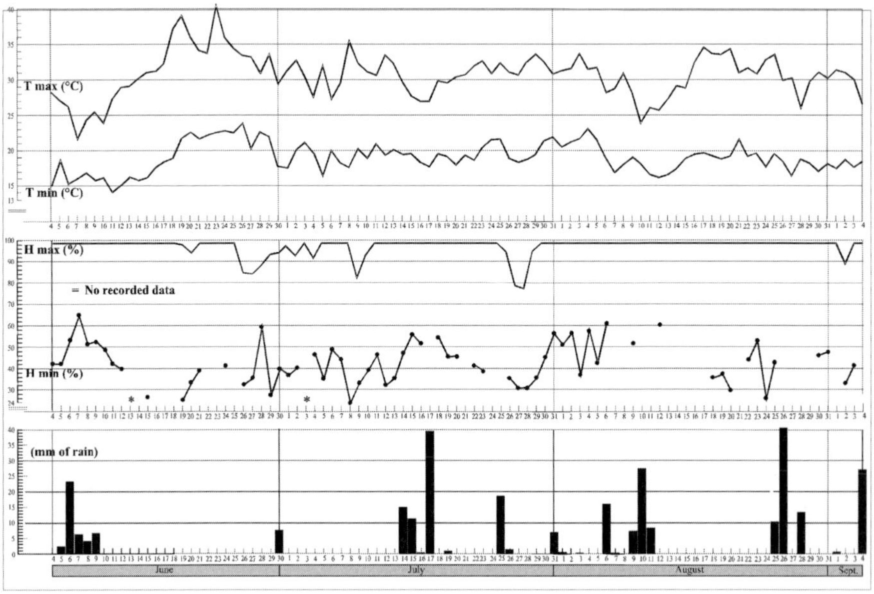


**Figure 1.** Temperatures, humidity and precipitations of the entire research period (4 June–4 September 2002).

and the high content of $Mg^{2+}$ in seawater restricted $Ca^{2+}$ uptake, enhancing, in contrast, the accumulation of P and Mn as well as of micronutrients in general [4]. Similar Ca restriction was observed in wheat roots irrigated with saline water, whereas a reduction of K occurred in shoots [5]. Previous research carried out in our lab [6–8] showed the activation of antioxidative mechanisms in sunflower seedlings treated with 10% and 20% seawater, which, although associated with different responses by shoots and roots, confirmed the adaptation of sunflower to salinity. The aim of the present research was to study the effects of 20% and 30% seawater on nutrient uptake and distribution in sunflower during the whole plant growth.

## Materials and methods

### *Plants growth and treatments*

A sandy loam soil, with neutral reaction, containing sufficient N, assimilable P and K as well as micronutrients, was air-dried and, after reduction of aggregates, distributed in 45 pots (7.5 kg each) divided in three sets of 15 pots each. In one set, the plants were irrigated with fresh water and in the other two sets the plants were irrigated with 20% or 30% seawater, respectively. Each pot was fertilized with urea, perphosphate and potash (1 g each).

**Table 1.** Chemical composition of irrigation waters (mg/l).

| Irrigation water | $Na^+$ | $K^+$ | $Ca^{2+}$ | $Mg^{2+}$ | $Cl^-$ | $HCO_3^-$ | $SO_4^{2-}$ |
|---|---|---|---|---|---|---|---|
| Seawater | 13300 | 437 | 1624 | 500 | 22950 | 158 | 2750 |
| Fresh water | 19 | 1 | 12 | 95 | 28 | 225 | 96 |
| 20% solution | 2675 | 87 | 334 | 176 | 4612 | 212 | 627 |
| 30% solution | 4005 | 131 | 496 | 217 | 6905 | 205 | 892 |

The main climatic parameters of the whole research period (4 June–4 September 2002) are reported in Figure 1. Achenes of a precocious sunflower (cv. Katharina, Piacenza ecotype), provided by Verneuil Italia, were soaked for 16 h in running water, and ten achenes were sown in each pot. After seeding, plants were irrigated with fresh water for a week and from the first leaf stage with 20% or 30% seawater, or maintained on fresh water. The solutions were added to each pot every 1–3 days, depending on the climatic conditions. Table 1 shows the chemical composition of the irrigation waters.

*Plant material and growth parameters*

Five plants from each group, in three replicates, were collected randomly, at full vegetation (I stage), floral bud appearance (II stage), seed formation (III stage), and seed full maturation (IV stage). Plant height was measured every 3 days, whereas leaf area was measured only at the first two stages, using the leaves of the fourth branch that represent an intermediate stage of plant vegetative development and are the last to fall due to the natural senescence processes.

*Nutrient determinations*

Total nitrogen determination was carried out by a micro-Kijeldhal method, after mineralization with $H_2SO_4$ at 350°C. Phosphorus and chloride quantifications were performed according to [9] and [10], respectively. Macro- and micronutrients were quantified in aliquots of plant material after wet digestion with concentrated $HNO_3$ with a 373 Perkin-Elmer atomic absorption spectrophotometer.

*Statistical analysis*

A two-way analysis of variance was applied to the data to evaluate the effect of growth, seawater and their interaction. Mean values and significance were determined by Duncan's multiple range test.

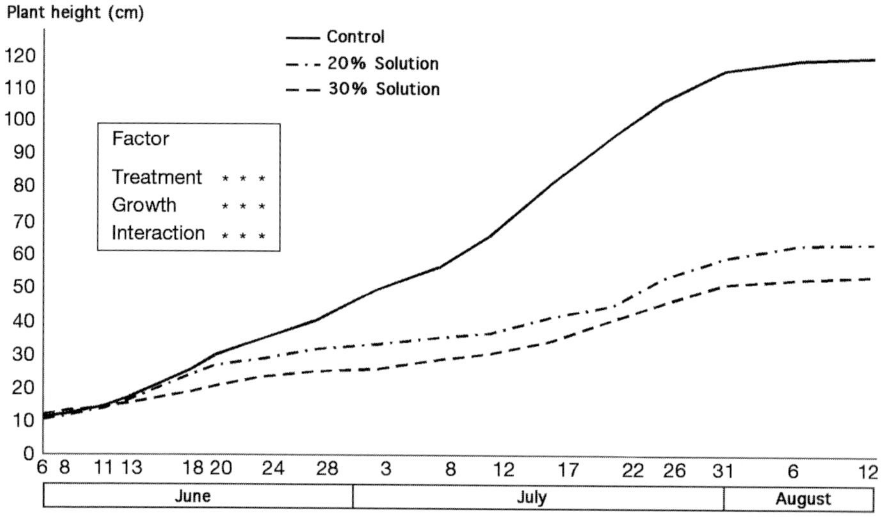


**Figure 2.** Sunflower plant height (cm) during the growth period. Significance: ns = not significant; *, ** and *** = significant for p ≤ 0.05, 0.01 and 0.001, respectively.

## Results and discussion

### *Plant height and leaf area*

One of the effects induced in plants by salt stress is the growth reduction, revealed by plant height reduction, as observed in maize plants treated with NaCl [11] and in wheat plants treated with seawater [12]. The growth of control and treated sunflower plants was similar up to 11 June, increasing thereafter with high rate in the control (more than 2.5 cm/day) and reaching about 120 cm at the end of the growth (Fig. 2). Plants irrigated with 20% or 30% seawater grew with a slower rate and reached at the end of the growth about half the height of the control plants. Leaf area reduction is also a visible index of growth reduction, as observed in tomato [2], wheat [13] and marigold [14] plants. A smaller the leaf area lowers the ability of the plant to capture light and thereby adversely affects the process of photosynthesis [15], which ultimately has a direct effect on biomass production. The leaf area in control plants increased from 40 to 53 cm$^2$, but this was significantly reduced by the two treatments, especially by 30% seawater, with the strongest reduction (more than 80%) at the first stage, independent of the treatment (Tab. 2).

This drastic reduction in leaf area and photosynthetic activity provokes a lower NADPH use in the Calvin cycle and a consequent decrease in NADP$^+$/NADPH ratio, so that the final acceptor of electrons becomes $O_2$ generating $O_2^{\cdot-}$ and other "activated" oxygen forms, potentially harmful

**Table 2.** Area ($cm^2$) of the fourth branch leaves at the first two growth stages of sunflower plants. Significance as in Figure 2.

| Treatments | I stage 4 July 2002 | II stage 18 July 2002 | Factor | |
|---|---|---|---|---|
| 0 | 40.3 | 53.0 | Treatment (TRT) | *** |
| 20% | 7.7 | 27.4 | Growth (G) | *** |
| 30% | 5.2 | 16.1 | TRT × G | ns |

for cellular components [16]. These “activated” oxygen forms might be the cause of the appearance of necrotic zones in the leaves.

## *Nutrients concentration*

In general, under stress conditions plants try to regulate ion movements into tissues to maintain growth. Seawater is a mixture of salts, which leads to more complex interactions of $Cl^-$ and $Na^+$ with the other nutrients. In treated plants, the $Cl^-$ contents increased significantly, especially with 30% seawater. In contrast to treatment with 20% seawater, the $Cl^-$ level in plants treated with 30% seawater decreased slightly with growth, but nonetheless remained at the highest levels (Tab. 3). Since salinity generally leads to an increase of all cations, the increase of $Cl^-$ during sunflower growth could be used to neutralize the excess of cationic charge, as observed in marigold [14]. At the same time, the cation increase could contribute to the osmotic adjustment to counteract secondary water stress, as this behavior is energetically less extravagant than the synthesis of organic osmolytes [17].

Seawater irrigation also resulted in a significant increase of Na content, which was higher at the first two stages with 30% seawater concentration. Salinity induces leaf necrosis in many species [18–20], so the high Na content found in sunflower irrigated with seawater could be responsible for the appearance of the rot collar and leaf necrosis, which was often followed by desiccation of the leaves and, in some cases, of the whole plant, especially with the 30% seawater treatment.

N content was significantly affected by growth, treatments and their interaction. At the first stage, N content was reduced by both seawater concentrations. Subsequently, in contrast, an increase was observed in treated plants compared with control plants. Under salt conditions, the $Cl^-/NO_3^-$ competition [21, 22] suggests that in the first phases of sunflower growth the lack of N was due to the high concentration of $Cl^-$ in the substrate that reduced $NO_3^-$ uptake, favoring also the appearance of chlorosis in the leaves of the first four branches. The leaves of the upper branches did not show any form of chlorosis, probably because of the accelerated reduction of $NO_3^-$ to $NH_4^+$ by salt stress [15], overcoming $Cl^-/NO_3^-$ competition.

**Table 3.** Nutrients content of sunflower during plant growth. Significance as in Figure 2.

| Nutrient | Treat-ment | I stage 4 July | II stage 18 July | III stage 19 August | IV stage 4 Sept. | Factor | |
|---|---|---|---|---|---|---|---|
| Cl | 0 | 15.4c | 13.4c | 12.1c | 11.6c | Treatment (TRT) | *** |
| (mg/g | 20% | 41.9b | 46.2b | 63.6ab | 62.2ab | Growth (G) | * |
| d.m.) | 30% | 76.7a | 60.5ab | 55.7ab | 53.6ab | TRT × G | *** |
| Na | 0 | 1.6d | 1.1d | 1.3d | 1.2d | Treatment (TRT) | *** |
| (mg/g | 20% | 7.7cd | 9.1bcd | 24.4a | 24.5a | Growth (G) | *** |
| d.m.) | 30% | 23.2ab | 18.6abc | 16.3abc | 20.2abc | TRT × G | *** |
| N | 0 | 37.7ab | 14.6cd | 28.7bc | 3.9d | Treatment (TRT) | *** |
| (mg/g | 20% | 28.6bc | 27.4bc | 50.7a | 7.6d | Growth (G) | *** |
| d.m.) | 30% | 27.7bc | 21.5bcd | 48.1a | 9.3d | TRT × G | *** |
| P | 0 | 1.7bc | 1.7bc | 1.2cde | 0.5e | Treatment (TRT) | ns |
| (mg/g | 20% | 1.4bcd | 2.5a | 0.8de | 0.5e | Growth (G) | *** |
| d.m.) | 30% | 1.8abc | 1.9ab | 1.3bcd | 0.5e | TRT × G | * |
| K | 0 | 36.0ab | 23.4cd | 13.6d | 12.4d | Treatment (TRT) | *** |
| (mg/g | 20% | 43.3a | 36.2ab | 25.8c | 24.5c | Growth (G) | *** |
| d.m.) | 30% | 30.9bc | 28.2bc | 25.1c | 23.9c | TRT × G | *** |
| Ca | 0 | 23.2de | 18.1efg | 16.6fg | 14.4g | Treatment (TRT) | *** |
| (mg/g | 20% | 29.6bc | 29.4bc | 26.0bcd | 22.7de | Growth (G) | *** |
| d.m.) | 30% | 37.4a | 31.5ab | 24.8cd | 20.8def | TRT × G | *** |
| Mg | 0 | 8.0a | 5.4b | 3.8b | 3.4b | Treatment (TRT) | * |
| (mg/g | 20% | 5.3b | 4.8b | 4.6b | 4.6b | Growth (G) | *** |
| d.m.) | 30% | 4.9b | 4.8b | 4.2b | 4.2b | TRT × G | *** |
| Fe | 0 | 11.7abcd | 7.0d | 15.2ab | 6.4d | Treatment (TRT) | * |
| (mg/100 g | 20% | 11.8abcd | 9.0cd | 17.2a | 14.5abc | Growth (G) | ** |
| d.m.) | 30% | 8.8cd | 9.0cd | 9.9 bcd | 9.5bcd | TRT × G | ns |
| Cu | 0 | 1.7bc | 1.2cd | 1.0d | 0.8d | Treatment (TRT) | *** |
| (mg/100 g | 20% | 2.0ab | 2.4a | 1.6bc | 1.6bc | Growth (G) | *** |
| d.m.) | 30% | 1.9ab | 2.2ab | 1.7bc | 1.9ab | TRT × G | ** |
| Zn | 0 | 4.4cde | 3.3efg | 3.2fg | 2.5g | Treatment (TRT) | *** |
| (mg/100 g | 20% | 6.4ab | 6.7a | 5.2cd | 3.7ef | Growth (G) | *** |
| d.m.) | 30% | 6.8a | 6.5ab | 5.5bc | 4.2def | TRT × G | * |
| Mn | 0 | 3.5bcdef | 2.5ef | 3.9abcde | 1.7f | Treatment (TRT) | *** |
| (mg/100 g | 20% | 4.6abcd | 4.7abc | 3.0cdef | 2.8def | Growth (G) | *** |
| d.m.) | 30% | 5.6a | 5.0a | 3.5bcdef | 3.3bcdef | TRT × G | *** |

In agreement with [4], seawater irrigation did not induce significant variations in P contents, but during growth a slight decrease occurred, showing the lowest content at the last stage. The decrease observed during growth might be due to an increase in $Cl^-$ in the soil. The interaction between growth and treatments played a role only at the lowest significance level (Tab. 3).

During the growth, K and Ca contents in the control plants decreased from 36 and 23 mg/g dry matter (d.m.) to 12 and 14 mg/g d.m., respectively. A similar decreasing trend was observed in the plants treated with both seawater concentrations, which, nonetheless, showed higher amounts of K and Ca in

**Table 4.** $K^+/Na^+$ and $Ca^{2+}/Na^+$ ratios expressed as meq/g d.m. in sunflower plants

| | $K^+/Na^+$ | | | | $Ca^{2+}/Na^+$ | | | |
|---|---|---|---|---|---|---|---|---|
| **Treat-ment** | **I stage 4 July** | **II stage 18 July** | **III stage 19 August** | **IV stage 4 Sep-tember** | **I stage 4 July** | **II stage 18 July** | **III stage 19 August** | **IV stage 4 Sep-tember** |
| 0 | 13.1 | 11.8 | 5.8 | 6.4 | 16.6 | 18.0 | 13.8 | 14.4 |
| 20% | 3.4 | 2.3 | 0.6 | 0.6 | 4.4 | 3.8 | 1.2 | 1.1 |
| 30% | 0.8 | 0.9 | 0.9 | 0.7 | 1.8 | 1.9 | 1.7 | 1.2 |

comparison to the control. In contrast, in salt-treated wheat [4, 23], maize [24], and marigold [14] a reduced uptake and translocation to leaves was observed. An increase in K in maize [11, 25] and cotton plants [26] with increasing salinity has been reported. In sunflower the interaction between growth and treatments also affected K and Ca significantly.

In spite of an increase in K content following the two treatments, a reduction of $K^+/Na^+$ ratio (Tab. 4), responsible for metabolic injuries in non-halophyte plants [11], was observed. The low $K^+/Na^+$ ratio has been correlated to a reduced photosynthetic ability of the plants under saline conditions and to inhibition of enzymatic activities. In spite of the higher uptake and translocation of Ca in the plant, a reduction of $Ca^{2+}/Na^+$ ratio occurred (Tab. 4) due to the very high increase in Na content. The reduction of this ratio, also observed in maize seedlings grown under NaCl [23], may indicate a replacement of $Ca^{2+}$ by $Na^+$, which, in turn, can result in alteration of membrane integrity, increase in $K^+$ efflux and decrease in $K^+/Na^+$ ratio [27].

Seawater treatments led to a reduction of $Mg^{2+}$ content only at the first stage, although growth and interaction between growth and treatments showed a very significant decreasing effect, in agreement with that observed in maize [23] and wheat [5] after NaCl treatments. In sunflower, the higher uptake of Ca, in addition to Na excess, likely contributed to reducing the Mg content. This reduction in seawater-treated sunflower could contribute to the chlorosis of the leaves of the first four branches and, at least in part, to growth reduction. Thereafter, no variations were observed during growth. This indicates that the plant was able to overcome the initial salt shock.

Generally, the contents of micronutrients are less affected by salinity than macronutrients and their variations have not received much attention in plants under salt conditions, so that their uptake and distribution in plants are not clear yet [28, 29]. In sunflower, Fe was less affected by treatments and growth, unless we consider the 20% seawater for which, at the last stage, Fe content was more than double in comparison with that in control plants.

Cu, Zn and Mn were strongly affected by growth and treatment, but less by the interaction of the two factors. In particular, the two seawater concentrations produced a similar and significant increase of Cu and Zn, found also in seawater-treated wheat [4]. The contents of all three cations were very low and decreased in control and treated plants during the growth, so the increase due to salinity had no significant relevance for osmotic adjustment, but may be important to increase, or at least to maintain, the efficiency of the enzymes delegated to regulate the antioxidant defenses of the plants under salt conditions [5]. Both Cu and Zn are part of chloroplastic, cytosolic and mitochondrial superoxide dismutase (SOD), while Mn is a part of mitochondrial and peroxisomal forms of the enzyme, with a pivotal role for superoxide anion elimination. The concomitant increase of Cu and Zn might be due to the observed synergism between these two ions [28] and may be correlated to the higher need of antioxidant defenses [30, 31].

In conclusion, although 20% seawater leads to some injuries on stems and leaves, Katharina Piacenza ecotype completes its cycle producing achenes, likely favored by micronutrients which increase the antioxidative defenses, so avoiding oxidative stress and by the fact that Na concentration in leaves and stems is quite similar to, or slightly higher than those of control plants.

## References


1 Reddy MP, Iyenger ERR (1999) Crop response to salt sea water application and prospects. In: M Pessarakli (ed): *Handbook of plants and crop stress*, 2nd edn. Marcel Dekker, New York, 1041–1068

2 D'Amico ML, Izzo R, Tognoni F, Pardossi A, Navari-Izzo F (2003) Application of diluted sea water to soilless culture of tomato (*Lycopersicon esculentum* Mill.): Effects on plant growth, yield, fruit quality and antioxidant capacity. *Food Agric Environ* 1: 112–116

3 Saure MC (2001) Blossom-end rot of tomato (*Lycopersicon esculentum* Mill.) – A calcium- or a stress-related disorder. *Sci Hort* 90: 193–208

4 D'Amico ML, Navari-Izzo F, Izzo R (2004) Alternative irrigation waters: Uptake of mineral nutrients by wheat plants responding to sea water application. *J Plant Nutr* 27: 1043–1059

5 Meneguzzo S, Navari-Izzo F, Izzo R (2000) NaCl effects on water relations and accumulation of mineral nutrients in shoots, roots and cell sap of wheat seedling. *J Plant Physiol* 156: 711–716

6 Di Baccio D, Navari-Izzo F, Izzo R (2000) Effetti dell'irrigazione del girasole con acqua di mare e meccanismi adattativi. *Atti del XVIII Convegno Nazionale della Società Italiana di Chimica Agraria* – Catania, 20–22 September 2000

7 Di Baccio D, Navari-Izzo F, Izzo R (2001) Meccanismi di risposta antiossidativi in girasole irrigato con acqua di mare al 20%. *Atti del XIX Convegno Nazionale della Società Italiana di Chimica Agraria* – Reggio Calabria, 25–28 September 2001

8 Di Baccio D, Navari-Izzo F, Izzo R (2004) Sea water irrigation: Antioxidant defence responses in leaves and roots of a sunflower (*Helianthus annuus*, L.) ecotype. *J Plant Physiol* 161: 1359–1366

9 Allen RJL (1940) The estimation of phosphorus. *Biochem J* 34: 858–865

10 Binzel ML, Hasegawa PM, Rhodes D, Handa S, Handa AK, Bressan RA (1987) Solute accumulation in tobacco cells adapted to NaCl. *Plant Physiol* 84: 1408–1415

11 Izzo R, Scagnozzi A, Belligno A, Navari-Izzo F (1993) Influence of NaCl treatment on Ca, K and Na interrelations in maize shoots. In: MAC Fragoso, ML van Beusichem (eds): *Optimization of Plant Nutrition*. Kluwer, Dordrecht, 557–582

12 D'Amico ML, Navari-Izzo F, Izzo R (2001) Efficienza produttiva e meccanismi di adattamento del grano irrigato con acqua di mare al 10%. *Atti del XIX Convegno Nazionale della Società Italiana di Chimica Agraria* – Reggio Calabria, 25–28 September 2001

13 Saneoka H, Shiota K, Kurban H, Chaudhary MI, Premachandra GS, Fujita K (1999) Effect of salinity on growth and solute accumulation in two wheat lines differing in salt tolerance. *Soil Sci Plant Nutr* 45: 873–880

14 Chaparzadeh N, Khavari-Nejad RA, Navari-Izzo F, Izzo R (2003) Water relations and ionic balance in *Calendula officinalis* L. under salinity conditions. *Agrochimica* XLVII: 69–79

15 Cramer GR, Alberico GJ, Schmidt C (1994) Salt tolerance is not associated with the sodium accumulation of two maize hybrids. *Aust J Plant Physiol* 21: 675–692

16 Asada K, Takahashi M (1987) Production and scavenging of active oxygen in photosynthesis. In: DJ Kyle, CB Osmond, CJ Arntzen (eds): *Photohinibition*. Elsevier, Amsterdam, 227–287

17 Munns R (2002) Comparative physiology of salt and water stress. *Plant Cell Environ* 25: 239–250

18 Hernandez J, Zerega L, Ordosgoitti A (2000) Necrosis del apice y del borde de las hojas y retraso en el crecimiento de la cana de azucar (*Saccharum* sp. *Hibrido*) en el Bajo Yaracuy, Venezuela. *Rev Fac Agron Univ Zulia* 17: 226–238

19 El-Siddig K, Ludders P (1994) N-nutrition and salinity tolerance in apple trees. *Angew Bot Berichte* 5: 184–188

20 Wang-Rui G, Chen-Shao L, Liu-Li Y, Hao-Zhi Y, Weng-Hai J, Li-He J, Yang-Shiang H, Duan-Shan L (2005) Genotypic differences in antioxidative ability and salt tolerance of three poplars under salt stress. *J Beijing Forestry Univ* 27: 46–52

21 Schroppel-Meier G, Kaiser WM (1988) Ion homeostasis in chloroplasts under salinity and mineral deficiency. I. Solute concentrations in leaves and chloroplasts from spinach plants under NaCl or $NaNO_3$ salinity. *Plant Physiol* 87: 822–827

22 Walker RR, Blackmore DH, Qing S (1993) Carbon dioxide assimilation and foliar ion concentrations in leaves of lemon (*Citrus lemon* L.) trees irrigated with NaCl or $Na_2SO_4$. *Aust J Plant Physiol* 20: 173–185

23 Izzo R, Navari-Izzo F, Quartacci MF (1991) Growth and mineral adsorption in maize seedlings as affected by increasing NaCl concentrations. *J Plant Nutr* 14: 687–699

24 Botella MA, Martinez V, Pardines J, Cerdà A (1997) Salinity induced potassium deficiency in maize plants. *J Plant Physiol* 150: 200–205

25 Izzo R, Belligno A, Muratore G, Navari-Izzo F (1996) Seedlings growth and $Ca^{2+}$, $K^+$ and $Na^+$ accumulation in maize roots as affected by NaCl. *Agrochimica* 40: 25–32

26 Brugnoli E, Björkman O (1992) Growth of cotton under continuous salinity stress: Influence on allocation pattern, stomatal and non-stomatal components of photosynthesis and dissipation of excess light energy. *Planta* 187: 335–347

27 Cramer GR, Epstein E, Läuchli A (1989) Na-Ca interactions in barley seedlings: Relationship to ion transport and growth. *Plant Cell Environ* 12: 551–558

28 Luo Y, Rimmer DL (1995) Zinc-copper interaction affecting plant growth on a metal-contaminated soil. *Environ Pollut* 88: 79–93

29 Cornillon P, Palloix A (1997) Influence of sodium chloride on the growth and mineral nutrition of pepper cultivars. *J Plant Nutr* 20: 1085–1094

30 Meneguzzo S, Navari-Izzo F, Izzo R (1998) Stromal and thylakoid-bound ascorbate peroxidases in NaCl-treated wheat. *Physiol Plant* 104: 735–740

31 Meneguzzo S, Navari-Izzo F, Izzo R (1999) Antioxidative responses of shoots and roots of wheat to increasing NaCl concentrations. *J Plant Physiol* 155: 274–280

Biosaline Agriculture and High Salinity Tolerance
Edited by Chedly Abdelly, Münir Öztürk, Muhammad Ashraf and Claude Grignon


# Diversity in the response of two potential halophytes (*Batis maritima* and *Crithmum maritimum*) to salt stress

Karim Ben Hamed[1], Dorsaf Messedi[1], Annamaria Ranieri[2] and Chedly Abdelly[1]

[1] *Laboratory of Plant Adaptation to Abiotic Stresses, Center for Biotechnology in Borj Cedria Ecopark. BP 901, Hammam Lif 2050, Tunisia*

[2] *Department of Agricultural Chemistry and Biotechnology, University of Pisa, I-56124 Pisa, Italy*

## Abstract

In this study, we compared the response to NaCl of *Batis maritima* and *Crithmum maritimum*, two potential halophytes with a different range of salinity tolerance. At high NaCl concentrations (800 mM for *B. maritima* and 300 mM for *C. maritimum*), the growth of both plants was significantly reduced. A split root experiment aimed at determining whether high NaCl conditions limit growth of plants through toxic effects of excessive salt accumulation in shoots or through impairment of some essential nutrient acquisition. The split root experiment was performed with three treatments. In the first treatment (B/S), half of the roots were immersed in a basal medium (B) and the other half in the same medium supplemented with NaCl (S). In the two other treatments, the two halves of the root system were immersed either in salt-free medium (B/B) or in the basal medium containing salt (S/S). Under split-root conditions, *B. maritima* and *C. maritimum* accumulated Na in their shoots, and displayed improved growth as compared to control plants. In *C. maritimum*, the B/S treatment partially restored K provision to the shoots but not that of Ca, suggesting that the inhibition of $K^+$ uptake by salt could only limit its growth under high salinity. In *B. maritima* (B/S plants), the concentration of $K^+$ and $Ca^{2+}$ were diluted by growth. The inhibition of $K^+$ and $Ca^{2+}$ uptake by salt did not seem to limit growth of *B. maritima* growth under high salinity. The growth of *B. maritima* and *C. maritimum* could be also limited by the restriction imposed by NaCl on N uptake.

## Need for halophytes as experimental plants

One of the least successful areas of crop science has been the search for salt-tolerant cultivars [1]. Traditionally, studies directed towards understanding salt tolerance have investigated the responses of glycophytes to salt stress [2]. A typical approach has included a comparison of growth and physiological parameters (ion, content, photosynthesis, respiration, water relations) in glycophytes grown with or without salt stress, looking for responses to added

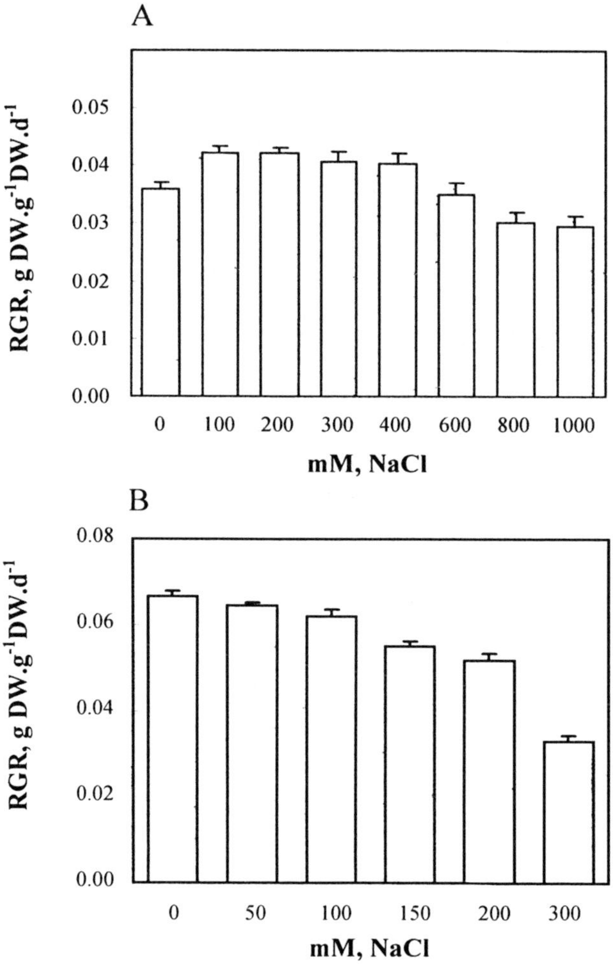


**Figure 1.** Effect of NaCl on the relative growth rate (RGR) in (A) *Crithmum maritimum* and (B) *Batis maritima*. RGR = M/M̲ Δt, where Δ is the difference between the values at the final and initial harvests, t the salt treatment duration (days), and M̲ the logarithmic mean of M, the whole plant dry weight (DW) is given in g; M̲ = ΔM/Δ ln M. RGR (g $DW.g^{-1}DW.d^{-1}$).

salinity [3]. The difficulty with such an approach is that it is not possible to distinguish those responses that are truly adaptive from those that are results of metabolic lesions or other types of damage. In many halophytes, the optimum salinity for growth is shifted to levels of salt at which most plants and all crop plants experience severe reduction in growth and yield [4]. As a result, halophytes provide the opportunity to compare growth and physiological parameters at low and high levels of salinity, without the confounding effects of $Na^+$ toxicity.

## Diversity in the response of halophytes to NaCl constraint: Case study of *Crithmum maritimum* and *Batis maritima*

In most plants, growth is gradually reduced as salinity increases above the threshold level of salinity tolerance (defined as the salt concentration leading to a growth depression of 50% [5]), which varies in different species [6]. An exception to this generalization can be found in some halophytic succulent plants in which growth is stimulated by low to moderate salinities [7]. For *Batis maritima*, a succulent halophytic shrub commonly habituating salt marshes and salt flats, the addition of 100–400 mM NaCl to the growth medium increased growth activity (estimated by the relative growth rate, RGR), which was insensitive to the addition of 600 mM NaCl. The RGR decreased only at higher NaCl concentrations (800 and 1000 mM) by 36% of the control RGR (Fig. 1A). The response to salinity displayed by *Crithmum maritimum*, a succulent halophyte thriving on salty cliffs, differed from that of *B. maritima*. The growth activity was decreased progressively as NaCl levels were raised in the medium from 100 to 300 mM (Fig. 1B). The range of salinity tolerance in *Crithmum* (50–300 mM) was less broad than that in *Batis* (100–1000 mM).

These results showed that *Batis* is an obligatory halophyte as it requires NaCl to express maximum growth capacities, while *C. maritimum* behaved like a facultative halophyte. At optimal salinities, the relative growth rate of plants of *B. maritima* was approximately 0.04/day over 50 days of NaCl treatment, which was slower than that (0.07/day) in *C. maritimum* grown for the same duration of salt treatment, and much slower than that (0.14/day) in *Atriplex halimus* seedlings for 35 d, 0.1–0.28/day in annual halophyte *Suaeda maritima* grown for 28–56 days [8] and 0.11/day in *Cakile maritima* for 42 days [9]. These comparisons with other halophytic species show that *B. maritima* is a slow growing species with a high degree of NaCl tolerance.

## Growth restriction under severe salinity conditions

### *Major constraints responsible for growth restriction in saline environments*

It is generally recognized that three adverse effects could simultaneously be responsible for growth reduction in plants, the osmotic, the toxic and the nutritional effects. However, the relative contribution of these three major constraints to growth inhibition at high substrate salinity is difficult to assess [10]. Munns [11] suggested a hypothesis of a biphasic model in which the external osmotic potential could be the main growth inhibitory factor in the first days to weeks of growth. In the vegetative growth stage accumulation and/or specific ion effects are increasingly important in the leaves. Other reviews stress the nutritional effects of salinity [12] as the immediate effect of excess Na on the Ca homeostasis of root and leaf cells [13]. Rengel [13] suggested that the Na-related changes of the normal pattern of Ca fluxes at the

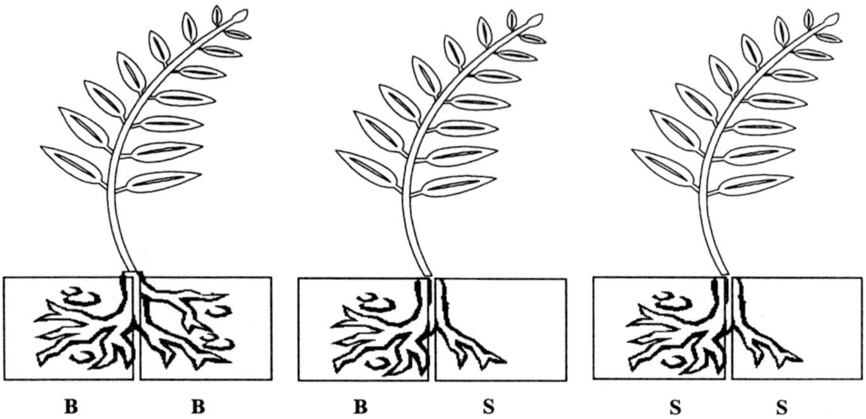


**Figure 2.** A schematic representation of the three treatments used in the split root experiment. B, Basal medium; S, basal medium supplied with NaCl. For each treatment, five replicates corresponding to single plants were used for both species.

plasma membrane is the primary signal of salt stress perceived by roots and translated into osmotic changes.

### *Split root system: To assess the contribution of salt toxicity and nutritional disruption to growth inhibition*

The split root experiment aimed at determining whether high NaCl conditions limit growth of plants through toxic effects of excessive salt accumulation in shoots or through impairment of some essential nutrient acquisition.

After rooting in liquid environment, plants were transferred individually in pots filled with appropriate nutrient solution. The split root experiment was performed with three treatments (Fig. 2). In the first treatment, one half of the roots were immersed in a basal medium (B) and the other half in the same medium supplemented with NaCl (B/S) (S refers to 300 mM for *C. maritimum* and 800 mM NaCl for *B. maritima*). In the two other treatments, the two halves of the root system were immersed either in salt-free medium (B/B) or in the basal medium containing salt (S/S).

### *Nutritional disruption as a major constraint in reducing the growth of halophytes*

Salt in the growth medium imposes both ionic and osmotic stress on plants [14] and disturbance of nutrient uptake [15]. The first effect appears as leaf necrosis (from the accumulation of sodium in leaf tissues) and as a significant

**Figure 3.** *C. maritimum* and *B. maritima* under split root system. B, Basal medium, S1, basal medium supplied with 300 mM NaCl; S2, basal medium supplied with 800 mM NaCl.

reduction in biomass production [11]. This effect was not found in *C. maritimum* and *B. maritima* even in plants having their whole root system in salty medium (S/S) (Fig. 3). The absence of necrosis confirms the efficiency of $Na^+$ compartmentalization systems in the leaf cells. The ability of *C. maritimum* and *B. maritima* to accumulate $Na^+$ inside the cells was also suggested by the absence of leaf dehydration. At higher salinities (300 and 800 mM NaCl for *C. maritimum* and *B. maritima*, respectively), the mean $Na^+$ accumulation in the leaves of *C. maritimum* amounted to 460 mM and in *B. maritima* 1.0 M. The maintenance of leaf hydration (14 and 6.5 ml $H_2O$ $g^{-1}$ dry weight (DW) in *B. maritima* and *C. maritimum*, respectively) in spite of such a high $Na^+$ accumulation (Fig. 4) indicates that much of $Na^+$ in leaves was osmotically active.

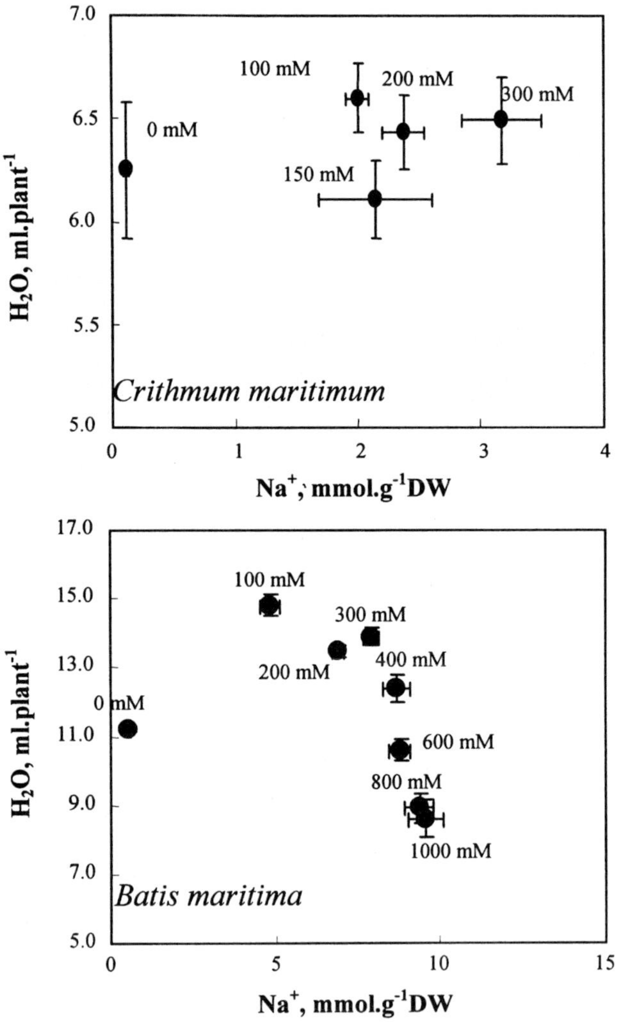


**Figure 4.** Relationship between water content and $Na^+$ content in shoots. The contents were measured at the end of the salt treatment.

Despite the high accumulation of $Na^+$ in their shoots, split root plants (B/S) displayed improved growth (more significant in *B. maritima* than in *C. maritimum*) as compared to control plants (B/B) (Fig. 5). Our results show that large accumulation of salt in the shoots of *B. maritima* and less in *C. maritimum* is compatible with a high growth rate when a part of the roots is maintained in a salt-free medium. For both halophytes, the detrimental effect of salt treatment on growth seems to correspond to a nutritional disruption rather than a toxic effect of the salt in shoots.

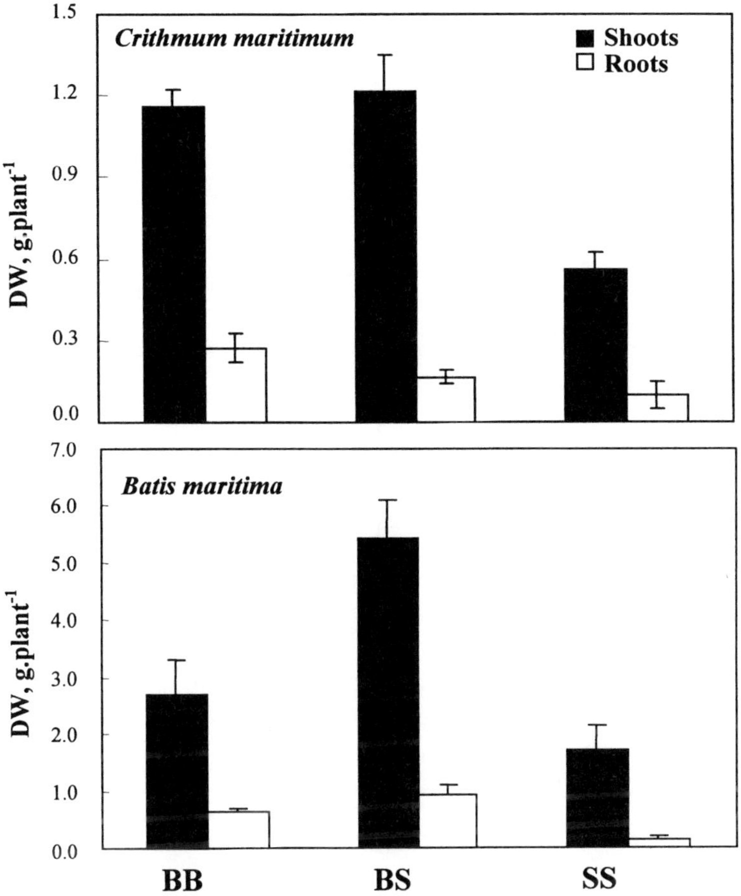


**Figure 5.** Biomass of *C. maritimum* and *B. maritima* plants grown for 50 days in a split root system. B, Basal nutrient solution; S, basal nutrient solution supplied with NaCl (300 and 800 mM, for *C. maritimum* and *B. maritima*, respectively).

For *C. maritimum*, S/S plants clearly suffered from potassium shortage, as indicated by the low $K^+$ concentration in their shoots (Fig. 6A). Probably, the salt in the medium inhibited $K^+$ absorption by roots as already shown in *Sesuvium portulacastrum* [15]. The B/S treatment partially restored $K^+$ provision to the shoots, as indicated by the high values of $K^+$ amounts and concentrations in theses organs (Fig. 6). Thus, it is reasonable to hypothesize that B roots were able to transport $K^+$ in B/S plants. As compared to the S/S treatment, B/S treatment led to large calcium amounts that were not associated to higher concentration in the tissues (Fig. 6). Therefore, the low calcium concentration in B/S plants resulted from dilution of $Ca^{2+}$ by growth rather than from restriction of Ca transport to the shoots.

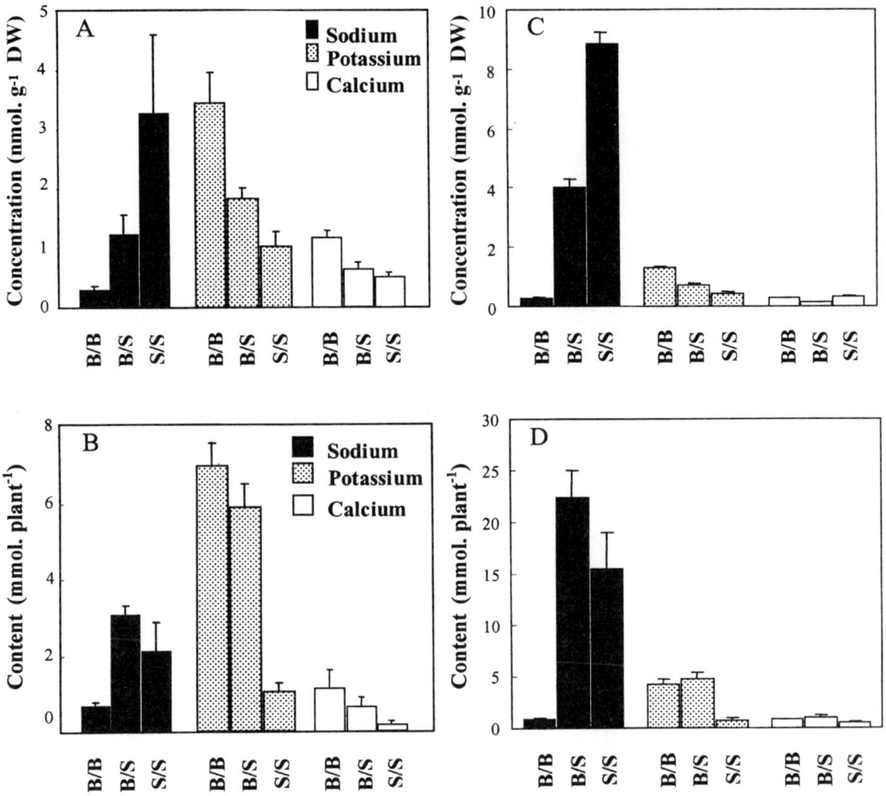


**Figure 6.** Concentrations and contents of Na, K and Ca in the shoots of *C. maritimum* (A, B) and *B. maritima* (C, D).

From these results, it seems that the inhibition of $K^+$ uptake by salt, only limits *C. maritimum* growth under high salinity. In *B. maritima* (B/S plants), the concentration of $K^+$ and $Ca^{2+}$ were diluted by growth, which suggest that the transport of these two nutrients to shoots was not restricted (Fig. 6). The inhibition of $K^+$ and $Ca^{2+}$ uptake by salt did not seem to limit growth of *B. maritima* growth under high salinity.

For *B. maritima* and *C. maritimum*, it can be suggested that the restriction of $NO_3^-$ uptake and transport to the aerial parts is responsible for the growth inhibition under excessive salinity conditions.

## Summary

*B. maritima* and *C. maritimum* are perennial halophytes with potential utility for the valorization and the rehabilitation of saline soils. Both plants are investigated in different NaCl concentrations (0–800 mM NaCl for *B. maritima* and 0–300 mM NaCl for *C. maritimum*) to compare their response to salt stress. Our results show that both species are able to produce biomass under all NaCl concentrations. *B. maritima* expressed maximal growth up to 400 mM NaCl, whereas *C. maritimum* seemed to be less tolerant to salt treatment. At higher NaCl concentrations (800 mM for *B. maritima* and 300 mM for *C. maritimum*), the growth of both plants is reduced. To determine the factors limiting the plant growth in the presence of salt excess, an sutdy using a split-root system was carried out. Our results show that large accumulation of sodium in the shoots of *B. maritima* and less in *C. maritimum* is compatible with a high growth rate when a part of the roots is maintained in a salt-free medium. For both halophytes, the detrimental effect of salt treatment on growth seems to correspond to nutritional disturbances rather than toxic effects of sodium in the shoots. The growth of *C. maritimum* could be limited at high salinity by the restrictions imposed by NaCl on K and N uptake. For *B. maritima*, the restriction of N uptake by shoots may be the major factor responsible for the growth inhibition at high NaCl concentration.

## Future research

The next step of this research will be directed to the comparison of the effect of salt on $K^+$, $Ca^{2+}$ and N acquisition by plants and their corresponding effects. Thereby, it should be possible to determine which nutrient(s) is the limiting factor causing growth reduction and the capacities of these species to accumulate NaCl in their aerial parts. On a larger scale, the addition of such nutrients to salty soils should enhance the potential of halophytes to extract salt and desalinize theses soils, which could constitute the fundament of new desalination technologies.

## References


1 Kerstiens G, Tych W, Robinson MF, Mansfield TA (2002) Sodium-related partial stomatal closure and salt tolerance of *Aster tripolium*. *New Phytol* 153: 509–515

2 Parks GE, Dietrich MA, Schumaker KS (2002) Increased vacuolar $Na^+/H^+$ exchange activity in *Salicornia bigelovii* Torr. in response to NaCl. *J Exp Bot* 53: 1055–1065

3 Balasuramanian R (2004) Investigation of salt tolerance mechanisms in the halophytes *Aster tripolium* L. and *Sesuvium portulacastrum* L. through physiological, biochemical, and molecular methods. Ph D thesis, Institut für Botanik, Universität Hannover

4 Greenway H, Munns R (1980) Mechanisms of salt tolerance in nonhalophytes. *Annu Rev Plant Physiol* 31: 149–190

5 Koyro HW (2003) Study of potential cash crop halophytes by a quick check system. In: Lieth H, Mochtchenko M (eds): *Cash crop halophytes: Recent studies*. Kluwer, Dordrecht, 5–17

6 Maas EV (1987) Salt tolerance of plants. In: Christie BR (ed): *CRC handbook of plant sciences in agriculture*, vol II. CRC Press, Boca Raton, 57–75

7 Gorham J (1996) Mechanisms of salt tolerance of halophytes. In: Choukrallah R, Malcolm CV, Hamdy A (eds): *Halophytes and biosaline agriculture*. Marcel Dekker, New York, 31–53
8 Short DC, Colmer TD (1999) Salt tolerance in the halophyte *Halosarcia pergranulata* subsp. *Pergranulata*. *Ann Bot* 83: 207–213
9 Debez A, Ben Hamed K, Chibani F, Abdelly C (2003) Some physiological and biochemical aspects of salt tolerance in two oleaginous halophytes: *Cakile maritima* and *Crithmum maritimum*. In: Lieth H, Mochtchenko M (eds): *Cash crop halophytes: Recent studies.* Kluwer, Dordrecht, 31–39
10 Marschner H (1995) *Mineral nutrition of higher plants*, 2nd edn. Academic Press, London
11 Munns R (2002) Comparative physiology of salt and water stress. *Plant Cell Environ* 25: 239–250
12 Grattan SR, Grieve CM (1999) Salinity-mineral nutrient relation in horticultural crop. *Sci Hort* 78: 127–157
13 Rengel Z (1992) The role of calcium in salt toxicity. *Plant Cell Environ* 15: 625–632
14 Tester M, Davenport R (2003) $Na^+$ tolerance and $Na^+$ transport in higher plants. *Ann Bot* 91: 503–527
15 Messedi D, Labidi N, Grignon C, Abdelly C (2004) Limits imposed by salt to the growth of the halophyte *Sesuvium portulacastrum*. *J Plant Nutr Soil Sci* 167: 720–725

Biosaline Agriculture and High Salinity Tolerance
Edited by Chedly Abdelly, Münir Öztürk, Muhammad Ashraf and Claude Grignon


# Differential effects of sodium salts on the germination of a native halophytic species from South America: *Prosopis strombulifera* (Lam.) Benth

Virginia M. Luna[1], Analía S. Llanes[1], Laura R. Sosa[2], Mariana A. Reginato[1] and Herminda E. Reinoso[3]

[1] *Laboratorio de Fisiología Vegetal, Departamento de Ciencias Naturales, Universidad Nacional de Río Cuarto, 5800 Río Cuarto, Argentina*
[2] *Laboratorio de Fisiología Vegetal, Facultad de Química, Bioquímica y Farmacia, Universidad Nacional de San Luis, 5600, San Luis, Argentina*
[3] *Laboratorio de Morfología Vegetal, Departamento de Ciencias Naturales, Universidad Nacional de Río Cuarto, 5800, Río Cuarto, Argentina*

## Abstract

*Prosopis strombulifera* is a halophytic shrub frequently found in the salinized areas of central Argentina. Interactions between temperature, ionic and osmotic components of salinity, and seed germination in this species are discussed in this chapter. Besides the osmotic effect, specific ion effects of salts play an important role in seed germination causing toxicity to the embryo. In saline soils where *P. strombulifera* is frequent, NaCl and $Na_2SO_4$ proportions are similar. Germination experiments with both salts, their iso-osmotic anionic and cationic mixtures and polyethylene glycol (PEG) were performed at 30°C and 35°C; the germination percentages registered with PEG were lower than those obtained with iso-osmotic Na-based monosaline solutions at osmotic potential ($\psi o$) of –1.2 MPa and lower, but greater than those in the salt mixtures, indicating that seeds were mainly affected by an osmotic effect rather than by ionic toxicity at 35ºC. The salt mixture accentuated ion toxicity showing that germination is inhibited by a combination of osmotic and ionic effects, the latter having greater influence at very high salt concentrations. The excess of $Cl^-$ or $SO_4^{2-}$ anions in both cationic mixtures produced equal magnitude of toxicity on the seeds. Although a deleterious effect of potassium was also observed, the anionic effects were evidently much more marked. From $\psi o$ of –1.2 MPa and lower, germination inhibition increased when salt concentration increased as the ionic effects were additive to osmotic effects. The germination percentages obtained with monosaline solutions at 35°C were superior to those obtained at 30°C, indicating that temperature played an important role in the germination response of this species by diminishing the osmotic effect of salt only in the case of monosaline solutions; however, the toxic effect of ions was accentuated when they were combined. Nevertheless, a partial reversion of sulfate toxicity was observed when seeds were placed in anionic salt mixtures at 30°C, demonstrating the differential effects of temperature on the osmotic and ionic components of salinity.

## Introduction

While all major crops, as well as most wild species, are glycophytes, i.e., sensitive to relatively low salt concentrations, there are also plants naturally adapted to conditions with high salinity in the soil. These plants, known as halophytes, include a large taxonomic variety and occupy diverse habitats, from extremely dry to temporarily waterlogged sites or salt marshes.

## Seed germination in saline environments

For the successful establishment of plants in saline environments, seeds must remain viable at high salinity in an imposed secondary dormancy and germinate when salinity decreases [1]. Salinity causes harmful effects on plant physiology that affect germination and seedling establishment, among others, by avoiding the water entrance to the embryo due to an external osmotic potential, and/or by death of the embryo caused by ion toxicity [2].

Many researchers have reported a decrease in germination percentage and seedling growth in high salinity [3–5]. Nevertheless, different species differ in their sensitivity to salt and salt tolerance [6, 7]; in the various stages of seedling development, organs, tissues and cells show a different degree of tolerance to this environmental condition [8, 9].

Very different salt tolerance limits have been reported among the halophytic species. For example, *Sporobolus virginicu* [10], *Halopyrum mucronatum* [11] and *Briza maxima* [12] germinate in NaCl concentrations up to 350 mM. *Urochondra setulosa* [13], *Aeluropus legopoides* [2] and *Puccinella festucaeformis* [14], among others, are able to germinate at 500 mM NaCl; *Salicornia bigelovii* germinates well at 856 mM NaCl [15], and others can still germinate in 1000 mM NaCl, e.g., *Cressa crefica* [16] and *Arthrocnemun indicum* [17]. It has been reported that inorganic ions were not more inhibitory than mannitol and polyethylene glycol (PEG) in several halophytes, indicating that seeds are mainly affected by osmotic stress rather than specific ion toxicities [18]. The same author stated that the distribution of halophytes in the United States indicates that there are species tolerant to a wide range of combinations of anions such as $Cl^-$, $SO_4^{2-}$ and $CO_3^{2-}$ and cations such as $Na^+$, $K^+$ and $Mg^{2+}$ [19], suggesting that for these species, specific ionic effects would have less influence on seed germination than the soil water potential. More recent results [20] on different sodium and potassium salts effects on *Atriplex prostrata* germination reveal that inhibition of seed germination and early growth was primarily due to an osmotic effect and not to a specific ion toxicity of either chloride or sulfate salts. Similarly, when comparing the effect of NaCl, $MgSO_4$ and $Na_2SO_4$ on germination and early growth of *Puccinella festucaeformis*, salt concentration had greater effect than the type of salt [21]. No differences in germination percentage were reported when seeds were exposed to different osmotic agents as in the case of *Sarcobatus vermiculatus* treated with NaCl, KCl and PEG [22], and *Securigera securidaca* treated with $Na_2SO_4$ and NaCl [23].

Nevertheless, some studies indicate that besides the osmotic effect, specific ion effects of different salts play an important role in germination of various species causing toxicity to the embryo [24–26]. Other have proposed that salinity effect on germination is fundamentally due to specific ions [27].

The relative importance of osmotic effects *versus* ionic effects on seed germination are, therefore, still a matter of discussion, and probably depends on the species under study.

## *Prosopis strombulifera*, a halophyte widely distributed in South America

The genus *Prosopis* includes several economically important species commonly found in saline zones in America [28]. Some of them are fundamental shrub-like members of native communities such as *P. strombulifera*, a spiny shrub frequently found in the salinized areas of Córdoba and San Luis (central area of Argentina). Previous results with *P. strombulifera* seedlings hydroponically grown under increasing NaCl concentrations showed that this is a salt-excluder halophytic species that shows stem and root growth stimulation up to 500 mM in 80% of treated seedlings in relation to non-treated plants [29]. This response is different to that obtained in other *Prosopis* species [30–32] in which growth was reduced in response to increasing NaCl salinity [32–34]. In saline soils where *P. strombulifera* is frequent, NaCl and $Na_2SO_4$ proportions are generally similar, although in some soil samples $Na_2SO_4$ is more abundant [35].

There have been only few recent studies on the effects of $Na_2SO_4$ on germination and plant growth. Most salinity studies have been performed using monosaline solutions, thus limiting the possibility of extrapolating the results to field conditions. Potassium toxicity has seldom been reported, probably due to the fact that $K^+$ is one of the main macronutrients essential for plant growth and metabolism [36]. There has been increasing interest in the comparison of NaCl effects on germination and seedling growth with other salts usually present in the soil, to determine if the observed inhibitory results are due to an osmotic effect or ion toxicity.

Previous results with *P. strombulifera* seeds indicated that optimum temperature limits for germination are 25° to 35°C [35]. Recently, it has been demonstrated that the germination percentage of this species was influenced not only by the salt concentration but also by the nature of the ions in the saline solutions. The germination percentage diminished when salinity increased, being substantially inhibited from osmotic potential ($\psi o$) of –1.9 MPa and lower, with no germination at $\psi o$ of –2.2 MPa [37, 38]. Other authors observed that *Prosopis alba*, *P. chilensis* and *P. tamarugo* reached a germination percentage similar to that of controls at $\psi o$ of –0.4 and –0.8 MPa NaCl [39].

*P. strombulifera* seeds showed a final germination percentage similar to that of controls when they were exposed to ψo of –0.4, –0.8 MPa NaCl, $Na_2SO_4$ monosaline solutions, NaCl + $Na_2SO_4$ bisaline solutions or PEG at 35°C [38] (Fig. 1). These results, together with those obtained by other authors [2, 10–12], suggest that a ψo of –1.2 MPa marks a limit from which germination is markedly affected in most halophytes. The germination percentage registered in *P. strombulifera* in the presence of PEG was lower than that obtained with iso-osmotic Na-based monosaline solutions at a ψo of –1.2 MPa and lower, but greater than that in the salt mixtures, which could indicate that seeds are mainly affected by an osmotic effect rather than by ionic toxicity when they are in Na-based monosaline solutions at 35°C [38]. The entrance of inorganic ions from monosaline solutions into the seeds could alleviate the inhibiting osmotic effect of the external medium by diminishing the internal osmotic potential, which cannot occur in the presence of PEG [40]. From a ψo of –1.2 MPa or lower, the germination percentage in Na- and K-based sulfate monosaline solutions decreased in comparison with those obtained with Na- or K-based chloride monosaline solutions, showing $SO_4^{2-}$ anion toxicity for this species that was accentuated at lower potentials. These results agree with other reports on the effect of NaCl, $CaCl_2$ and $Na_2SO_4$ at s ψo of –0.3 to –1.5 MPa on the germination of *P. juliflora* [41] where the germination percentage was more affected by $Na_2SO_4$ than NaCl from ψo of –0.9 MPa or lower. This response reflects germination dependency on the $Na^+$-associated anion. A similar response was also observed in *Hordeum vulgare* [42] and *Medicago sativa* [43], where $Na_2SO_4$ was more inhibitory than NaCl.

In the presence of KCl, mainly when experiments were carried out at 30°C, a lower germination percentage was registered with *P. strombulifera* seeds in comparison with NaCl treatment, showing a greater $K^+$ than $Na^+$ toxicity (Fig. 2). This is in agreement with the data obtained for other halophytes where $K^+$ toxicity was explained to be due to its incapability of replacing $Na^+$ for osmotic adjustment [36, 44, 45]. In contrast, it was reported that in *A. prostrata* the germination percentage was more inhibited by NaCl and $Na_2SO_4$ in comparison with KCl and $K_2SO_4$ at a ψo of –1.5 MPa. [46]. In the experiments with *P. strombulifera*, Na- or K-based sulfate monosaline solutions behaved similarly to each other, but germination percentage was lower than that obtained with Na- or K-based chloride monosaline solutions throughout the whole experiment, indicating again a greater toxicity of $SO_4^{2-}$ anions, which delayed and lowered germination (Figs 1 and 2). In addition, experiments performed at 35°C (Fig. 1) showed that the toxic effect exerted by $SO_4^{2-}$ anion on germination was independent of the accompanying cation, whereas, in the case of $Cl^-$ anion, differences were observed according to the accompanying cation. In the presence of bisaline solutions the germination percentage was lower than that in monosaline solutions, being very low at ψo of –1.5 MPa and almost null at ψo of –1.9 MPa [38]. Thus, the salt mixture accentuated ion toxicity in these experiments, in agreement with the concept that germination is inhibited by a combination of osmotic and ionic effects, the latter having a

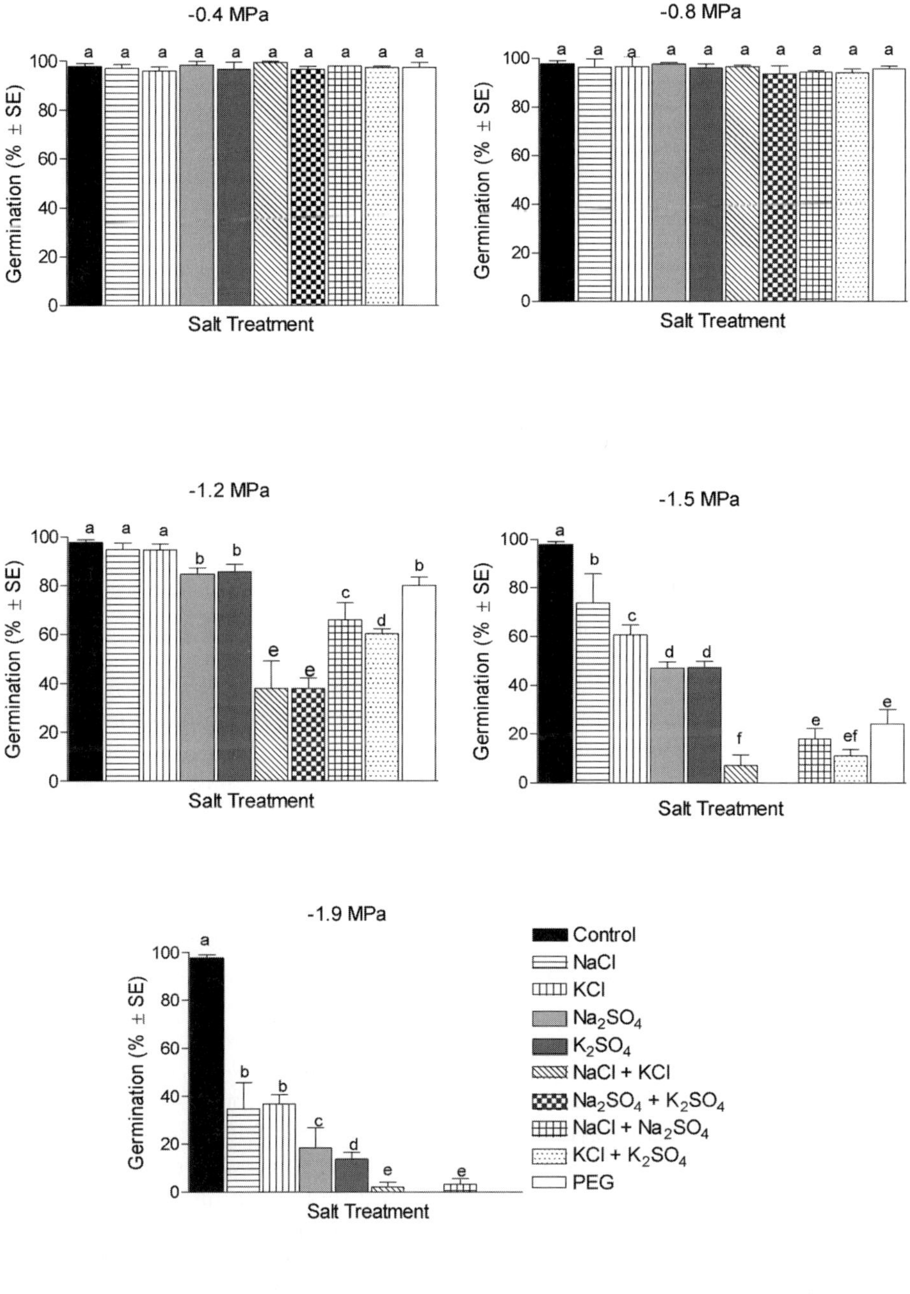


**Figure 1.** Percentage germination (mean ± SE) of *P. strombulifera* seeds at ψo –0.4, –0.8, –1.2, –1.5 and –1.9 MPa with monosaline and bisaline solutions and PEG at 35 ± 1°C after 168 h of treatment. Values with the same letter are not significantly different at $p > 0.05$. Further details of the experiments can be found in Llanes et al. [38].

greater influence at very high salt concentrations [47]. Since using either cationic bisaline solutions ($Na_2SO_4$ + $K_2SO_4$ and NaCl + KCl) or anionic bisaline solutions ($Na_2SO_4$ + NaCl and $K_2SO_4$ + KCl) yielded similar results (final germination percentage at ψo of –1.2 MPa was lower than that registered in the presence of $Cl^-$ and $SO_4^{2-}$ monosaline solutions in both cases), in this species the excess of $Cl^-$ or $SO_4^{2-}$ anions in both cationic mixtures produced equal magnitude of toxicity on the seeds. Although a deleterious effect of potassium on *P. strombulifera* seeds was also observed, the anionic effects were evidently much more marked (Fig. 1) [38]. Thus, in these experiments, from a ψo of –1.2 MPa and lower, germination inhibition increased when salt concentration increased due to ionic effects that were additive to osmotic effects.

On the other hand, a partial reversion of sulfate toxicity was reported when *P. strombulifera* seeds were placed in the presence of anionic salt mixtures at 30°C [37].

Further evidence indicates that the salt effect in monosaline solutions can be considerably alleviated in nature by synergistic interactions among them. For example, *Securigera securidova* seeds placed to germinate in $Na_2SO_4$ + $MgSO_4$ and NaCl + $MgSO_4$ showed an increase in their germination tolerance limit to both bisaline solutions in relation to monosaline solutions [48]. Similar improvements of monosaline solutions effects were obtained with $MgCl_2$ + NaCl and $CaCl_2$ + $MgCl_2$ [19].

Comparing both experiments with *P. strombulifera*, one performed at 30°C (Fig. 2) [37] and the other at 35°C (Fig. 1) [38], the germination percentage obtained with monosaline solutions at 35°C were quite superior to those obtained at 30°C. This would indicate that temperature played an important role in the germination response of this species by diminishing the osmotic effect of salt. However, when salt mixtures were used, the higher temperature did not favor germination. This is because, by increasing the water diffusion capacity in the solution, a higher temperature could alleviate the osmotic effect only in the case of monosaline solutions, but the toxic effect of ions was accentuated when they were combined, leading to a greater inhibition of germination. These results confirm previous reports on the key role that temperature plays in salinity responses. In fact, it is known that temperature influences the germination response in many halophytes [13, 49], affecting a great number of biochemical processes including membrane-bound protein activities and cytosolic enzymes [11, 50].

In summary, here we have discussed the interactions between temperature, ionic and osmotic components of salinity, and seed germination in the halophyte *P. strombulifera*. The diversity of results obtained by different authors trying to explain temperature-salinity interactions may be due not only to the intrinsic characteristics of each species, but also to the differential effects of temperature on the osmotic and ionic components of salinity. In addition, ionic effects depend on the chemical nature of the ions involved, the interactions of which may change from synergistic to antagonistic or *vice versa*. We have demonstrated that the germination response is different when the

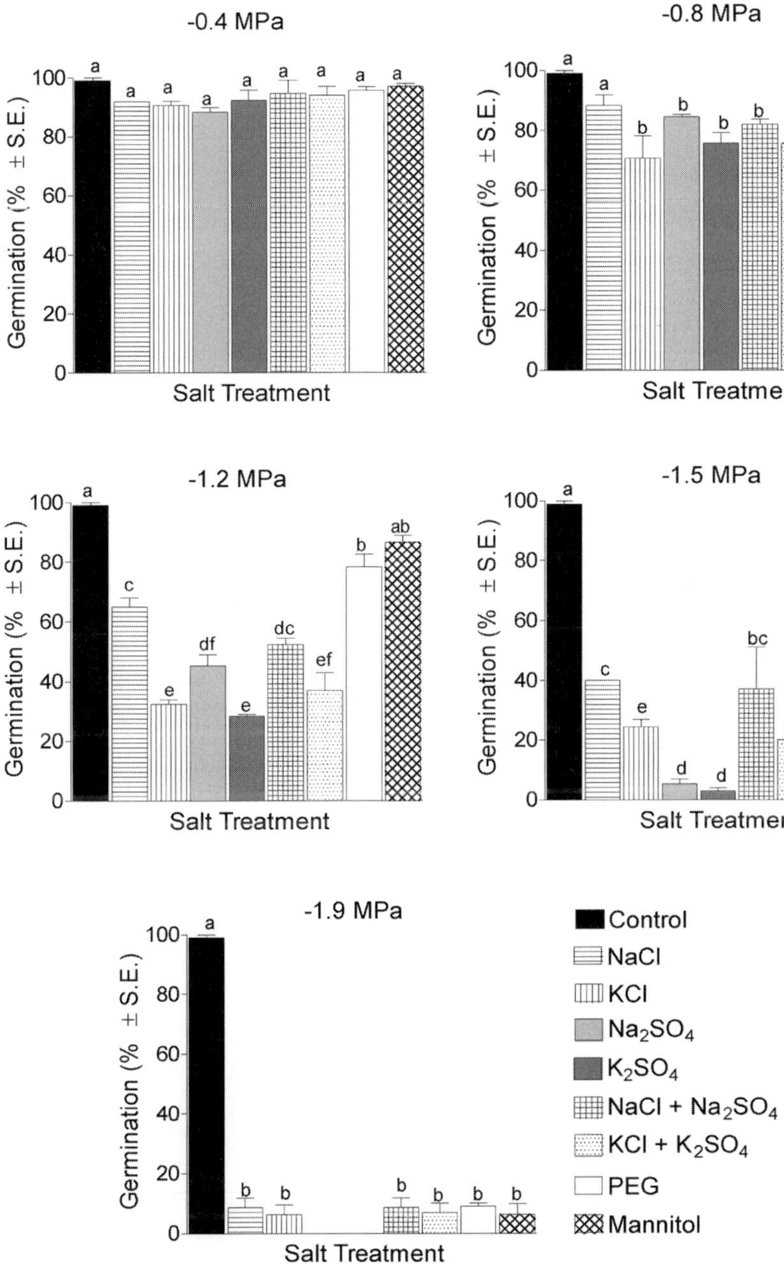


**Figure 2.** Percentage germination (mean ± SE) of *P. strombulifera* seeds at ψo –0.4, –0.8, –1.2, –1.5 and –1.9 MPa with monosaline and bisaline solutions, PEG and mannitol at 30 ± 1°C after 168 h of treatment. Values with the same letter are not significantly different at $p > 0.05$. Further details of the experiments can be found in Sosa et al. [37].

major salts present in most salinized soils of America are used separately than when they are used in iso-osmotic mixed solutions, a situation that can mimic what actually occurs in the field (particularly for the saline soils from central Argentina). However, once the definite mechanisms of salt tolerance at the germination stage are defined, they will be extremely valuable for improving plant resistance to salinity.

### *Acknowledgments*

The study was supported with funds from SECYT-Universidad Nacional de Río Cuarto, Argentina (research grant no. 038/02-04 and research program no. 023/04-06), and Consejo Nacional de Investigaciones Científicas y Tecnológicas (CONICET) de la República Argentina (PIP 2005–2006), to V. Luna.

## References

1 Woodell SRJ (1985) Salinity and seed germination patterns in coastal plants. *Vegetatio* 61: 223–229
2 Gulzar S, Khan MA (2001) Seed germination of a halophytic grass *Aeluropus lagopoides*. *Ann Bot* 87: 319–324
3 Bernstein L (1975) Effect of salinity and sodicity on plant growth. *Annu Rev Phytopathol* 13: 295–312
4 Ghoulam C, Fares K (2001) Effect of salinity on seed germination and early seedling growth of sugar beat (*Beta vulgaris* L.). *Seed Sci Technol* 29: 357–364
5 Van Hoorn JW (1991) Development of soil salinity during germination and early seedling growth and its effect on several crops. *Agric Water Manag* 20: 17–28
6 Troech FR, Thompson LM (1993) *Soil and soil fertility*. Oxford University Press, New York
7 Brady NC, Weil RR (1996) *The nature and properties of soils*, 11th edn. Prentice Hall, Simon and Shuster, New Jersey
8 Munns R (1993) Physiological processes limiting plant growth in saline soils: Some dogma and hypotheses. *Plant Cell Environ* 16: 15–24
9 Ashraf M (1994) Breeding for salinity tolerance in plants. *Crit Rev Plant Sci* 13: 17–42
10 Breen CM., Everson C, Rogers K (1997) Ecological studies on *Sporobolus virginicus* (L.) Kunth. with special reference to salinity and inundation. *Hydrobiología* 54: 135–140
11 Khan MA, Ungar IA (2001) Alleviation of salinity stress and the response to temperature in two seed morphs of *Halopyrum mucronatum* (Poaceae). *Aust J Bot* 47: 777–783
12 Lombardi T, Fochetti T, Onnis A (1998) Germination of *Briza maxima* L. Seeds effect of temperature, light, salinity and seed harvesting time. *Seed Sci Technol* 26: 463–470
13 Gulzar S, Khan MA, Ungar IA (2001) Effect of salinity and temperature on the germination of *Urochondra setulosa*. *Seed Sci Technol* 29: 21–29
14 Onnis A, Miceli P (1975) *Puccinellia festucaeformis* (Host) Parl.: Dormienza e influenza della salinita sulla germinazione. *G Bot Ital* 109: 27–37
15 Rivers WG, Weber DJ (1971) The influence of salinity and temperature on seed germination in *Salicornia bigelovii*. *Physiol Plant* 24: 73–75

16 Khan MA (1991) Studies of germination of *Cressa cretica* L. Seed Pak. *J Weed Sci Res* 4: 89–98
17 Khan MA, Gul B (1998) High salt tolerance in the germinating dimorphic seeds of *Arthrocnemun indicum*. *Int J Plant Sci* 159: 826–832
18 Ungar I (1978) Halophyte seed germination. *Bot Rev* 44: 233–264
19 Ungar I (1991) *Ecophysiology of vascular halophytes*. CRC Press, Boca Raton
20 Egan TP, Ungar I, Meekins JF (1997) The effect of different salts of sodium and potassium on the germination of *Atriplex prostrata* (Chenopodiaceae). *J Plant Nutr* 20: 1723–1730
21 Onnis A (1981) Le specie come indicatori ambientali. Estratto da Colloquio su "Inquinqmento eindicatori biologici, Roma, 3–4 giugno 1980. Collana del programmol/La finalizzato" Promozione delta germinazione in relazione alle variazioni di temperature e salinita del substrato nel periodo estivo-autunnale. *G Bot Ital* 110: 127–136
22 Romo JT, Haferkamp MR (1978) Effects of osmotic potential, potassium chloride, and sodium chloride on germination of Greasewood (*Sarcobatus vermiculatus*). *Great Basin Nat* 47: 110–116
23 Al-Jibury L, Clor M, Talabany D (1986) Effect of certain salts and their combinations on germination and seedling development of *Securigera securidaca*. *Arab Gild J Sci Res* 4: 5–11
24 Strogonov BP (1969) *Physiological basis of salt tolerance of plants*. Akad Mauk USSR (translation by Israel Prog. Sci Trans Jerusalem)
25 Hardegree SP, Emmerich WE (1990) Effect of polyethylene glycol exclusion on the water potential of solution-saturated filter paper. *Plant Physiol* 92: 462–465
26 Zekri M (1993) Salinity and calcium effects on emergence, growth and mineral composition of seedlings of eight citrus rootstocks. *J Hort Sci* 68: 53–62
27 Tobe K, Zhang L, Omasa K (1999) Effects of NaCl on seed germination of five nonhalophytic species from a Chinese desert environment. *Seed Sci Technol* 27: 851–863
28 Burkart A (1976) A monograph of the genus *Prosopis* (Leguminosae subfamily Mimosoideae). Catalogue of the recognized species of *Prosopis*. *J Arnold Arbor Univ* 57: 450–523
29 Reinoso H, Sosa L, Ramírez L, Luna V (2004) Salt-induced changes in the vegetative anatomy of *Prosopis strombulifera* (Leguminosae). *Can J Bot* 82: 618–628
30 Jarrell W, Virginia R (1990) Response of mesquite to nitrate and salinity in a simulated freatic environment: Water use, dry matter and mineral nutrient accumulation. *Plant Soil* 125: 185–196
31 Catalán L, Balzarini M, Taleisnik E, Sereno R, Karlin U (1994) Effects of salinity on germination and seedling growth of Prosopis flexuosa (D.C). *Forest Ecol Manag* 63: 347–357
32 Cony M, Trione S (1998) Inter-and intraspecific variability in *Prosopis flexuosa* and *P. chilensis*: Seed germination under salt and moisture stress. *J Arid Environ* 40: 307–317
33 López Villagra G, Galera F (1992) Soil salinity-sodicity effects on germination, survival and development in four populations of *Prosopis strombulifera* (Lam) Benth (Fabaceae: Mimosoideae). In: RW Dutton (ed): Prosopis *species. Aspects of their Value, Research and Development.* FAO-UNESCO-IPA, 219–234
34 Villagra PE (1997) Germination of *Prosopis argentina* and *P. alpataco* seeds under saline conditions. *J Arid Environ* 37: 261–267
35 Sosa L, (2005) Adaptaciones fisiológicas de *Prosopis strombulifera* a condiciones de salinidad por cloruros y sulfatos. PhD Thesis, Universidad Nacional de Río Cuarto, Córdoba, Argentina
36 Egan T, Ungar I (1998) Effect of different salts of sodium and potassium on the growth of *Atriplex prostrata* (Chenopodiaceae). *J Plant Nutr* 21: 2193–2205
37 Sosa L, Llanes A, Reinoso H, Reginato M, Luna V (2005) Osmotic and specific ion effects on the germination of *Prosopis strombulifera* (Lam.) Benth. *Ann Bot* 96: 261–267

38 Llanes A, Reinoso H, Luna V (2005) Germination and early growth of *Prosopis strombulifera* seedlings in different saline solutions. *World J Agric Sci* 1: 120–128
39 Arce P, Balboa O (1987) Factores que inciden en la propagación por estacas en *Prosopis chilensis. Cien Invest Agr* 14: 51–62
40 Tobe K, Li X, Omasa K (2000) Seed germination and radicle growth of a halophyte, *Kalidium capsicum* (Chenopodiaceae). *Ann Bot* 85: 391–396
41 Perez A, Tambelini C (1995) Effect of saline and water stress and early aging on the "Algarroba" seed germination. *Pesq Agropec Bras* 30: 1289–1295
42 Huang J, Redmann RE (1995) Physiological responses of canola and wild mustard to salinity and contrasting Ca supply. *J Plant Nutr* 18: 1931–1949
43 Redmann R (1974) Osmotic and specific ion effects on the germination of alfalfa. *Can J Bot* 52: 803–808
44 El-Haddad E, O'Leary J (1994) Effects of salinity and K/Na ratio of irrigation water on growth and solute content of *Atriplex amnícola* and *Sorghum bicolor. J Appl Irrig Sci* 14: 127–133
45 Kefu, Z, Hai F, Harris P (1995) The physiological basis of growth inhibition of halophytes by potassium. In: MA Khan, IA Ungar (eds): *Biology of salt tolerant plants*. Department of Botany, University or Karachi, Pakistan. Book Crafters, Chelsea, MI, 221–227
46 Todd PE, Ungar I, Forrest Meekins J (1997) The effect of different salts of sodium and potassium on the germination of *Atriplex prostrata* (Chenopodiaceae). *J Plant Nutr* 20: 1723–1730
47 Poljakoff-Mayber A, Somers GF, Werker E, Gallagher JL (1994) Seeds of *Kosteletzkya virginica* (Malvaceae): Their structure, germination, and salt tolerance. II. Germination and salt tolerance. *Am J Bot* 81: 54–59
48 Al-Jibury L, Clor M (1986) Interaction between sodium, calcium and magnesium chlorides affecting germination and seedling growth of *Securigera securidaca. Ann Arid Zone* 25: 105–111
49 Delesalle VA, Blum S (1994) Variation in germination and survival among families of *Sagittaria latifolia* in response to salinity and temperature. *Int J Plant Sci* 155: 187–195
50 Khan MA, Ungar IA (1999) Seed germination and recovery of *Triglochin maritima* from salt stress under different thermoperiods. *Great Basin Nat* 59: 144–150

Biosaline Agriculture and High Salinity Tolerance
Edited by Chedly Abdelly, Münir Öztürk, Muhammad Ashraf and Claude Grignon


# Growth and nitrogen-fixing performances of *Medicago truncatula - Sinorhizobium meliloti* symbioses under salt (NaCl) stress: Micro- and macro-symbiont contribution into symbiosis tolerance

Haythem Mhadhbi and Mohamed Elarbi Aouani

*Laboratoire Interactions Légumineuses Microorganismes (LILM), Centre de Biotechnologie de Borj Cedria (CBBC), BP 901, 2050 Hammam lif, Tunisie*

## Abstract

The effect of salt (NaCl) stress on plant growth and nitrogen-fixing apparatus was studied for symbiotic associations established between three *Medicago truncatula* lines inoculated and two *Sinorhizobium meliloti* strains. Salinity modulated all parameters analyzed as indices of growth performance, nitrogen-fixing capacity (acetylene-reduction assay) and nodule antioxidant system. Under stressful conditions, symbioses showed variability of response to salt application. Contrasting symbiotic associations were identified for nodulation, nitrogen-fixing capacity and salt tolerance. The total variance of analyzed performance indices under stressful conditions depended essentially on plant genotype factor. In nodules, the NaCl application had a decreasing effect on the rates of protein content and some antioxidant enzymes mainly catalase. Other enzymes such as guaiacol peroxidase increased in stressed nodules.

## Introduction

Accentuated by agricultural practices such as irrigation, salinity represents today the major cause of land degradation and plant yield decrease over all regions in the world [1]. This constraint affects most of the agricultural lands in the arid and semi-arid regions [2]. In these bioclimatic stages, legume production that depends on nitrogen fixation is often limited by severe environmental stress, mainly salinity and drought [3]. To overcome this situation, many research programs are involved in the enhancement of legume production under severe conditions [4]. These programs have focused on the selection of effective and tolerant rhizobial strains and legume genotypes to improve symbiotic nitrogen fixation under stress [5, 6]. Nevertheless, there have only been few reports on the relationships of specific interactions between symbiotic partners and on the environment of symbiosis [7, 8]. Salinity causes various forms of stress, such as an osmotic effect and ionic toxicity, and affects the overall cell metabolic activities [9]. One of the first results of this metabolic disturbance is the over-production of reactive oxygen species (ROS), inducing the oxidative stress [1]. In nodules, ROS are reported to have deleterious effects on tissue integrity and nodule functioning [10]. Symbiosis response to this constraint includes morphological modifications, such as the change of nodule cortex structure [10], and biochemical adaptations, such as the modulation of antioxidant enzyme expression in nodules [11].

## Effect of salt stress on performance and symbiotic partner's involvement

The complexity of the study of plant response to salt stress has been well documented. This process is further complicated in the case of symbiosis by the number of involved partners and interactions. The use of models such as *Medicago truncatula* and *Sinorhizobium meliloti* can be a solution for the dissection of certain mechanisms of tolerance and/or sensitivity.

In the present study, experiments were performed with three *M. truncatula* lines, a reference line Jemalong (J6) and two local lines TN8.20 and TN6.18 from different regions in Tunisia. Plants were inoculated with two *S. meliloti* strains, one reference RCR2011 and one native strain, TII7 [12]. Symbioses were grown in a green house, under controlled conditions.

### *Statistic assessment of the factor influencing the symbiotic performance*

Results showed a high influence of stress application (75 mM NaCl) on the total variance of growth and nitrogen-fixing capacity. The symbiotic interactions under salt stress were assessed by the evaluation of the contribution of each symbiotic partner and their interaction on the total variance of different parameters (Tab. 1). Under salt stress, the effect of the plant genotype factor averaged 88% and 79% for shoot and root dry weight, respectively, and it reached 64% for nodule fresh weight. The same level of contribution was noted under non-stressful conditions. For the acetylene-reduction assay (ARA) and the nodule number, the strain factor effect remained at high levels (47.5% and 33%, respectively), despite the fact that stress application led to an increase in plant line factor. The interaction factor effect was generally decreased under stress condition, mainly on the shoot and root biomass production and nitrogen-fixing activity (ARA). These results confirm other reports [7, 8, 12] concerning the dependence of plant biomass production on plant genotype. This could be explained by the sensitivity of plants, even the tolerant lines, in comparison to the bacterial partner (*Sinorhizobium*), which is known to grow in highly stressful medium [3, 13]. Thus, changes in symbiotic performances under salt stress are caused by response of plant partner. Nevertheless, it has been reported that the performance of

**Table 1.** Contribution of each partner of the symbiosis and the interaction between the two partners in the total variance of shoot dry weight (SDW), root dry weight (RDW), nodule number (NN), nodules fresh weight (NFW) and nitrogen fixation ARA of the *M. truncatula-S. meliloti* symbiosis under non-stressed (control, T) and salt stress (75 mM NaCl, S) conditions. Percentages are calculated from two-way variance analysis.

| | | Parameters | | | | | | | | | |
|---|---|---|---|---|---|---|---|---|---|---|---|
| | | SDW | | RDW | | NN | | NFW | | ARA | |
| | | T | S | T | S | T | S | T | S | T | S |
| Strain effect | % | 13.7 | 6.2 | 22 | 19 | 45 | 33 | 16 | 15 | 53 | 47.5 |
| | F | 8.91 * | 5.7 * | 107.11 ** | 27.48 * | 50.67 ** | 30.86 ** | 60.38 * | 30.14 * | 96.12 *** | 78.81 *** |
| Line effect | % | 74.48 | 88.6 | 73 | 79 | 30 | 51 | 67 | 64 | 11.7 | 25 |
| | F | 70.38 *** | 118.7 *** | 346.25 *** | 110.8 *** | 49.25 ** | 72.6 ** | 253.7 *** | 136.06 *** | 18.8 * | 38.62 ** |
| Interaction | % | 11.95 | 5.2 | 5 | 2 | 25 | 16 | 18 | 21 | 34 | 27.5 |
| | F | 6.87 * | 5.51 * | 9.6 * | 3.12 ns | 15.16 ** | 17.42 * | 61.00 * | 43.86 * | 59.37 ** | 44.85 ** |

*$p \leq 0.05$; **$p \leq 0.01$; ***$p \leq 0.001$; ns, not significant

bacterial partner could lead to an enhancement of legume production under stress [3, 11].

The important effect of the interaction factor on the variance of the nitrogen-fixing capacity under salt stress confirms the necessity of optimal interaction between the two symbiotic partners for the enhancement of symbiotic performance [12], and to overcome the stress effect on nodule metabolism.

## *Salt effect on symbiotic performance and nodule metabolism indices*

The effect of salt stress on changes in symbiotic interactions was analyzed for each association on all parameters used to estimate plant growth, nodulation and nitrogen-fixing capacity (Fig. 1). We noted variability of salt effect between symbioses with respect to the inoculated rhizobial strain and the genotype of host plant. Indeed, symbiosis involving the line TN8.20 with the two inoculated strains and symbiosis involving Jemalong with the local TII7 *S. meliloti* strain seemed to have the highest level of tolerance to salt stress. These symbioses maintained high levels of aerial biomass production, with only 16–19% inhibition and reduction in ARA values of 19–30% under stressful conditions. The symbioses implicating TN6.18 line were the most affected by salt application with inhibition rate respectively of 47–50% and 40–52%,

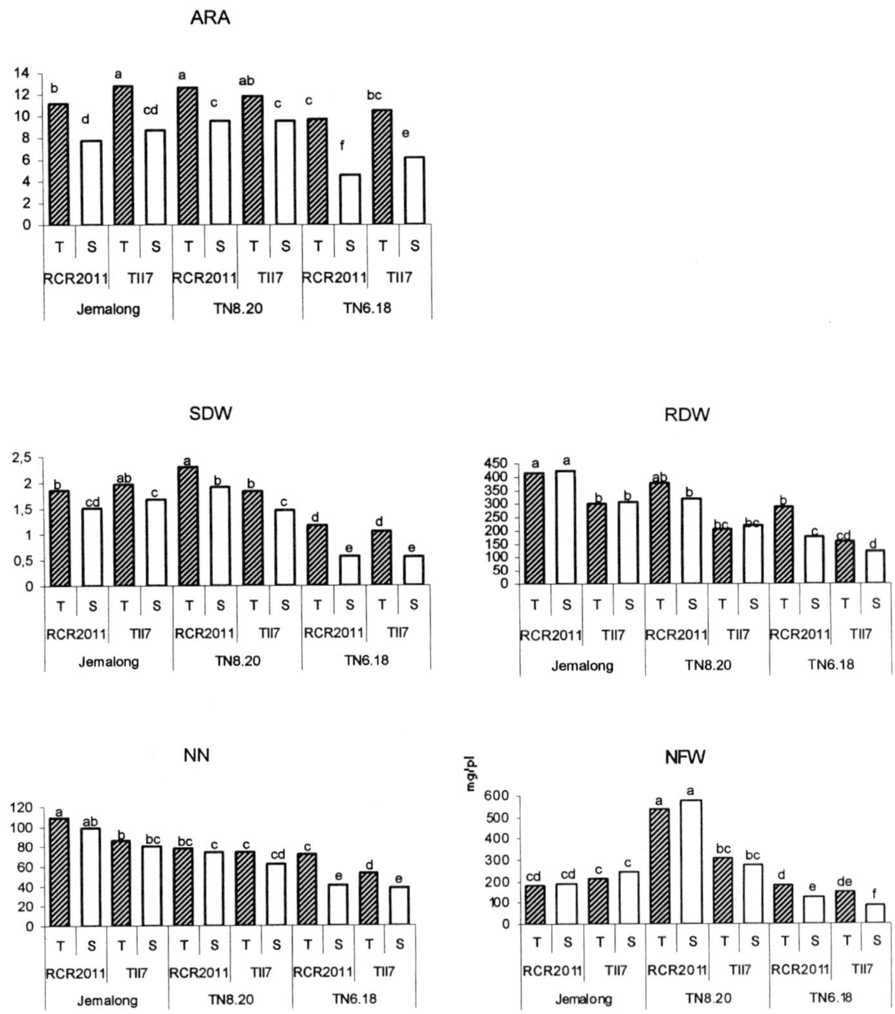


**Figure 1.** Effect of salinity (75 mM NaCl) on nitrogen-fixing capacity (ARA, μmol/h per plant) shoot dry biomass (SDW, g/plant), root dry biomass (RDW, mg/plant), number of nodules (NN), nodule fresh biomass (NFW, mg/plant) and of *M. truncatula* lines Jemalong, TN8.20 and TN6.18 inoculated by RCR2011 and TII7 *S. meliloti* strains. Values are means of nine to ten repetitions. Means denoted with different letters differ significantly at $p \leq 0.05$ based on Duncan's Multiple-Range test. T, control; S, salt stressed.

respectively. Plants of Jemalong line inoculated with RCR2011 reference strain also showed a higher sensitivity to salt stress than those associated to TII7 strain, with inhibition rates averaging 22% for shoot dry weight and 31% for ARA. Theses results could be attributed to the different level of response among the *M. truncatula* lines, implying different structural metabolic changes

[14]. The root biomass and the nodule number and biomass were less affected by salt stress. Indeed, NaCl application was performed 15 days after inoculation, and so could not affect the first generations of nodules. Concerning roots, under stressful conditions implying nutriment deprivation, plants could develop their root system to the maximum in search of needed nutriments. This strategy was further facilitated in our experimental conditions since plants were grown in hydroponic medium.

## Nitrogen-fixation protection against salt-generated oxidative stress

The application of salt stress had a significant effect on the variance of nodule antioxidant activities. The salinity effect was more pronounced on guaiacol peroxidase (POX, EC 1.11.1.11) and catalase (CAT EC, 1.11.1.6) activities ($p \leq 0.001$). The variability of antioxidant enzyme expression was analyzed under salt stress and for control symbioses in relation to the symbiotic partners and their interactions.

### *Modulation of changes in antioxidant activities under salt stress*

The total variance of antioxidant activities was highly influenced by plant line effect in stressed nodules, where it was increased compared to non-stressful conditions (Tab. 2). The contribution of line factor was 36–54% under salt stress and it was most evident for CAT (54%) and POX (53%) activities. The effect of plant line was decreased for nodule protein rate, which was essentially dependant on strain factor (62%) under salt stress. Moreover, the effect of rhizobial strain was manifested essentially on the variance of the ascorbate peroxidase (APX, EC 1.11.1.11; 44%) and the superoxide dismutase (SOD, EC 1.15.1.1; 32%) activities, despite the decrease of its contribution for SOD under stressful conditions. The interaction factor effect was clearly pronounced into the total variance of these antioxidant enzymes under salt stress, reaching 40% for CAT, 20% for APX, 25% for SOD and 32% for POX activities. The important contribution of interaction factor to changes in ARA values and antioxidant enzymes under stress imply a high dependence between symbiosis tolerance, nodule protection and functioning under such conditions. It has been shown that nodule ionic status and nodule-cortex oxygen permeability are very sensitive to salt application [14].

### *Antioxidative response in nodules of salt stressed nodules*

The application of salt stress had an inductive effect on nodular SOD activity for symbioses involving TN8.20 (13%) and mainly TN6.18 (19% and 32%)

**Table 2.** Contribution of each partner of the symbiosis and the interaction between the two partners in the total variance of total soluble protein rate, superoxide dismutase (SOD), catalase (CAT), guaiacol peroxidase (POX) and ascorbate peroxidase (APX) in the nodules of the *M. truncatula-S. meliloti* symbiosis under non-stressed (control, T) and salt stress (75 mM NaCl, S) conditions. Percentages are calculated from two-way variance analysis.

| | | Parameters | | | | | | | | | |
|---|---|---|---|---|---|---|---|---|---|---|---|
| | | Proteins | | SOD | | CAT | | POX | | APX | |
| | | T | S | T | S | T | S | T | S | T | S |
| Strain effect | % | 43 | 62.5 | 62 | 32 | 20 | 16 | 18 | 25 | 30 | 44 |
| | F | 35.12 *** | 70.3 *** | 57.83 *** | 20.01 *** | 26.24 * | 11.21 * | 5.18 * | 6.53 * | 54.1 *** | 21.05 *** |
| Line effect | % | 45 | 25 | 18.59 | 43 | 48 | 54 | 42 | 53 | 34 | 36 |
| | F | 36.92 *** | 23.78 * | 17.48 * | 42.55 *** | 76.56 ** | 24.84 *** | 13.95 *** | 10.87 *** | 61.31 *** | 17.02 *** |
| Interaction | % | 12 | 12.5 | 20 | 25 | 32 | 40 | 40 | 32 | 36 | 20 |
| | F | 8.39 * | 8.81 * | 18.4 * | 14.36 * | 62.61 ** | 20.23 *** | 11.29 *** | 8.00 *** | 63.5 *** | 8.36 * |

$^{*}p \leq 0.05$; $^{***}p \leq 0.001$

**Table 3.** Summary of percent changes of the nodule protein content and antioxidant activities; SOD, CAT, POX and APX of *M. truncatula-S. meliloti* symbioses under salt stress (75 mM NaCl).

| | Jemalong | | TN8.20 | | TN6.18 | |
|---|---|---|---|---|---|---|
| | RCR2011 | TII7 | RCR2011 | TII7 | RCR2011 | TII7 |
| Proteins inhibition % | 16.41 | 14.73 | 11.12 | 8.00 | 16.57 | 18.11 |
| CAT inhibition % | 32.28 | 31.28 | 22.86 | 24.49 | 15.82 | 15.18 |
| SOD stimulation % | $1.48^{ns}$ | $1.00^{ns}$ | 13.47 | 11.32 | 32.89 | 18.32 |
| POX stimulation % | 38.18 | 23.01 | 29.24 | 50.03 | 19.36 | 72.81 |
| APX stimulation % | 18.76 | 30.02 | $4.76^{ns}$ | $10.80^{ns}$ | $7.47^{ns}$ | $11.87^{ns}$ |

with the two *S. meliloti* strains (Tab. 3). However, this induction was not significant for symbioses with the Jemalong line. SOD has been reported to play a key role in the process of nodule establishment and functioning [15, 16]. Nevertheless, these have been many different and contradictory reports concerning the involvment of this enzyme in the response to environmental stresses such as salinity and drought [11, 17, 18]. This also applies to the role of APX in the symbiosis functioning and protection. In the present study, APX activity showed an opposite effect between symbioses to that of SOD.

APX was not significantly affected by salt stress for symbioses involving TN8.20 and TN6.18 *M. truncatula* lines, but it was induced in nodules of Jemalong. CAT activity was decreased under stress conditions for all symbioses. This enzyme is reported for its correlation with nitrogenase (ARA) [12, 17] and thus its involvement in the symbiotic process. Nevertheless, despite being a free hydrogen peroxide scavenger, CAT did not seem to be a key enzyme of the antioxydative defense of stressed nodules. Defense against oxidative stress generated by salinity or other environmental stresses such as drought and strong light is well managed by POX activity, which was clearly enhanced by salt stress in nodules of all analyzed associations and has been reported as a defense tool for several salt- and drought-stressed plant [11, 18].

## Conclusion

It is well known that salt stress affects symbiotic association mainly at the level of nodule functioning, estimated in our study by nitrogen fixation. This study identified contrasting associations among symbioses of *S. meliloti* and *M. truncatula* for some traits of symbiotic performance and salt tolerance. The application of salt (NaCl) decreased the activities of key enzymes in nodules, such as CAT and nitrogenase, as well as the protein content. However, some antioxidant enzymes were stimulated, mainly POX that seemed to be involved in the process of symbiosis defense under salt stress.

### *Acknowledgements*

This work was supported by the PICS Project, 2004–2006: 'Projet International de Cooperation Scientifique' (France). We thank Ms. Jamila HAMMAMI for her technical assistance.

## References

1 Zhu J-K (2001) Plant salt tolerance. *Trends Plant Sci* 6: 66–71
2 Graham PH, Vance CP (2003) Legumes: Importance and constraints to greater use. *Plant Physiol* 131: 872–877
3 Zahran HH (1999) Rhizobium legume symbiosis and nitrogen fixation under severe conditions and in arid climate. *Microbiol Mol Biol Rev* 63: 968–989
4 Drevon JJ, Abdelly C, Amarger N, Aouani ME, Aurag J, Gherbi H, Jebara M, Lluch C, Payre H, Schump O et al (2001) An interdisciplinary research strategy to improve symbiotic nitrogen fixation and yield of common bean (*Phaseolus vulgaris*) in salinised areas of the Mediterranean basin. *J Biotechnol* 91: 257–268
5 Serraj R., Sinclair TR, Purcell LC (1999) Symbiotic N2 fixation response to drought. *J Exp Bot* 50: 143–155

6 Sadiki M, Rabih K (2001) Selection of chickpea (*Cicer arietinum*) for yield and symbiotic nitrogen fixation ability under salt stress. *Agronomie* 21: 659–666
7 Fesenko AN, Provorov NA, Orlova IF, Simarov BV (1994) The role of the pea (*Pisum sativum* L.) cultivar genotype and the *Rhizobium leguminosarum* strain in the effectiveness of symbiosis. *Russian J Genet* 30: 725–729
8 Robinson KO, Beyene DA, van Berkum P, Knight-Mason R, Bhardwaj HL (2000) Variability in plant-microbe interaction between *Lupinus lines* and *Bradyrhizobium strains. Plant Sci* 159: 257–264
9 Tonon G, Kevers C, Faivre-Ranpant O, Graziani M, Gaspar T (2004) Effect of NaCl and mannitol iso-osmotic stresses on proline and free polyamine levels in embryogenic *Fraxinus angustifolia* callus. *J Plant Physiol* 161: 701–708
10 Matamoros MA, Dalton DA, Ramos J, Clemente MR, Rubio MC, Becana M (2003) Biochemistry and molecular biology of antioxidants in the rhizobia-legume symbiosis. *Plant Physiol* 133: 449–509
11 Mhadhbi H, Jebara M, Limam F, Aouani ME (2004) Rhizobial strain involvement in plant growth, nodule protein composition and antioxidant enzyme activities of chickpea-rhizobia symbioses: modulation by salt stress. *Plant Physiol Biochem* 42: 717–722
12 Mhadhbi H, Jebara M, Limam F, Huguet T, Aouani ME (2005) Interaction between *Medicago truncatula* lines and *Sinorhizobium meliloti* strains for symbiotic efficiency and nodule antioxidant activities. *Physiol Plant* 124: 4–11
13 Jebara M, Mhamdi R, Aouani ME, Ghrir R, Mars M (2001) Genetic diversity of *Sinorhizobium* populations recovered from different *Medicago* varieties cultivated in Tunisian soils. *Can J Microbiol* 47: 139–147
14 Aydi S, Drevon JJ, Abdelly C (2004) Effect of salinity on root-nodule conductance to the oxygen diffusion in the *Medicago truncatula-Sinorhizobium meliloti* symbiosis. *Plant Physiol Biochem* 42: 833–840
15 Santos R, Hérouart D, Puppo A, Touati D (2000) Critical protective role of bacterial superoxide dismutase in *Rhizobium*-legume symbiosis. *Mol Microbiol* 38: 750–759
16 Jamet A, Sigaud S, Van de Sype G, Puppo A, Herouart D (2003) Expression of the bacterial catalase genes during *Sinorhizobium meliloti-Medicago sativa* symbiosis and their crucial role during the infection process. *Mol Plant Microbe Interact* 16: 217–225
17 Tejera NA, Campos R, Sanjuan J, Lluch C (2004) Nitrogenase and antioxidant enzyme activities in *Phaseolus vulgaris* nodules formed by *Rhizobium tropici* isogenic strains with varying tolerance to salt stress. *J Plant Physiol* 161: 329–338
18 Lee DH, Kim YS, Lee CB (2001) The inductive responses of the antioxidant enzymes by salt stress in the rice (*Oryza sativa* L.). *J Plant Physiol* 158: 737–745

Biosaline Agriculture and High Salinity Tolerance
Edited by Chedly Abdelly, Münir Öztürk, Muhammad Ashraf and Claude Grignon


# Physiological responses of two *Arabidopsis thaliana* isolates, *N1438* and *Col*, to different salts

Hela Mahmoudi, Houneida Attia, Imen Tarchoun, Zeineb Ouerghi and Mokhtar Lachaâl

*Unité de Physiologie et Biochimie de la Tolérance au Sel des Plantes, Département de Biologie, FST, Campus Universitaire, 2092 Tunis, Tunisie*

## Abstract

Growth inhibition by salt stress in glycophytes like *Arabidopsis thaliana* is associated with a significant accumulation of $Na^+$ and $Cl^-$ in rosette leaves and to a reduction of their supply of essential nutrients such as potassium and calcium. In the present work, we attempted to evaluate the contribution of each of these factors to changes in the physiological functions of the species. The experiments were carried out under greenhouse conditions. Three-week old plants of *A. thaliana* from *Col* and *N1438* isolates were cultivated for 17 days in a basal medium supplemented with either 12.5 mM $Na_2SO_4$, 12.5 mM $K_2SO_4$, 25 mM NaCl, or 25 mM KCl. A salt-free medium was used as control. On harvesting, plants were cut into rosette leaves, bolts, and roots, and their fresh and dry weights, water contents, and major nutrient contents were determined. In *Col*, growth was decreased more by $K^+$ salts (KCl and $K_2SO_4$) than by $Na^+$ salts (NaCl and $Na_2SO_4$), and more by $K_2SO_4$ than by KCl, whereas in *N1438*, no difference was observed between the different salt treatments. Sodium was primarily accumulated in shoots, where it was probably compartmentalized into vacuoles, since it seemed to participate in osmotic adjustment. The growth sensitivity of *Col* to $K^+$ salt treatments was associated with excessive accumulation of $K^+$ in plant tissues. In conclusion, the variability of salt responses in *A. thaliana* was more dependent on cations ($Na^+$ or $K^+$) than on their associate anions ($Cl^-$ or $SO_4^{2-}$).

## Introduction

Plant growth reduction under saline conditions can be due to osmotic effects, salt ion toxicity, and/or ion imbalance imposed by the salt excess [1]. These kinds of stress affect plant growth by inducing changes in water status, membrane structure, cellular enzymatic activities, protein synthesis, and gene expression [2]. Two experimental approaches are usually used to distinguish between the effects of high accumulation of toxic ions in tissues and those due to nutrient deficiencies. The first approach is based on the comparison of plant responses to various salt types [3] and the second uses a split-root system, with a part of the root system in control medium and the other part in the presence of salt [4]. In the present work, we studied the implication of $Na^+$ and $Cl^-$ accumulation in rosette leaves, and of limitation of essential nutrients acquisition, using different salts (NaCl, KCl, $Na_2SO_4$, and $K_2SO_4$).

## Material and methods

Seeds of *A. thaliana* (isolates *Col* and *N1438*) were sown in pots containing a 1:2 (vol:vol) mixture of sand and peat, and grown in culture chamber with a 12-h photoperiod [150 μmol $m^{-2}$ $s^{-1}$ photosynthetically active radiation (PAR)]. The temperature and average relative humidity were 22°C and 60% during the day and 18°C and 80% during night. The seedlings were irrigated with distilled water during the first 9 days then with a nutrient solution [5]. At the age of 17 days, the plants were transferred into individual pots containing a complete nutrient solution. A first harvest was performed after 7 days. The remaining plants were divided into five lots corresponding to control (without salt), and four salt treatments: 12.5 mM $K_2SO_4$, 12.5 mM $Na_2SO_4$, 25 mM KCl, and 25 mM NaCl. All plants were harvested 17 days later. Fresh and dry weigh of roots, rosette leaves and bolts were separately measured. Ions were assayed in acid extracts, with flame photometry for cations and coulometry for $Cl^-$.

## Results

### *Salt effect on growth*

In *Col*, biomass production was more affected by potassium than by sodium salts (Fig. 1). Bolt growth was particularly sensitive to potassium salts; it was

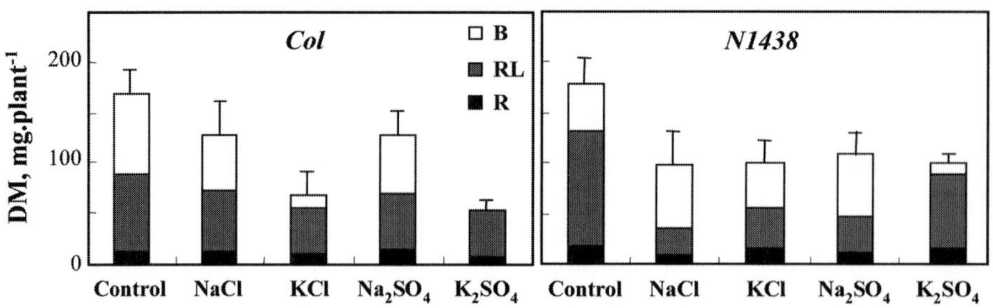


**Figure 1.** Biomass (DM= dry matter) of two *Arabidopsis thaliana* isolates (*Col* and *N1438*). The plants were grown for 17 days in the presence of the indicated salts: salt-free medium (control), 25 mM NaCl, 25 mM KCl, 12.5 mM $Na_2SO_4$, or 12.5 mM $K_2SO_4$. Means of ten replicates ± SE. R, roots; RL, rosette leaves; B, bolts.

completely inhibited in $K_2SO_4$. In *N1438*, however, plant growth was reduced by 40% in comparison to the control regardless of the kind of salt treatment. Only $K_2SO_4$ inhibited bolt growth. Rosette leaves were the most adversely affected organs in plants treated with sodium salts.

### *Sodium uptake and distribution*

Independently of the treatments, $Na^+$ was primarily accumulated in shoots of *Col* and *N1438* (Fig. 2). The accompanying anion did not exert significant effect on $Na^+$ uptake. Sodium concentration reached high values [about 3 mmol $g^{-1}$ dry matter (DM)] in leaves of NaCl-treated plants.

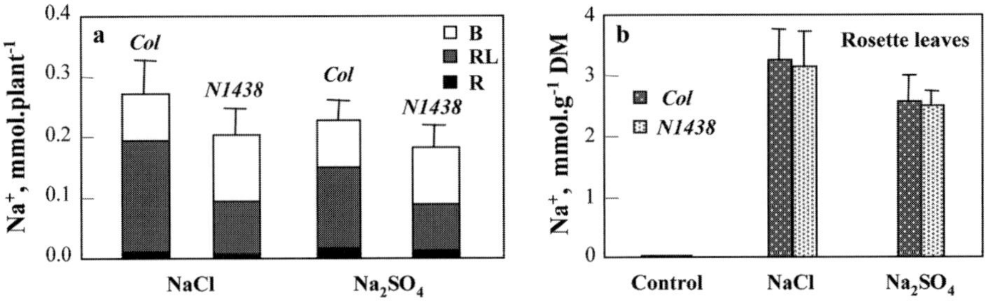


**Figure 2.** $Na^+$ content of the different organs (a) and $Na^+$ concentration in rosette leaves (b). Plants were grown for 17 days in the presence of the indicated salts: salt-free medium (control), 25 mM NaCl or 12.5 mM $Na_2SO_4$. Means of ten replicates ± SE. R, roots; RL, rosette leaves; B, bolts.

### *Potassium uptake and distribution*

When grown in the presence of NaCl or $Na_2SO_4$, the plants did not take up potassium during the entire treatment period, since the quantities per plant found at the initial and the final harvests were virtually the same (Fig. 3). These plants used their initial reserves for growth, and redistributed $K^+$ from the leaves to the bolts. In treatments with high $K^+$ concentration in the medium, $K^+$ uptake was stimulated only when the accompanying anion was $Cl^-$. In the presence of $K_2SO_4$, the plants absorbed the same amount of $K^+$ as in the control. The mean $K^+$ concentration in leaf tissues was dependent on the type of salt in the medium (Fig. 4). It was largely increased in the presence of potassium salts. This enhancement was more important in the presence of KCl than in $K_2SO_4$, probably because of the low mobility of $SO_4^{2-}$. In contrast, it was decreased in response to sodium salts, indicating that $K^+$ transport into leaves was more inhibited than leaf growth.

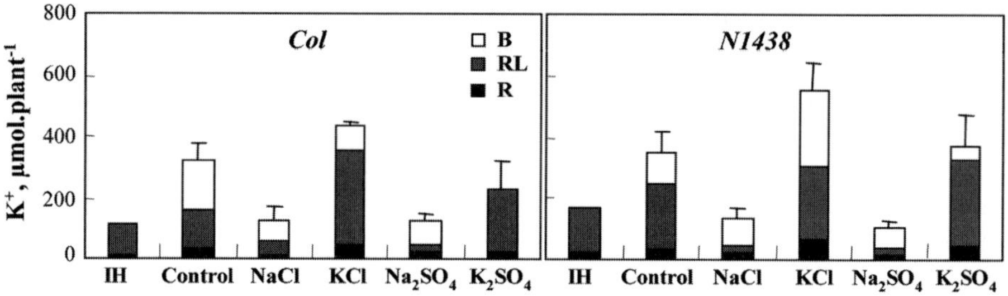


**Figure 3.** Potassium content in the different organs of two *A. thaliana* isolates (*Col* and *N1438*). R, roots; RL, rosette leaves; B, bolts. IH, initial harvest. Treatments: 17 days on salt-free medium (control), 25 mM NaCl, 25 mM KCl, 12.5 mM $Na_2SO_4$, or 12.5 mM $K_2SO_4$. Means of ten replicates ± SE.

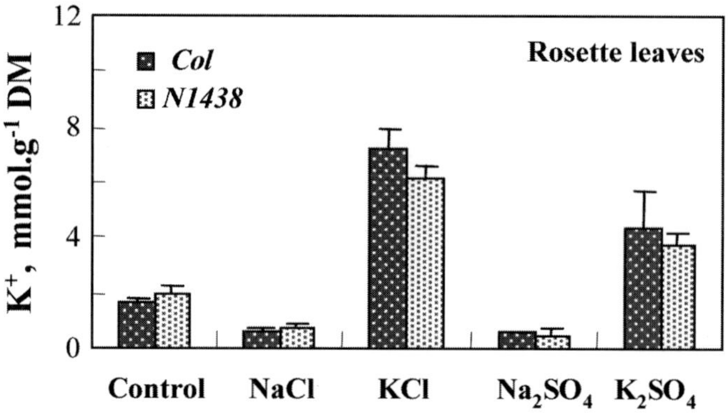


**Figure 4.** Potassium concentration in rosette leaves in two *A. thaliana* isolates (*Col* and *N1438*). The plants were grown for 17 days in the presence of the indicated salts: salt-free medium (control), 25 mM NaCl, 25 mM KCl, 12.5 mM $Na_2SO_4$, or 12.5 mM $K_2SO_4$. Means of ten replicates ± SE.

## *Water relationships*

The maintenance of cell turgor by osmotic adjustment is an important physiological adaptation to minimize the detrimental effects of salt stress [6]. $K^+$, $Na^+$ and $Cl^-$, as main soluble inorganic intracellular ions, participate in the osmotic adjustment in glycophytes [7, 8]. In *N1438*, the leaf water content was not modified by $Na^+$ salt treatments (Fig. 5). Thus, the observed large $Na^+$ accumulation was probably associated to ion compartmentalization into the leaf cells, permitting water retention. On the contrary, $K^+$ salt treatments (especially $K_2SO_4$ treatment) led to dehydration of the leaves of some plants. The latter behavior was much more accentuated in *Col*. In this isolate, the

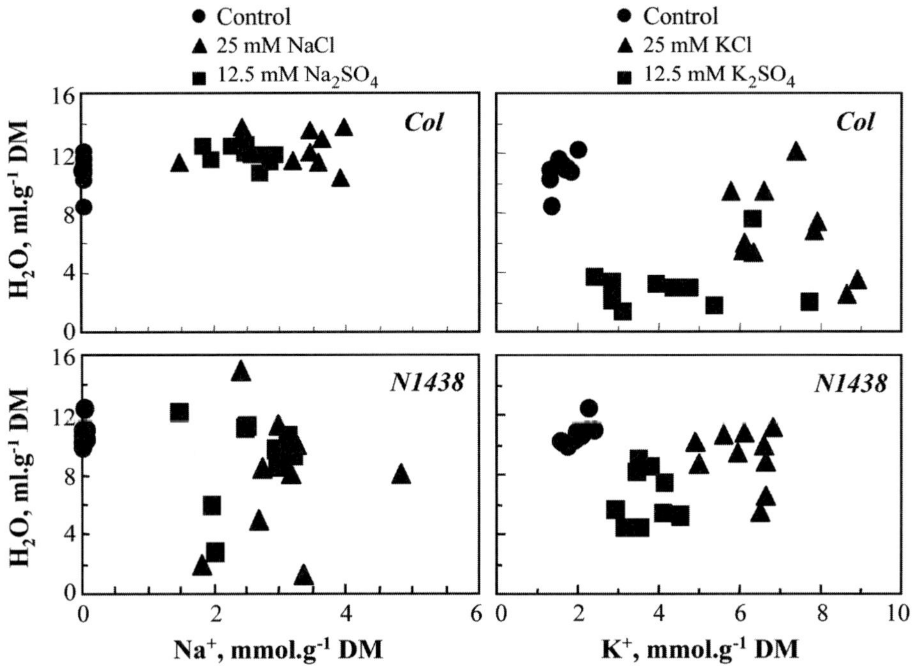


**Figure 5.** Relationship of leaf hydration to total ionic concentration in two isolates of *A. thaliana*. The total ionic concentration in leaf tissues is estimated as 2 ($Na^+$ + $K^+$) where parentheses indicate ion content (mmol g$^{-1}$ DM) divided by water content (ml $H_2O$ g$^{-1}$ DM).

leaf dehydration event was more frequent, more pronounced, and occurred in both $K^+$ and $Na^+$ salt treatments.

We indirectly estimated the participation of these ions in the leaf tissue osmolarity as O=2 ($K^+$ + $Na^+$) where parentheses indicate ion content (mmol g$^{-1}$ DM) divided by water content (ml $H_2O$ g$^{-1}$ DM), and the factor 2 stands for accompanying anion, supposed to be univalent. In control plants of both isolates, O was ca. 0.3–0.4 M (Fig. 6). It was considerably augmented in the presence of potassium salts in *Col*, reaching 2.5–3.5 M, but it did not exceed 0.5 M in the presence of sodium salts. In *N1438*, the osmolarity was in the 1–1.5 M range for all salt treatments. These results indicate that the capacity to compartmentalize ions provided at high concentration differed between the two isolates.

## Discussion

Sodium and potassium salts at high concentration reduced the growth of the two *Arabidopsis thaliana* isolates, *Col* and *N1438*, in different ways according

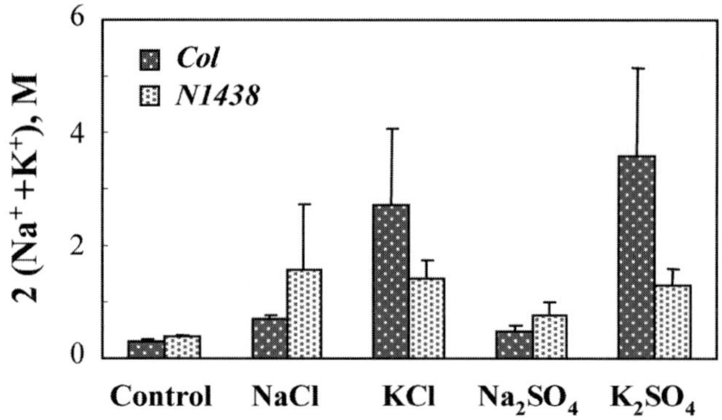


**Figure 6.** Total ionic concentration in leaf tissues or osmolarity estimated as 2 ($Na^+$ + $K^+$) where parentheses indicate the amount (mol) of ions per liter of tissue water. The factor 2 stands for hypothesized univalent accompanying anion. The plants were grown for 17 days in salt-free medium (control), or in the presence of salt (25 mM NaCl, 25 mM KCl, 12.5 mM $Na_2SO_4$, or 12.5 mM $K_2SO_4$). Means of ten replicates ± SE.

to isolates and salt type. Potassium salts (KCl and $K_2SO_4$) inhibited the DM production of *N1438* similarly to sodium salts. Although typical $Cl^-$ toxicity has been observed in some species such as bean and alfalfa [9, 10], this did not seem to be the case for *N1438*. In *Col*, the growth was relatively tolerant to sodium salts (DM production being reduced by only 25% of control), but highly sensitive to potassium salts. This sensitivity to potassium salts was associated with necroses and leaf and bolt desiccation, which were more severe in the presence of $K_2SO_4$ than in KCl (not shown). Zid and Grignon [3] observed that at low salt concentration, the growth of *Citrus* was more reduced by KCl than by NaCl and $Na_2SO_4$. These results are in agreement with those observed in the wild-type and the mutant *stl2* of *Ceratopteris richardii* in which KCl and $K_2SO_4$ appeared more toxic for growth than sodium salts [11]. We observed a similar sensitivity to potassium salts in *Col*. In contrast; the four salt types induced the same sensitivity in *N1438*. Thus, the effects of salts on *Col* growth seemed mainly due to the cations ($Na^+$ or $K^+$) rather than to their accompanying anions ($Cl^-$ or $SO_4^{2-}$). In *Citrus*, the presence of $Na^+$ in the leaves was the cause of an ionic disturbance in the foliar apoplast, which probably contributed to the appearance of characteristic necrotic spots [12].

Concerning the leaves, the growth sensitivity of the two isolates depended closely on the type of salt treatment. *Col* was less sensitive to sodium salts than *N1438*, and more sensitive to potassium salts. The sensitivity of *Col* leaves to potassium salts and of *N1438* leaves to sodium salts was associated with leaf dehydration, which led to total withering of some plants. A major point in our results was that the differences between *Col* and *N1438* concerned only the growth response, the water relations, and the mean concentration of ions in leaf tissue water. The ion accumulation in leaves, expressed as mmol $g^{-1}$ DM, was virtually the same in both isolates. Thus, the capacity to absorb $K^+$ and $Na^+$ and to transport these ions into the leaves were the same, and only the capacity to compartmentalize these ions differentiated *Col* from *N1438*, the former isolate being unable to compartmentalize $K^+$ at high concentration, and the latter isolate being unable to compartmentalize $Na^+$. These differences could be due to limitations of the plasma membrane transporters or of vacuolar ion compartmentalizing systems, which would leave large accumulation of $K^+$ (*Col*) or $Na^+$ (*N1438*) in leaf apoplast, thereby dehydrating leaf cells.

In summary, our work identified a natural variability of univalent cation compartmentalization in *Arabidopsis thaliana*. Our results suggest that this variability concerns the efficiency and the $K^+$-$Na^+$ selectivity of leaf cell transport systems.

## References

1 Hasegawa PM, Bressan RA, Zhu JH (2000) Plant cellular and molecular responses to high salinity. *Annu Rev Plant Physiol Plant Mol Biol* 51: 463–499

2 Blomberg A, Adler L (1992) Physiology of osmotolerance in fungi. *Adv Microbiol* 33: 145–212

3 Zid E, Grignon C (1986) Effets comparés de NaCl, KCl et Na2SO4 sur la croissance et la nutrition minérale de jeunes *Citrus aurentium L. Oecol Plant* 7: 407–416

4 Hajji M (1980) La responsabilité de la racine dans la sensibilité du Laurier-rose au chlorure de sodium. *Physiol Vég* 18: 505–515

5 Gay AP, Hauck B (1994) Acclimation of *Lolium temulentum* to enhanced carbon dioxide concentration. *J Exp Bot* 45: 1133–1141

6 Munns R, Greenway H, Kirst GO (1983) Halotolerant eukaryotes. In: OL Lang, PS Nobel, CB Osmond, H Ziegles (eds): *Encyclopaedia of plant physiology*, vol. 12C. New Series. Springer, New York, 59–135

7 Greenway H, Munns R (1980) Mechanisms of salt tolerance in nonhalophytes. *Annu Rev Plant Physiol* 31: 149–190

8 Pitman MG (1984) Transport across root and shoot-root interactions. In: RC Staples (ed): *Salinity tolerance in plants: Strategies for crop improvement.* Wiley, New York, 93–123

9 Cachorro P, Ortiz A, Cerda A (1994) Implications of calcium nutrition on the response of *Phaseolus vulgaris* L. to salinity. *Plant Soil* 159: 205–212

10 Chaudhary MT, Merrett MJ, Wainwright SJ (1997) Growth, ion content and proline accumulation in NaCl-selected and non-selected lines of Lucerne cultured on sodium and potassium salts. *Plant Sci* 127: 71–79

11 Vogelien DL, Hickpk LG, Warne TR (1996) Differential effects of $Na^+$, $Mg^{2+}$, $K^+$, $Ca^{2+}$ and osmotic stress on the wild type and the NaCl-tolerant mutants Stl1 and Stl2 of *Ceratopteris richardii. Plant cell environ* 19: 17–23

12 Zid E, Grignon C (1985) Sodium-calcium interactions in leaves of *Citrus aurantium* grown in the presence of NaCl. *Plant Physiol* 23: 895–903

Biosaline Agriculture and High Salinity Tolerance
Edited by Chedly Abdelly, Münir Öztürk, Muhammad Ashraf and Claude Grignon


# Physiology of salt tolerance in *Atriplex halimus* L.

Hela Ben Ahmed, Dorra Ben Ammar and Ezzeddine Zid

*Unité d'Ecophysiologie et Nutrition des Plantes, Département de Biologie, Faculté des Sciences Mathématiques, Physiques et Naturelles, Université Tunis El Manar, Campus Universitaire, 1060 Tunis, Tunisie*

## Abstract

*Atriplex halimus* is a common shrub in Tunisia, which represents a palatable food for sheep and camels. Furthermore, its halophytic behavior makes it a model for the study of mechanisms of salt tolerance in plants. We present here results obtained on *Atriplex halimus* var. *halimus*. In this species, the germination of seeds is very sensitive to salinity, since low concentrations of sodium chloride (50 mM) in the medium delayed the germination and reduced the capacity of the seedlings to emerge. Germination was completely inhibited, but reversibly, by NaCl concentrations up to 200 mM. Nevertheless, after the development of the radicle and the emergence of the cotyledons, which occurred 5 days after the imbibition of the seeds, the seedlings were able to tolerate this high concentration of NaCl and their growth was stimulated by salt. Also, 1-month-old plantlets, grown in a hydroponic medium, showed an optimal growth on 50–200 mM NaCl, and tolerated NaCl concentrations up to 300 mM. Our results indicate that, in *Atriplex halimus*, salt tolerance is acquired at an early vegetative stage of the plant development, and is related to: (i) the absorption and transport to shoots of high quantities of $Na^+$ and $Cl^-$ and their use in the osmotic adjustment, (ii) the efficiency of the vacuolar compartmentation of these ions, which prevents the ionic damage of the cytoplasm, and (iii) the aptitude of whole plant to ensure a sufficient supply of $K^+$, by maintaining a high selectivity for this essential nutriment, in spite of large amount of $Na^+$ in the medium.

## Introduction

About 15 million hectares are affected by salinity in the Maghreb and the Middle East. In Tunisia, the (semi) arid Mediterranean bio-climatic regions are frequently irrigated with saline water. Halophytes, which are exposed to high salt levels in their native habitats, have evolved several mechanisms to deal successfully with these stressful conditions [1, 2]. *Atriplex halimus* is a common shrub in Tunisia, which represents palatable forage for sheep and camels [3]. Furthermore, its halophytic behavior makes it a model for the study of mechanisms of salt tolerance. *Atriplex* shrubs have adaptations enabling them to tolerate the effect of salts internally, or excrete salt from cells and tissues [2, 4, 5]. As a result they have an advantage over plant species that lack strategies to deal with salt in the soil and are thus excellent competitors in saline environments. Generally, low salinity levels do not appear to have a deleterious effect on the growth of *Atriplex* spp. and may actually stimulate growth [6, 7]. *Atriplex* spp. must tolerate the high salt concentrations of their habitats, and they must absorb water from a soil solution that has a low water potential. To maintain water uptake and turgor under these conditions, halophytes need to maintain a water potential that is more negative than that existing in the soil solution. It is advantageous if this is achieved mainly by the accumulation of inorganic ions that can easily be taken up from the soil [8]. The aim of this work was to study the response of germination, growth and the nutritional behavior of *A. halimus* to different NaCl concentrations varying from 50 to 300 mM.

## Material and methods

### *Germination study*

Seeds were removed from the bracts by hand, and were then sown on pots filled with inert sand irrigated with pure NaCl solutions (0, 50, 100, 150 and 200 mM). Every day, the emergence of seeds was counted. To study the aptitude of emergence recovery germination after exposure to salinity for 15 days, ungerminated seeds from the NaCl treatment were removed from the pots, rinsed with deionized water and transferred into new plastic seedling trays and germinated on distilled water for a further period for germination (15 days).

### *Growth study*

Seeds were sown in pots filled with inert sand irrigated with distilled water under artificial light (100 $\mu$mol $m^{-2}$ $s^{-1}$, photoperiod: 16 h), at 25°C (light)/20°C (dark) and relative humidity ranging between 60% and 85%. After 1 month, plantlets were transferred into nutrient solutions containing macro- and micronutrients, and NaCl (0–300 mM). After 3 weeks, plantlets were harvested and separated into roots, stems and leaves. Extraction of ions was achieved in $HNO_3$ (0.1 N). $Na^+$, $K^+$ and $Ca^{2+}$ were assayed by flame emission photometry, whereas $Cl^-$ was assayed by coulometry.

## Results

The germination of seeds is very sensitive to salinity, since low concentrations of sodium chloride (50 mM) in the medium delayed the germination and reduced the capacity of emergence of the seedlings (Fig. 1). Germination was completely inhibited, but in a reversible way, by NaCl concentrations ranging up to 200 mM. Nevertheless, after the development of the radicle and the

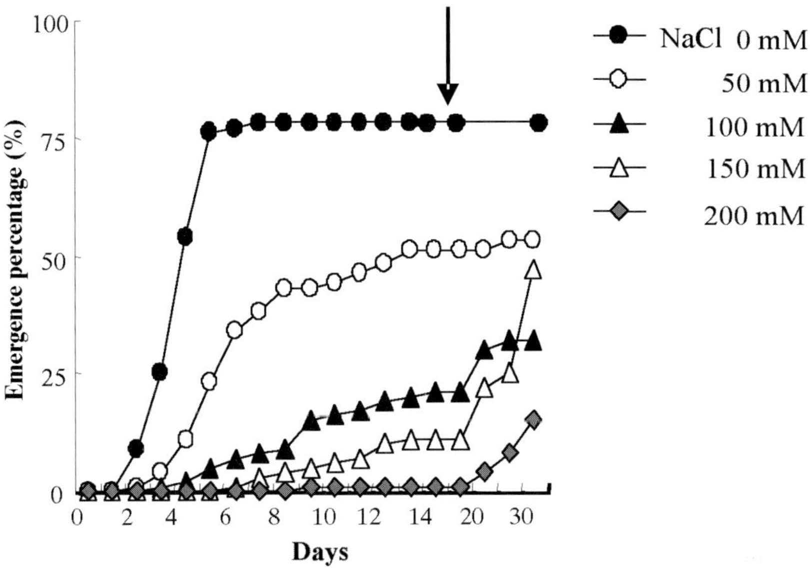


**Figure 1.** Effect of NaCl (0–200 mM) on the emergence percentage of *Atriplex halimus*. The arrow indicates the moment when the salted solutions were replaced by distilled water containing no NaCl.

emergence of the cotyledons, which occurred 5 days after the imbibition of the seeds, the seedlings became able to tolerate this high concentration of NaCl and their growth was stimulated by salt (results not shown).

One-month-old plantlets, grown in hydroponic medium, showed an optimal growth on 50–200 mM NaCl (Fig. 2), and tolerated NaCl concentrations up to 300 mM. In fact, salinity increments induced significant increases in total fresh weight and dry weight of the whole plant. Similar effects were evident on the growth of different parts of the plant. The growth stimulation in the roots was much less marked than in the stems and leaves, with major increases over the control in fresh and dry weights. The increase in shoot fresh weight over the control was largely due to an increase in water content (succulence). At 200 mM NaCl, a significant decrease in growth occurred compared to that at an optimum salinity range of 50–200 mM. In fact, at salinities above 200 mM NaCl, fresh and dry weights declined significantly, but remained slightly over the control.

The increase in dry weight production observed after 3 weeks in the presence of NaCl was associated with a high accumulation of $Na^+$ and $Cl^-$ in different organs. The $Na^+$ and $Cl^-$ content of leaves, stems and roots increased markedly with an increase in salinity, whereas $K^+$ and $Ca^{2+}$ contents decreased significantly (Fig. 3).

To assess the ability of *A. halimus* for selective uptake in the presence of high amount of $Na^+$, a K/Na selectivity ratio was calculated from leaf, stem

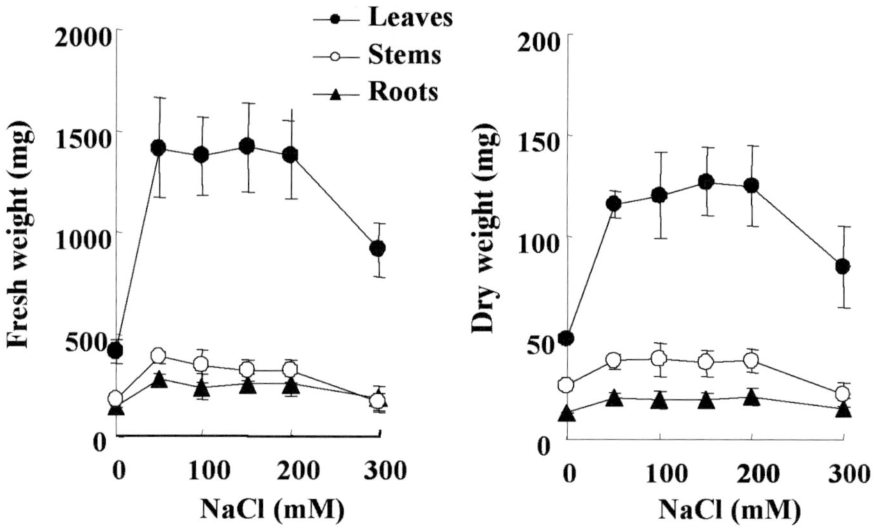


**Figure 2.** Effect of NaCl (0–300 mM) on fresh and dry weights of leaves, stems and roots in *A. halimus* plants after 3 weeks of growth. Each point is the mean of 30 replicates (error bars at 5%).

and root ion contents (Tab. 1). The results showed that the plant always remained selective to $K^+$. The K/Na ratio determined in the leaves, stems and roots was reduced by salinity, but remained higher than that calculated in the medium.

According to Oertli's hypothesis [9], *A. halimus* showed increased vacuolar accumulation of $Na^+$. Our results showed stability between the leaf water and their sodium concentration (Fig. 4a). Thus, sodium appears well sequestred inside the vacuole, which might play a role in osmotically adjusting the plant under saline conditions (Fig. 4b).

## Discussion

In general, both halophytes and glycophytes respond in a similar manner to increased salinity stress, with a reduction in the total number of seeds germinating and a delay in the initiation of the germination process [10, 11]. This agrees with our results; *A. halimus* showed a great sensitivity to salinity at the germination stage. The maximum germination was recorded in distilled water, whereas high concentrations of salt inhibited this germination, but did not kill the seed (Fig. 1). Nevertheless, after the development of the radicle and the emergence of the cotyledons, which occurred 5 days after the imbibition of the seeds, the seedlings were able to tolerate this high concentration of NaCl and their growth was stimulated by salt. Additionally, not only do halophytic

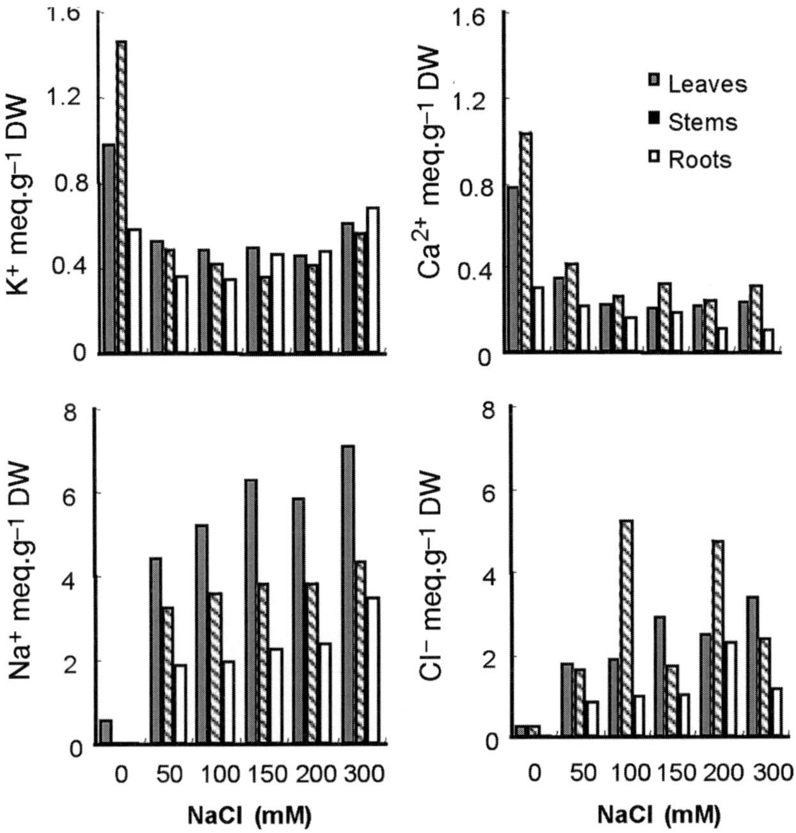


**Figure 3.** Effect of NaCl (0–300 mM) on $K^+$, $Ca^{2+}$, $Na^+$ and $Cl^-$ contents of roots, stems and leaves of *A. halimus* after 3 weeks of growth. Each point is the mean of 30 replicates.

**Table 1.** Effect of NaCl (0–300 mM) on the K/(K+Na) ratio in leaves, stems and roots of *Atriplex halimus* after 3 weeks of growth.

| NaCl (mM) | Medium | Roots | Stems | Leaves |
|---|---|---|---|---|
| 50 | 0.0029 | 0.160 | 0.132 | 0.107 |
| 100 | 0.0014 | 0.150 | 0.106 | 0.083 |
| 150 | 0.0009 | 0.147 | 0.087 | 0.077 |
| 200 | 0.0007 | 0.160 | 0.093 | 0.073 |
| 300 | 0.0004 | 0.158 | 0.115 | 0.071 |

plants tolerate salinity, they can also be stimulated by NaCl [7]. In agreement with other studies, we demonstrated that growth was stimulated by NaCl (Fig. 2). All salinity treatments from 50 to 200 mM NaCl caused an increase in fresh and dry weights of *Atriplex* compared to the control. The increase

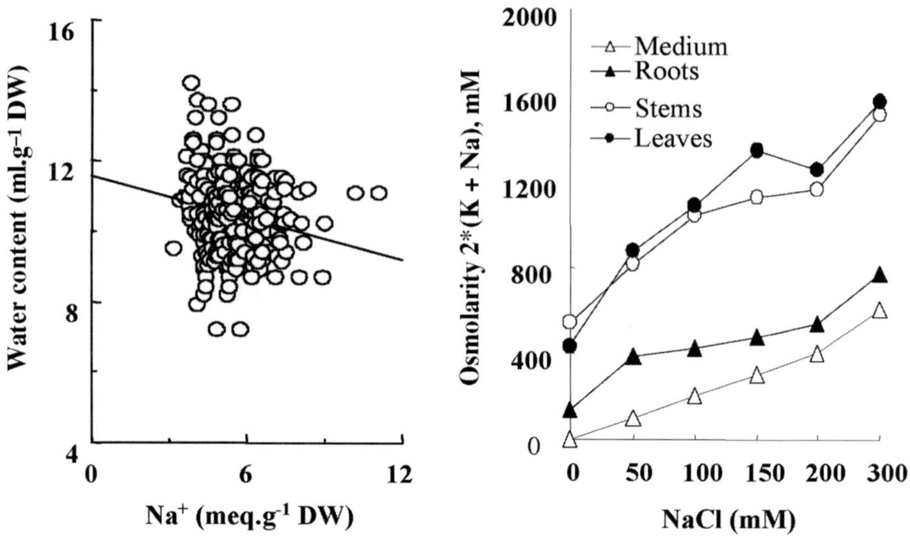


**Figure 4.** (a ) The relationship between sodium and water in the leaves of *A. halimus* cultivated for 3 weeks in the presence of NaCl (0–300 mM). (b) Estimation of the osmolarity of leaves, stems and roots of plants. The osmolarity, expressed in mM, was calculated from the mean concentrations of $K^+$ and $Na^+$ by the formula 2 × (Na+K).

in leaf fresh weight over the control was largely due to an increase in water content and a greater accumulation of ions. In fact, inorganic ions ($Na^+$ + $K^+$ + $Ca^{2+}$ + $Cl^-$) contributed to 30% of the dry weight (Fig. 5). In dicotyledonous halophytes, inorganic ions, especially $Na^+$ and $Cl^-$, may contribute to 30–50% of the dry weight [12]. In *Sarcocornia natalensis*, inorganic ions form 37% of the dry mass. Massive ion accumulation was considered to be of adaptive signification in the Chenopodiaceae [13]. *A. halimus* showed a good ability to accumulate massive quantities of sodium in leaf vacuole (Fig. 4). This sequestration of inorganic ions constitutes the basic mechanism by which *Atriplex* adjusts the osmotic potential of their internal tissues to the external salinity [14]. Another means of osmotic adjustment is the synthesis and concentration of non-toxic solutes in the cytoplasm. Species of Chenopodiaceae commonly accumulate glycinebetaine in the cytoplasm, which acts as an osmoprotectant [5, 15].

In conclusion, the salt tolerance in *A. halimus* seems to be related to: (i) the absorption and transport to shoots of high quantities of $Na^+$ and $Cl^-$ and their use in the osmotic adjustment, (ii) the efficiency of the vacuolar compartmentation of these ions, which prevents the ionic damage to the cytoplasm, (iii) the aptitude of whole plant to ensure a sufficient $K^+$ supply, by maintaining a high selectivity for this essential nutrient, in spite of large amount of $Na^+$ in the medium.

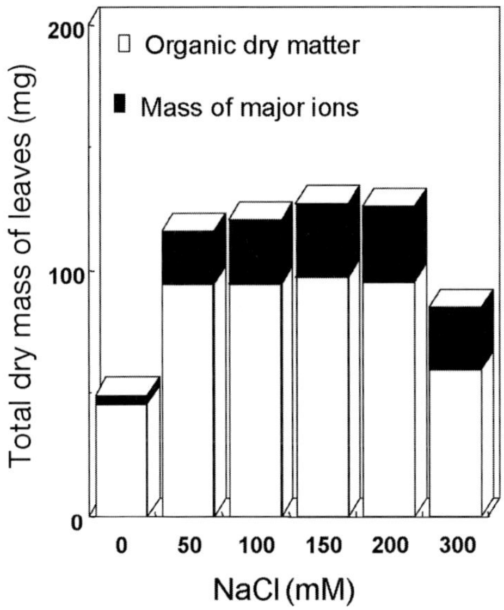


**Figure 5.** Effect of NaCl (0–300 mM) on organic dry matter and mass of major ions of total plants of *A. halimus* after 3 weeks of growth. Mean of 30 replicates.

## References


1 Hachicha M, Job JO, Mtimet A (1994) Les sols salés et la salinisation en Tunisie. *Sols de Tunisie* 5: 271–341

2 Popp M (1995) Salt resistance in herbaceous halophytes and mangroves. *Prog Bot* 56: 416–429

3 Le Houerou HN (1995) Forage halophytes in the Mediterranean basin. In: R Choukr-Allah, CV Malcolm, A Hamdy (eds): *Halophytes and biosaline agriculture*. Marcel Dekker, New York, 115–136

4 Flowers TJ, Troke PF, Yeo AR (1977) The mechanism of salt tolerance in halophytes. *Annu Rev Plant Physiol* 28: 89–121

5 Osmond CB, Björkman O, Anderson DJ (eds) (1980) *Physiological processes in plant ecology. Toward a synthesis with Atriplex*. Springer, Berlin

6 Ben Ahmed H, Zid E, El Gazzah M, Grignon C (1996). Croissance et accumulation ionique chez *Atriplex halimus* L. *Cah Agric* 5: 367–372

7 Khan MA, Ungar IA, Showalter AM (2000) The effect of salinity on the growth, water status, and ion content of a leaf succulent perennial halophyte, *Suaeda fruticosa* (L.) Forssk. *J Arid Environ* 45: 73–84

8 Koyro HW (2006) Effect of salinity on growth, photosynthesis, water relations and solute composition of the potential cash crop halophyte *Plantago coronopus* (L.). *Environ Exp Bot* 56: 136–146

9 Flowers TJ, Hajibagheri MA, Yeo AR (1991) Ion accumulation in the cell walls of rice plants growing under saline conditions: evidence for the Oertli hypothesis. *Plant Cell Environ* 14: 319–325

10 Ungar IA (1982) Germination ecology of halophytes. *Tasks Veg Sci* 2: 143–154

11 Ungar IA (1996) Effect of salinity on seed germination, growth and ion accumulation of *Atriplex patula* (Chenopodiaceae). *Am J Bot* 83: 604–607
12 Flowers TJ, Yeo AR (1986) Ion relations of plants under drought and salinity. *Aust J Plant Physiol* 13, 75–91
13 Naidoo G, Rughunanan R (1990) Salt tolerance in the succulent, coastal halophyte, *Sarcocornia natalensis*. *J Exp Bot* 41: 497–502
14 Bartels D, Nelson D (1994) Approaches to improve stress tolerance using molecular genetics. *Plant Cell Environ* 17: 659–667
15 Rozema J, Gude H, Bijl F, Wesselman H (1981) Sodium concentration in xylem sap in relation to ion exclusion, accumulation and secretion in halophytes. *Acta Bot Neerl* 30: 309–311

Biosaline Agriculture and High Salinity Tolerance
Edited by Chedly Abdelly, Münir Öztürk, Muhammad Ashraf and Claude Grignon


# Comparison of salinity tolerance of two related subspecies of *Beta vulgaris*: The sea beet (*Beta vulgaris* ssp. *maritima*) and the sugar beet (*Beta vulgaris* ssp. *vulgaris*)

Salma Daoud[1], Chérif Harrouni[2], Bernard Huchzermeyer[3] and Hans-Werner Koyro[4]

[1] *Faculté des Sciences, Université Ibn Zohr, Dakhla, Agadir, Morocco*
[2] *Inst.Agronomique et Vétérinaire Hassan II, Compl. Horticole d'Agadir, Agadir, Morocco*
[3] *University of Hanover, Institute for Botany, Hanover, Germany*
[4] *Justus-Liebig-University Giessen, Institute for Plant Ecology, Giessen, Germany*

## Abstract

Sea beet (*Beta vulgaris* ssp. *maritima*), which has a real potential to become a cash crop halophyte in the Mediterranean region, and sugar beet (*Beta vulgaris* ssp. *vulgaris*), were studied with the aim of investigating the physiological mechanisms involved in these species to overcome high salinity. Four-week-old plants of the two species were grown for 7 weeks under greenhouse conditions in an automated culture system "quick check system" irrigated with tap water (control) and four seawater concentrations (25%; 50%; 75% and 100% seawater). The five treatments were fertilized with half Hoagland nutrient solution. Plants of the two species had 100% survival in all treatments with optimal growth in the low salinity treatment (25% seawater). Salinity tolerance of the two beets is related to their ability to accomplish osmotic adjustment by regulating their ion and water uptake from the culture medium. To avoid toxicity due to excess ion accumulation, the two species adjust their osmotic potential by accumulating large amounts of ions, especially Na and Cl, in shoot vacuoles and by the synthesis of compatible solutes in the cell cytoplasm. The reduction of stomatal conductance and transpiration participates in maintaining the level of leaf turgescence, and this may contribute to a long-term survival in saline environments. The high level of photosynthesis of the beets in high saline conditions shows that the reduction of growth in these conditions is not the consequence of photosynthesis reduction but of ion toxicity.

## Introduction

The sea beet, *Beta vulgaris* ssp. *maritima* (L.), which survives in seawater salinity, is the halophyte ancestor of the sugar beet, *Beta vulgaris* ssp. *vulgaris* (L.), a salt tolerant glycophyte [1–3]. Salt tolerance of these species is a complex phenomenon requiring a series of integrated adaptations involving cellular systems, tissues and the whole plant. It is determined by a number of separate but interrelated mechanisms controlled by many genes [4–6]. Salt tolerance depends on the ability of the plant to control uptake, transport and compartmentation of salt through its tissues. All halophytes have well-developed mechanisms to control the uptake, transport and compartmentation of salt. Many glycophytes also exhibit these mechanisms to various degrees [7]. One approach towards understanding mechanisms of salt tolerance at the whole plant level is to follow the series of events that exposure to salinity initiates [8]. Much of the physiological research on salinity has concentrated on three topics: water relations, photosynthesis, or the accumulation of a particular metabolite, assuming that one or more of these processes would limit growth in saline conditions [8].

To compare the response of the sugar beet and its ancestor the sea beet in saline conditions, a detailed study was carried out to characterize the physiological changes generated by exposure to saline stress in the two species. The overall aim of the investigation was to promote cash crop halophytes for the reclamation of saline areas.

## Growth response of *Beta vulgaris* ssp. *maritima* (sea beet) and *Beta vulgaris* ssp. *vulgaris* (sugar beet) to seawater irrigation

The halophyte sea beet and the salt-tolerant glycophyte sugar beet survived at 100% seawater salinity with an optimum growth for shoot and roots of sea beet at 25% seawater salinity (Fig. 1). This salinity level is considered as the optimum for the growth of a number of dicotyledonous halophytes [7, 9, 10]. On the other hand, the optimum growth for sugar beet was in 10% seawater salinity. The 25% seawater salinity considerably reduced fresh biomass of sugar beet compared to the optimum (Fig. 1). Nevertheless, the plants developed actively and did not show any deficiency symptoms.

The two beet subspecies varied in salt concentration they can tolerate before their biomass starts to be reduced. Kinzel [11] defined the threshold of salinity tolerance as substrate concentration leading to a depression of 50% in fresh weight in comparison with optimal growth treatment (horizontal line in Fig. 1). This threshold was at 41% and 34% seawater salinity for the sea beet and the sugar beet, respectively (Fig. 1). This threshold is close to that found by Greenway and Munns [1] for the sugar beet: 260 mM of univalent salts. Even if the threshold of salinity tolerance of the sugar beet is lower than the sea beet, it is higher than many other glycophytic species. It is not surprising that sugar beet growth overlaps with that of halophytes. It has probably inherited some salt tolerance characters from its halophyte ancestor, the sea beet, which naturally grows in salt marshes.

Generally, the difference in the response of halophytes and glycophytes to saline conditions depends on the degree of salinity. The biomass of

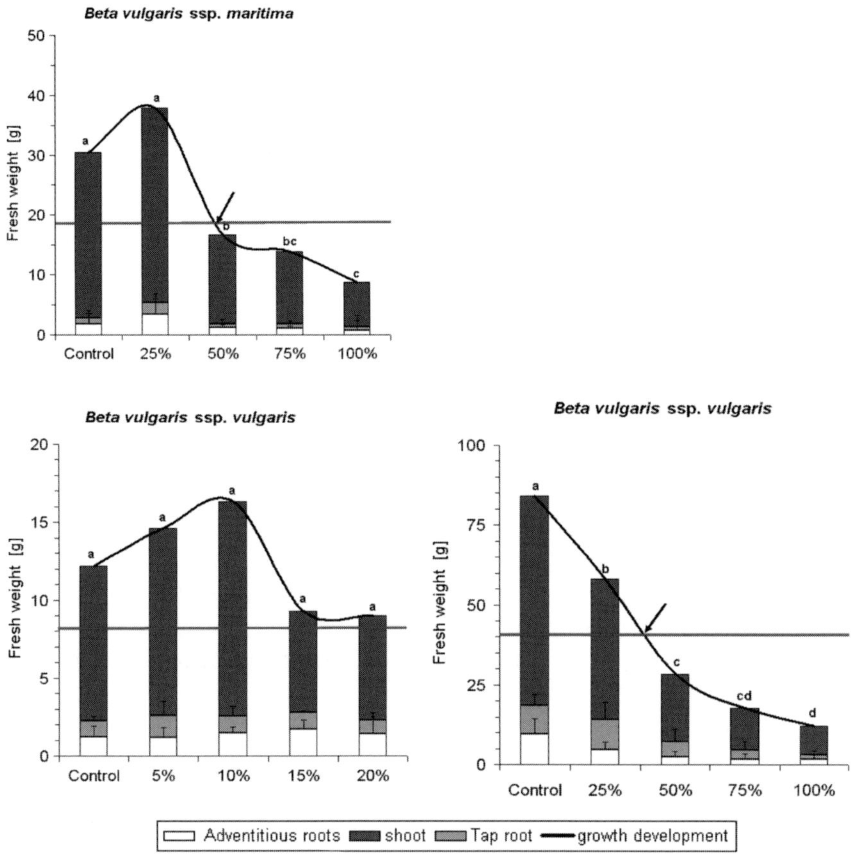


**Figure 1.** Effect of seawater salinity on fresh weight of different parts of *Beta vulgaris* ssp. *maritima* and *Beta vulgaris* ssp. *vulgaris*. The curve represents evolution tendency of fresh weight of the two subspecies in different seawater treatments, and the horizontal line indicates 50% of optimal fresh weight. The arrow indicates the intersection point of the two lines that represent the threshold level of salinity tolerance of each species. Bars with the same letter are not significantly different at $p \leq 0.05$.

dicotyledonous halophytes is stimulated to a higher degree by salt concentrations than dicotyledonous glycophytes [7]. However, the most used criterion to separate the two groups is the characteristic of halophytes to finish their life cycle in salt concentrations ranging between 100 and 200 mM NaCl [3].

High salinities affect sea beet biomass as well as that of sugar beet. Similar results were reported for other halophyte and glycophyte species [3, 12–18]. However, biomass reduction of sugar beet was more pronounced (81%) than that of sea beet (71%) in 100% seawater salinity compared to the control (tap water) (Fig. 1). In the sea beet the biomass decrease was more pronounced

**Table 1.** Salinity effect on total fresh weight; root/shoot ratio; tap root/adventitious roots ratio; and tap root diameter of *Beta vulgaris* ssp. *maritima* (sea beet) and *Beta vulgaris* ssp. *vulgaris* (sugar beet).

| **Treatment** | **Plant (g)** | | **Root/shoot** | | **Tap root/adventitious roots** | | **Tap root diameter (mm)** | |
|---|---|---|---|---|---|---|---|---|
| | Sea beet | Sugar beet | Sea beet | Sugar beet | Sea beet | Sugar beet | Sea beet | Sugar beet |
| Control | 30.43 a | 84.08 a | 0.10 a | 0.29 a | 0.58 a | 1.12 a | 5.48 a | 17.89 a |
| 25% Seawater | 37.87 a | 58.27 a | 0.17 a | 0.37 a | 0.56 a | 2.22 a | 7.23 a | 17.69 a |
| 50% Seawater | 16.71 a | 28.43 b | 0.13 a | 0.36 a | 0.58 a | 2.32 a | 5.20 a | 11.44 b |
| 75% Seawater | 13.95 a | 17.88 b | 0.15 a | 0.32 a | 0.77 a | 3.37 a | 4.45 a | 08.04 b |
| 100% Seawater | 8.86 a | 12.33 b | 0.14 a | 0.29 a | 1.25 a | 0.92 a | 3.72 a | 06.60 b |

For a given parameter, values that have the same letter are not significantly different at $p \leq 0.05$.

in leaves (73%) than in roots, especially tap root (28%), in 100% seawater treatment compared to the control (tap water). Hence, the root to shoot ratio increased with the increase of salinity in irrigation water (Tab. 1 and Fig. 1). Similar results were reported on a number of species by Munns and Termaat [19], on maize by Koyro and Huchzermeyer [20] and on sea beet by Koyro and Huchzermeyer [3]. Regarding sugar beet, biomass reduction in high salinity treatment was due to the reduction of fresh weight of both the shoot (86%) and roots, especially the tap root (83%), in 100% seawater treatment compared to the control (tap water) (Tab. 1 and Fig. 1).

Koyro and Huchzermeyer [3, 20] and Munns [7] reported that growth and survival of plants in saline conditions depend on three major constraints: water deficit arising from the low water potential in the culture medium, ion toxicity due to excessive uptake of Na and Cl to adapt to low water potential and high NaCl concentration, and nutrient imbalance by depression in uptake or internal distribution of mineral nutrients.

## Salinity and water relations in sea beet and sugar beet

The increase of seawater concentration in the culture medium generated a parallel decrease of water and osmotic potentials in different tissues of sea beet and sugar beet (Fig. 2). However, the water potential of adult leaves of the two species in seawater treatments was lower than their irrigation solutions (25%, 50%, 75% and 100% seawater). These results show that the water constraint, imposed to plants by salinity, was alleviated by osmotic regulation mechanisms allowing plants to take up water from saline medium to maintain turgescence.

The osmolalities of sea beet and sugar beet show an increasing gradient from adventitious roots to the shoot in all saline treatments, suggesting the

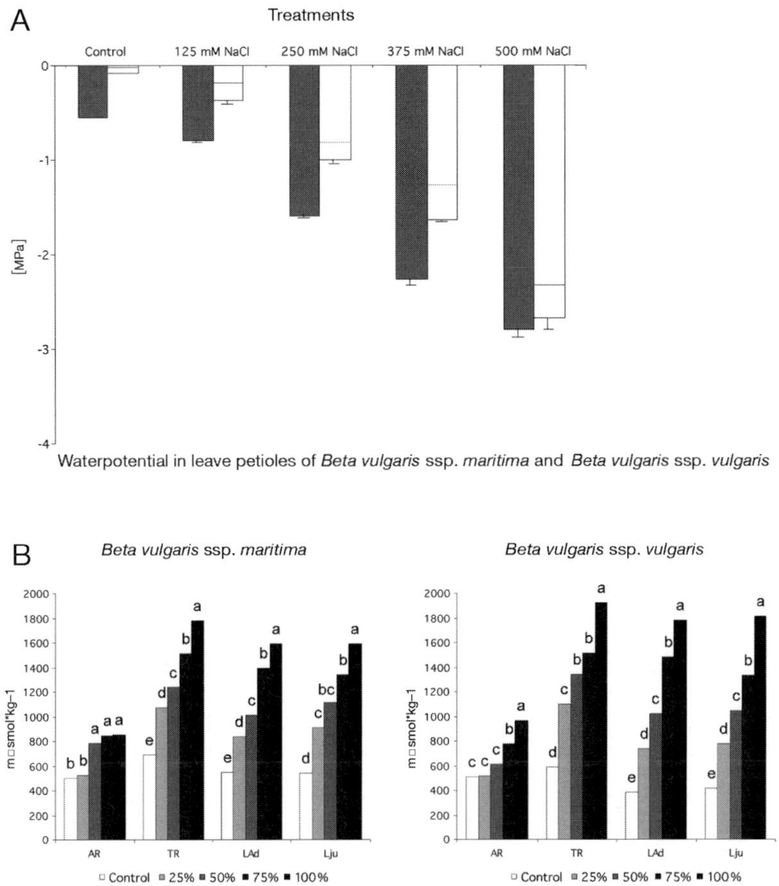


**Figure 2.** Effect of salt on water potential of leaf petioles (A) and on osmolality (mOsmol/kg) of adventitious roots (AR), tap root (TR), lamina of adult leaves (Lad) and lamina of juvenile leaves (Lju) of *Beta vulgaris* ssp. *maritima* and *Beta vulgaris* ssp. *vulgaris* (B). Bars with the same letter are not significantly different for different treatments at $p \leq 0.05$.

existence of continued water flux from roots to shoots (Fig. 2). This can only be achieved by the presence of an increasing concentration gradient of osmotically active solutes along the soil-plant-atmosphere continuum. Water relations and the aptitude of plants to osmotic adjustment are considered to be important determinants of dicotyledonous halophytes growth in saline conditions [7, 21, 22]. Indeed, osmotic adjustment – reduction of osmotic potential by the increase of intracellular solutes – is recognized as an adaptation mechanism to salt stress for a number of halophytes, and it is considered as a major component of mechanisms of tolerance to salinity [7].

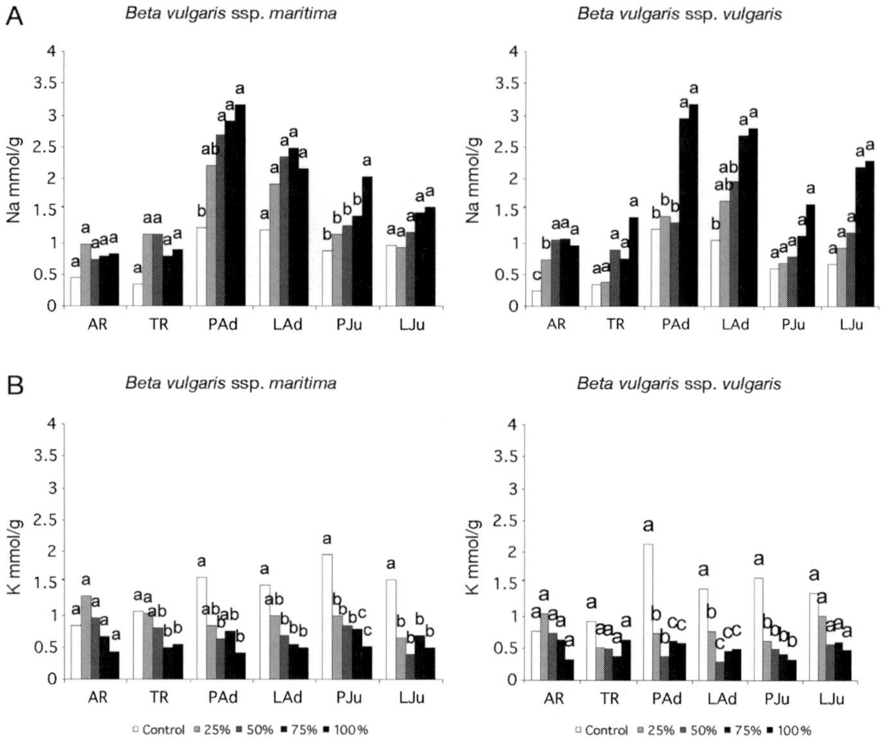


**Figure 3.** Effect of tap water (control), 25%, 50%, 75% and 100% seawater salinities on $Na^+$ (A) and $K^+$ (B) concentrations of adventitious roots (AR); tap root (TR); petiole and lamina of adult leaves (PAd, LAd) and petiole and lamina of juvenile leaves (Pju, Lju) of *Beta vulgaris* ssp. *maritima* and *Beta vulgaris* ssp. *vulgaris*. Bars represent the mean of three repetitions. Bars with the same letter are no significantly different at $p \leq 0.05$.

## Specific effect of salt on growth of the two beet subspecies

The sea beet and sugar beet were able to balance the low water potential of seawater treatments by high $Na^+$ concentrations in leaf tissues as a first osmoticum (Fig. 3). In the control (tap water) and in moderate salinity treatments, $Na^+$ content of juvenile and adult sea beet leaves were higher than those of the sugar beet. In high salinity treatments, $Na^+$ concentration in the two types of leaves of the sugar beet became higher than that in the sea beet, in which $Na^+$ content in 100% seawater was reduced compared to 75% seawater. $Na^+$ accumulation in leaves was associated with a reduction of $K^+$ content, suggesting that the $Na^+$ interferes with the assimilation and the transport of $K^+$, disturbing $K^+/Na^+$ selectivity and ionic balance [3, 20, 23, 24]. However, adult leaves accumulated more $Na^+$ than young leaves (Fig. 3) to prevent them from the toxic effect of NaCl [25]. In spite of the high amounts of $Na^+$ in dif-

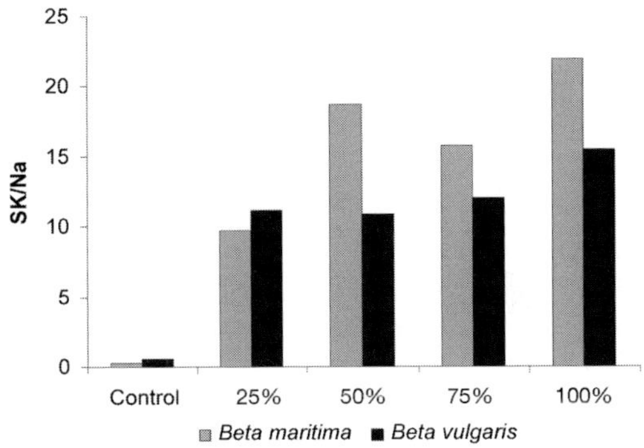


**Figure 4.** Effect of tap water (control), 25%, 50%, 75% and 100% seawater salinities on K to Na selectivity [SK/Na, i.e., (Na in irrigation solution/K in irrigation solution) × (K in roots/Na in roots)] in adventitious roots of *Beta vulgaris* ssp. *maritima* and *Beta vulgaris* ssp. *vulgaris*.

ferent parts of the plants of the two beet subspecies growing in high salinities, no ion deficiency or toxicity symptoms were observed.

$Na^+$ accumulation in the roots was much lower than in the shoots. The progressive accumulation of $Na^+$ in the roots was accompanied by $K^+$ reduction. The presence of exclusion mechanisms in halophytes roots was proposed to explain the low Na concentration at this level [26]. On the other hand, the level of $K^+$ to $Na^+$ selectivity ($SK^+/Na^+$) in adventitious roots of the two subspecies increased with the increase of seawater concentration in the culture medium, and it was relatively higher in the sea beet that was more tolerant to salinity than in sugar beet that is less tolerant (Fig. 4). The maintenance of an adequate net uptake of $K^+$ at high $Na^+$ concentrations is important, since the physiological functions of $K^+$ in plants cannot be substituted by $Na^+$, with the exception of the osmotic role of $Na^+$ in the vacuoles. It is, therefore, possible that $K^+/Na^+$ discrimination is associated with salinity tolerance. Sustained $K^+/Na^+$ selectivity is therefore proposed as a physiological marker for the ionic component of salt stress, providing ion homeostasis in growing roots [27].

On the other hand, a high selectivity was observed for the accumulation of $Ca^{2+}$ and $Mg^{2+}$ in all parts of the two subspecies (Fig. 5). $Ca^{2+}$ and $Mg^{2+}$ concentrations in the shoot and roots evolved proportionally with the increase of seawater in the culture medium since seawater is rich in these ions and $Na^+$ does not interfere with them. The maintenance of optimal $Ca^{2+}$ and $Mg^{2+}$ concentrations, especially in adventitious roots under high salinity conditions, allows the maintenance of a steady rate through the whole plant and prevents nutritional deficiency. $Ca^{2+}$ plays an important role in the increase of salinity tolerance of the plant, by maintaining membrane integrity in presence of excess NaCl, and thus controls the uptake and transport selectivity of ions [2].

Ion exchange in the tap root, which represents the carbohydrates storage organ, of the two beet subspecies was lower than in adventitious roots, which

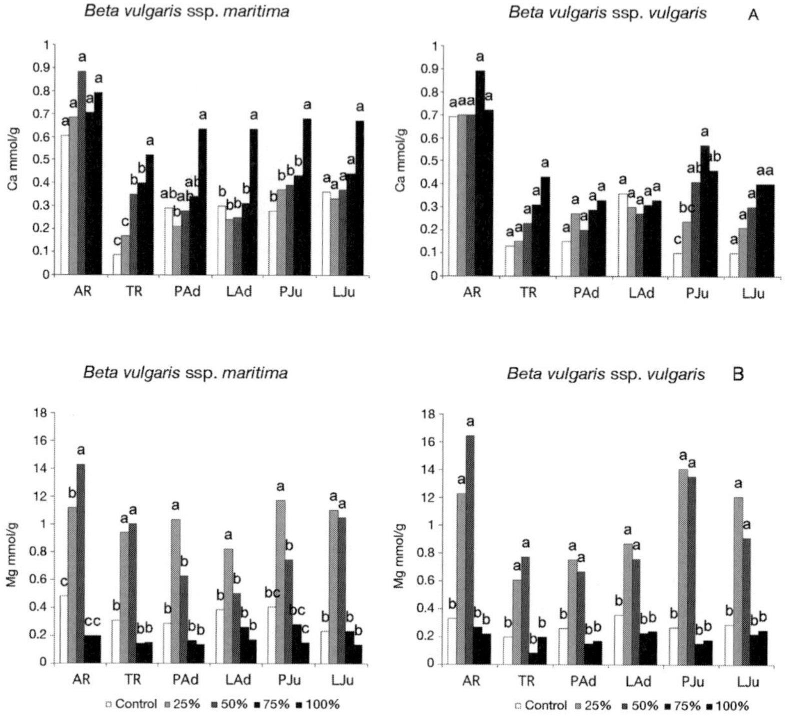


**Figure 5.** Effect of tap water (control), 25%, 50%, 75% and 100% seawater salinities on Ca (A) and Mg (B) concentrations of adventitious roots (AR); tap root (TR); petiole and lamina of adult leaves (PAd, LAd) and petiole and lamina of juvenile leaves (Pju, Lju) of *Beta vulgaris* ssp. *maritima* and *Beta vulgaris* ssp. *vulgaris*. Bars with the same letter are not significantly different for different treatments at $p \leq 0.05$.

are the absorption site of water and minerals by plants. Mineral analysis shows that the changes in water and osmotic potentials in leaves of sea beet and sugar beet plants in saline conditions are due to changes of their mineral composition. However, $K^+$, $Ca^{2+}$, $Na^+$, and $Mg^{2+}$ cannot solely explain the low osmotic potential values, especially in the tap root of the two beet subspecies (Figs 3–5); other solutes contribute to that, particularly organic solutes, which are mainly carbohydrates (Fig. 6). In the tap root of the two beet subspecies, carbohydrates concentrations, especially sucrose in saline treatments, were high enough to compensate the osmotic stress imposed by saline treatments. At the adventitious root level, the carbohydrates status was also improved but to a lower degree compared to the tap root. Salinity increased carbohydrates reserves, mainly glucose and fructose, in the leaves but at a lower degree compared to adventitious roots.

A positive relationship between the increase of salinity and amino acid accumulation was found in different parts of both beet subspecies (Fig. 7).

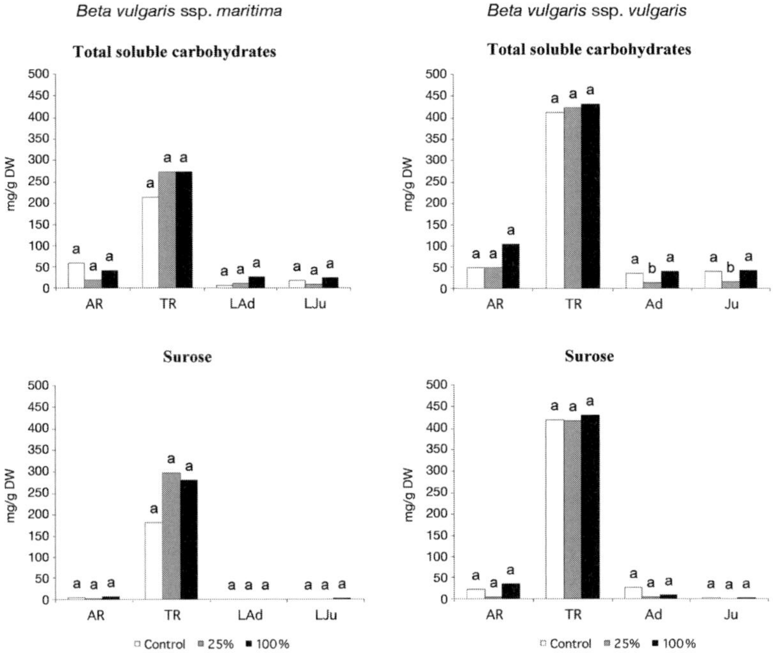


**Figure 6.** Total soluble carbohydrates and sucrose concentrations (mg/g dry weight) of adventitious roots (AR), tap root (TR), adult leaves (LAd) and juvenile leaves (LJu) of *Beta vulgaris* ssp. *maritima* and *Beta vulgaris* ssp. *vulgaris* irrigated with tap water (Control), 25% seawater and 100% seawater treatments. Bars represent the mean of three repetitions. Bars with the same letter are not significantly different for different treatments at $p \leq 0.05$.

Among a number of amino acids tested, glutamine and proline contributed the most to osmotic adjustment in roots. Their concentrations were 6.4 and 61 times higher, respectively, in the tap roots of sea beet plants growing in 100% seawater treatment compared to the control treatment without salt. However, these proportions were much lower for the sugar beet, in which the tap root accumulation of glutamine and proline were increased 2.4 and 16.6 times, respectively, in high salinity treatment (100%) compared to the control.

Colmer et al. [28], Hasegawa et al. [29] and Rhode et al [30] reported that the accumulation of proline in root tips can facilitate osmotic adjustment and sustain root growth in saline conditions, and thus contribute to salt tolerance. The difference in the response of sea beet and sugar beet to the synthesis and accumulation of carbohydrates and amino acids at the tap root level in saline conditions is probably due to the difference in their degree of salinity tolerance. While sea beet showed a considerable increase in carbohydrates and amino acids concentrations, sugar beet kept its carbohydrates levels

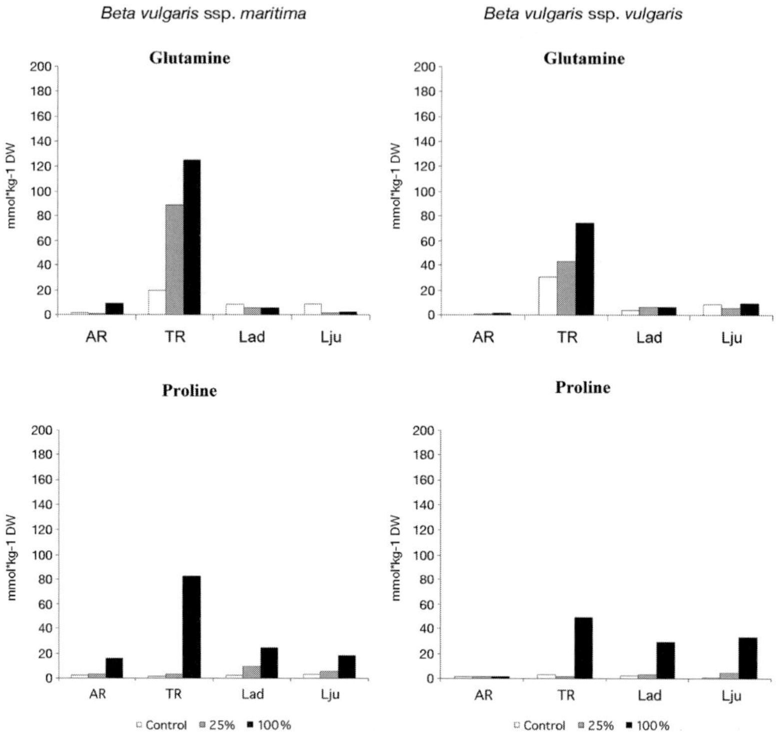


**Figure 7.** Glutamine and proline concentrations (mmol/kg dry weight) of adventitious roots (AR), tap root (TR), adult leaves (LAd) and juvenile leaves (LJu) of *Beta vulgaris* ssp. *maritima* and *Beta vulgaris* ssp. *vulgaris* irrigated with tap water (Control), 25% and 100% seawater treatments.

– mainly sucrose – unchanged and presented a relatively low increase of amino acids mainly glutamine and proline compared to sea beet. Morphologically, the osmotic adjustment of the two beet subspecies in saline treatments was expressed by the increase in leaf succulence, which was reached mainly by the lengthening of palisade parenchyma cells, and the stretch of spongy parenchyma cells. In these conditions the intercellular space shrinks and the resistance against water diffusion increases, which minimizes water loss during gas exchange through stomata. Adaptation of the two beet subspecies to seawater treatments is also expressed by reducing leaf stomatal conductance and consequently decreasing leaf transpiration. Stomatal conductance and transpiration of sea beet leaves were reduced to a higher degree than those of sugar beet (Fig. 8), suggesting that sea beet is more capable of regulating its transpiration to limit water losses in high salinity conditions. This was accomplished by increasing leaf stomata frequency relative to increasing salinity in the culture medium (Fig. 9) to minimize the opening time of stomata to reduce water loss [3, 31].

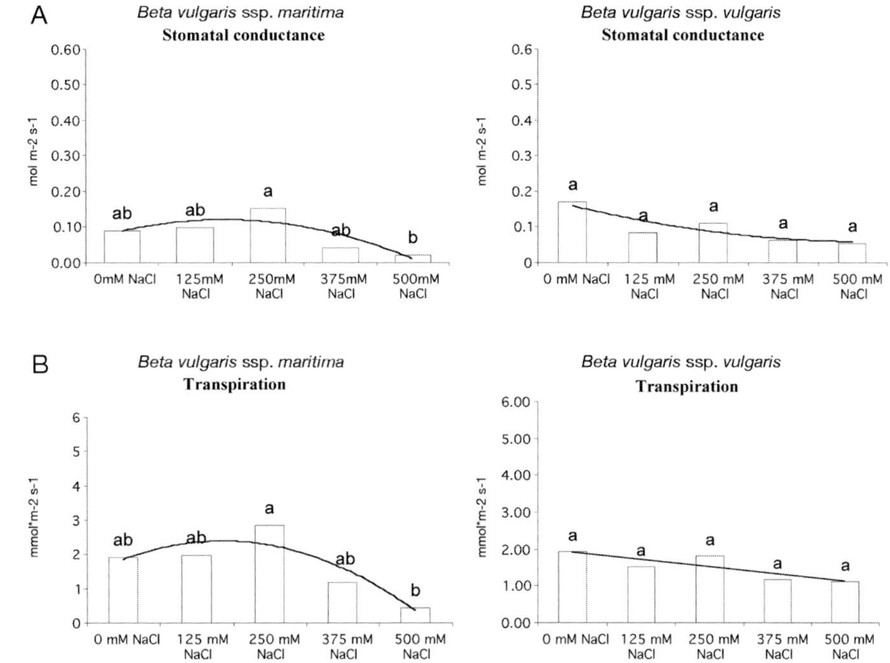


**Figure 8.** Stomatal conductance (mol $m^{-2}$ $s^{-1}$) (A) and transpiration (mmol $m^{-2}$ $s^{-1}$) (B) adult leaves of *Beta vulgaris* ssp. *maritima* and *Beta vulgaris* ssp. *vulgaris* of control (0 mM NaCl ) and 4 salinity treatments (125, 250, 375 and 500 mMol. $m^{-3}$ NaCl). Bars with the same letter are not significantly different for different treatments at $p \leq 0.05$.

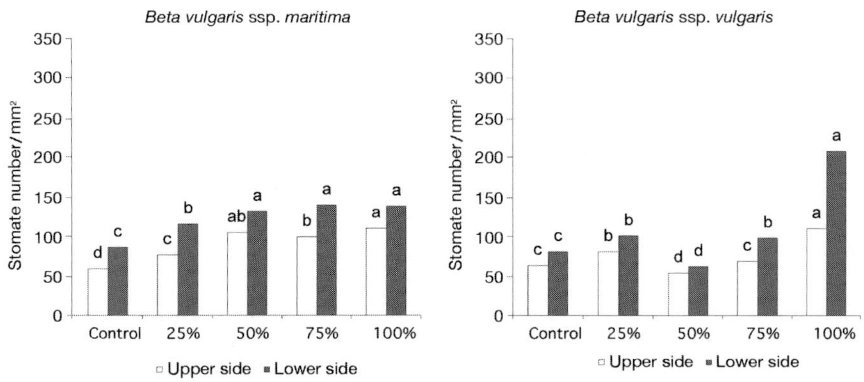


**Figure 9.** Effect of seawater salinity on stomata numbers per $mm^2$ of the upper and lower sides of adult leaves of *Beta vulgaris* ssp. *maritima* and *Beta vulgaris* ssp. *vulgaris*. Bars with the same letter are not significantly different for different treatments at $p \leq 0.05$.

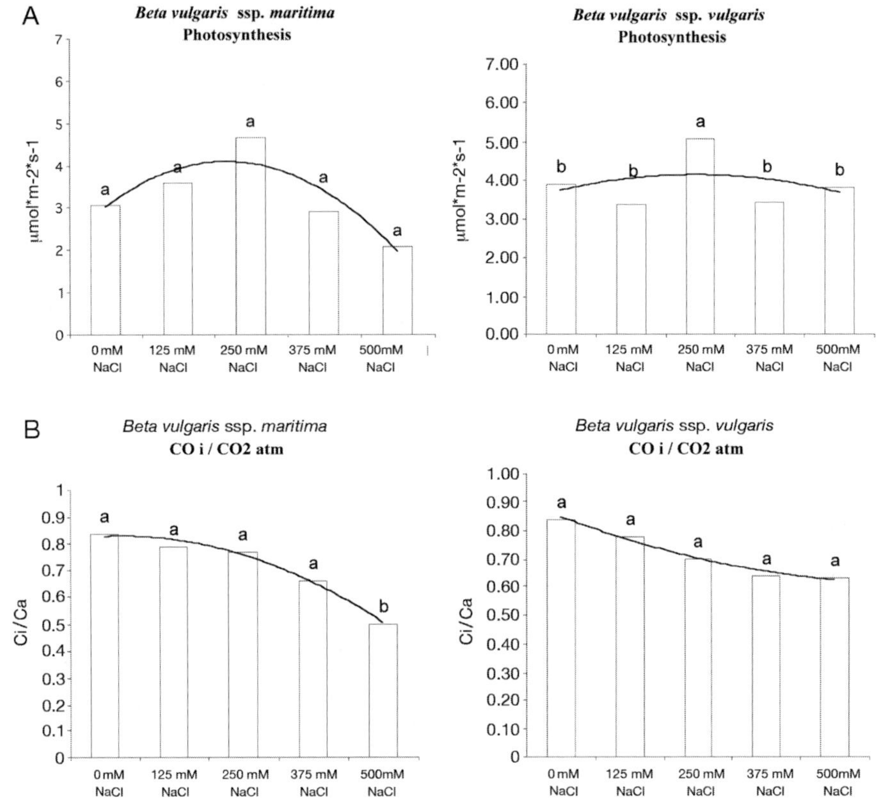


**Figure 10.** Photosynthesis (µmol $m^{-2}$ $s^{-1}$) (A) and $CO_2$ intracellular and $CO_2$ atmospheric ratio (Ci/Ca) (B) of adult leaves of *Beta vulgaris* ssp. *maritima* and *Beta vulgaris* ssp. *vulgaris* of control (0 mM NaCl) and four salinity treatment (125, 250, 375 and 500 mMol. $m^{-3}$ NaCl). Bars with the same letter are not significantly different for different treatments at $p \leq 0.05$.

Movements of stomata guard cells and photosynthesis rate are closely coordinated. Stomata opening is reduced to optimize water-use efficiency (WUE) to maintain water loss at a minimal value [32]. Photosynthetic activity of the two beet subspecies was stimulated by moderate salinity; it reached maximum values in 250 mM NaCl and was slightly reduced at high salinity (Fig. 10).

This leads to a high WUE in saline conditions. WUE conservation contributes to maintaining favorable ionic [33] and water [34] relations. Flowers et al. [9] and Osmond et al. [35] reported that salinity affects plant carbon assimilation through anatomical and morphological leaf changes, and through reduction of leaf area rather than the reduction of photosynthetic activity. However, the high stomata number in saline conditions enables the two beet subspecies to assimilate high amounts of $CO_2$, while keeping a low transpiration level.

During the day, low stomatal conductance helps the plant to minimize water loss. A high photosynthetic rate in high salinity levels is probably associated with a sufficient osmotic adjustment in leaves in response to low water potential in the irrigation solution and to the regulation of minerals distribution through cellular compartmentation [36]. Thus, $Na^+$ and $Cl^-$, which are toxic ions, are important in the vacuoles of leaf cells to maintain metabolic activities in cytoplasm of plants growing in saline habitats.

## Conclusion

Salt tolerance of sea beet and sugar beet is a response at the level of the whole plant, and a number of metabolic processes are involved. It is regulated by several interrelated parameters. The combination and the balance of these parameters can be used to describe salinity tolerance of each species, and it is not possible to evaluate the importance of one parameter without knowing the others.

### *Acknowledgements*

This work was carried out in the framework of the Concerted Action Project "Sustainable Utilization of Halophytes in the Mediterranean and Subtropical Dry Regions". The authors would like to acknowledge the facilities provided by the Justus-Liebig Institute of Giessen where some of the experiments were undertaken. The help of the Institute's technical staff is fully appreciated.

## References

1 Greenway H, Munns R (1980) Mechanisms of salt tolerance in non halophytes. *Annu Rev Plant Physiol* 31: 149–190

2 Marschner H (1995) *Mineral nutrition of higher plants*. Academic Press, London, 889

3 Koyro H-W, Huchzermeyer B (1999) Influence of high NaCl-salinity on growth, water and osmotic relations of the halophyte *Beta vulgaris* ssp. *maritima* – Development of a quick check system. In: A Hamdy, H Lieth, M Todorovic, M Moschenko (eds): *Halophyte uses in different climates*. Backhuys, Leiden, 89–103

4 Cheeseman JM (1988) Mechanisms of salinity tolerance in plants. *Plant Physiol* 87: 547–550

5 Breckle S-W (2002) Salinity, halophytes and salt affected natural ecosystems. In: A Läuchli, U Lüttge (eds): *Salinity: Environment – Plants – Molecules*. Kluwer Academic Publishers, Dordrecht, 53–77

6 Pitman M G, Läuchli A (2002) Global impact of salinity and agricultural ecosystems. In: A Läuchli, U Lüttge (eds): *Salinity: Environment – Plants – Molecules*. Kluwer Academic Publishers, Dordrecht, 3–20

7 Munns R (2002) Salinity, growth and phytohormones. In: A Läuchli, U Lüttge (eds): *Salinity: Environment – Plants – Molecules*. Kluwer Academic Publishers, Dordrecht, 271–290

8 Munns R (1993) Physiological processes limiting plant growth in saline soils: Some dogmas and hypothesis. *Plant Cell Environ* 16: 15–24

9 Flowers TJ, Hajibagheri MA, Clipson NJW (1986) Halophytes. *Q Rev Biol* 61: 313–337

10 Ungar IA (1991) *Ecophysiology of vascular halophytes*. CRC Press, Boca Raton

11 Kinzel H (1982) *Pflanzenökologie und Mineralstoffwechsel*. Eugen Ulmer, Stuttgart, 534

12 Yeo AR, Flowers TJ (1980) Salt tolerance in the halophyte *Suaeda maritima* (L.) Dum: Evaluation of the effect of salinity upon growth. *J Exp Bot* 31: 1171–1183

13 Kelley DB, Goodin JR, Miller DR (1982) Biology of Atriplex. *Tasks Veg Sci* 2: 79–107

14 Munns R, Greenway H, Delane R, Gibbs J (1982) Ion concentration and carbohydrate status of the elongating leaf tissue of *Hordeum vulgare* growing at high external NaCl. II. Cause of the growth reduction. *J Exp Bot* 33: 574–583

15 Flowers TJ, Läuchli A (1983) Sodium *versus* potassium: Substitution and compartmentation. *Encyclopaedia of Plant Physiology*, New series. 15B: 651–681

16 Harrouni MC, Daoud S, El Alami A, Debbagh B, Choukr-Allah R, Bengueddour R (1999) Responses of some halophytes to seawater irrigation in Morocco. *Biometeorology* 14: 57–75

17 Daoud S, Koyro H-W, Harrouni MC (2003) Salinity tolerance of *Beta vulgaris* ssp. *maritima*. Part I. Biomass production and osmotic adjustment. In: H Lieth (ed): *Cash crop halophytes: Recent studies.* Kluwer Academic Publishers, Dordrecht, 41–49

18 Harrouni MC, Daoud S, Koyro H-W (2003) Effect of seawater irrigation on biomass production and ion composition of seven halophytic species in Morocco. In: H Lieth (ed): *Cash crop halophytes: Recent studies.* Kluwer Academic Publishers, Dordrecht, 59–70

19 Munns R, Termatt A (1986) Whole-plant responses to salinity. *Aust J Plant Physiol* 13: 143–160

20 Koyro H-W, Huchzermeyer B (1999) Salt and drought stress effects on metabolic regulation in maize. In: M Pessarakli (ed): *Handbook of plant and crop stress.* Springer, Dordrecht, 843–878

21 Flowers TJ, Troke PF, Yeo AR (1977) The mechanism of salt tolerance in halophytes. *Annu Rev Plant Physiol* 28: 89–121

22 Munns R, Greenway, Kirst GO (1983) Halotolerant eucaryotes. In: OL Lange, P Nobel, CB Osmond, H Ziegler (eds): *Physiological Plant Ecology, II, Encyclopedia of Plant Physiology,* New Series 12C. Springer, Berlin, 59–135

23 Niu X, Bressan RA, Hasegawa PM, Pardo JM (1995) Ion homeostasis in NaCl stress environments. *Plant Physiol* 109: 735–742

24 Wyn Jones RG, Gorham J (2002) Intra-and inter-cellular compartmentation of ions, a study in specificity and plasticity. In: A Läuchli, U Lüttge (eds): *Salinity: Environment – Plants – Molecules.* Kluwer Academic Publishers, Dordrecht, 159–180

25 Koyro H-W (1999) Protocol for the installation of an improved quick check system in the I.A.V. Hassan II (Agadir, Morocco). *Third annual report: Sustainable Halophyte Utilisation in the Mediterranean and Subtropical Dry regions.*

26 Stelzer R, Laüchli A (1978) Salt and flooding tolerance of *Puccinellia peisonis*. III. Distribution and localisation of ions in the plant. *Z Pflanzenphysiol* 88: 437–448

27 Mühling KH, Läuchli A (2001) Physiological traits of sodium toxicity and salt tolerance. In: WJ Horst et al (eds): *Plant nutrition-Food security and sustainability of agro-ecosystems.* Kluwer Academic Publishers. Dordrecht, 378–379

28 Colmer TD, Fan TWM, Higashi RM, Läuchli A (1996) *Physiol Plant* 97: 421–424

29 Hasegawa PM, Bressan RA, Zhu J-K, Bohnert HJ (2000) Plant cellular and molecular responses to high salinity. *Annu Rev Plant Physiol Plant Mol Biol* 51: 463–499

30 Rhode D, Nadolska-Orczyk A, Rich PJ (2002) Salinity, osmolytes and compatible solutes. In: A Läuchli, U Lüttge (eds): *Salinity: Environment – Plants – Molecules*. Kluwer Academic Publishers, Dordrecht, 181–204

31 Daoud S, Koyro H-W, Harrouni MC, Schmidt A, Papenbrock J (2003) Salinity tolerance of *Beta vulgaris* ssp. *maritima*. Part II. Physiological and biochemical regulation. In: H Lieth (ed): *Cash crop halophytes: Recent studies*. Kluwer Academic Publishers, Dordrecht, 51–58

32 Ball M, Farquhar GD (1984) Photosynthetic and stomatal responses of two mangrove species, *Aegiceras corniculatum* and *Avicennia marina*, to long term salinity and humidity conditions. *Plant Physiol* 74: 1–6

33 Ball MC (1988) Salinity in the mangroves *Aegiceras corniculatum* and *Avicennia marina* I. water use in relation to growth, carbon partitioning and salt balance. *Aust J Plant Physiol* 15: 447–464

34 Sperry JS, Donnely JR, Tyree MT (1988) A method for measuring hydrolic conductivity and embolism in xylem. *Plant Cell Environ* 11: 35–40

35 Osmond CB, Winter K, Ziegler H (1982) Functional significance of different pathways of $CO_2$ fixation in photosynthesis. In: OL Lange, PS Nobel, CB Osmond, H Ziegler (eds): *Encyclopedia of plant ecology. Physiological plant ecology. II: Water relations and carbon assimilation.* New Series vol. 12. B. Springer, Berlin, 479–547

36 Gonasageran N, Jahnke J, Von Willert DJ (1995) Gas exchange responses of the C4 grass, *Sporobolus virginicus*, to salinity stress. In: M Ajmal Khan, IA Ungar (eds): *Biology of salt tolerant plants.* University of Karachi, Pakistan, 121–130

Biosaline Agriculture and High Salinity Tolerance
Edited by Chedly Abdelly, Münir Öztürk, Muhammad Ashraf and Claude Grignon


# Salinity influence on soil microbial population metabolism and enzymatic activities in lysimeter-grown *Olea europaea* and *Nicotiana glauca*

Adalgisa Belligno[1], Marco Russo[1], Vito Sardo[2] and Ju Ying Wu[3]

[1] *Dipartimento di Scienze Agronomiche, Agrochimiche e delle Produzioni Animali, University of Catania, Via S. Sofia 98, 95123 Catania, Italy*
[2] *Department of Agricultural Engineering, University of Catania, Via S. Sofia 100, 95123 Catania, Italy*
[3] *Beijing Research Center for Grass and Environment, Banjing Haidian District, 100097 Beijing, China*

## Abstract

Since soil microorganisms are quite sensitive to changes in their habitat, their response in terms of selected metabolic and enzymatic activities was investigated as a possible indicator of the effects of saline irrigation in lysimeter-grown plants of *Olea europea* and *Nicotiana glauca*. Water electrical conductivity ranged from 0.8 dS/m in the control to 8.9, 17.5, and 26.2 dS/m in treatments with 20%, 40% and 60% diluted seawater, respectively. While some results were non-significant, the following main conclusions could be drawn: Microbial biomass carbon, soluble soil carbon and respiration activity were not significantly correlated to salinity. In addition, the microbial biomass nitrogen was not correlated to salinity. Contrary to this, enzyme activities as shown by hydrolysis rate of fluorescein diacetate (FDA), acid and alkaline phosphatase, and β-glucosidase decreased consistently with salinity without any significant difference between the two plant species. FDA gave the most sensitive and consistent response.

## Introduction

The ever-growing pressure on existing water resources necessitates the use of water that was believed unfit for a sustainable long-term irrigation, particularly water rich in sodium chloride, which is widely available. This makes the problem of assessing the long-term sustainability complex.

Since microorganisms in the soil are responsible for nutrient and energy cycles [1] as well as for the decay of pollutants [2] and the elaboration of humus from organic matter [3], it is of interest to consider their response to salinity as they are particularly sensitive to salt stress [4]. Because of this sensitivity to salinity-induced changes in their habitat, monitoring variations in their metabolic and enzyme activities can be expected to be an effective

**Table 1.** Water electrical conductivity (EC) and total applied NaCl amounts.

| Treatment | Water EC (dS/m) | Added NaCl (g/plant) |
|---|---|---|
| 0% seawater (C) | 0.8 | 155 |
| 20% seawater (S1) | 8.9 | 1384 |
| 40% seawater (S2) | 17.5 | 2722 |
| 60% seawater (S3) | 26.2 | 4076 |

method for assessing the effects of salts on the soil-plant system; the response of microorganisms is also known to be influenced by the root system of plants present in the soil.

The above considerations prompted the present research, aimed at gaining a better insight into the impact of salinity on the soil-plant system and the possibility of using microbial activity variations as a warning tool in evaluating the sustainability of irrigating with saline water.

## Material and methods

The experimental site is located south of Catania, Sicily, along the coast of the Ionian Sea (37° 20' N) and is equipped with a meteo station including a thermohygrograph, an anemometer, a class A evaporimeter, a pluviograph and a heliophanograph.

Eighteen 1-year old plants of olive (*Olea europea* L.) and nicotiana (*Nicotiana glauca* G.) were planted in 20-L microlysimeters, filled with inert quartz sand and freely draining into 20-L tanks. Plants were submitted to four irrigation treatments with various levels of diluted seawater: 0% (control, C), 20% (S1), 40% (S2), and 60% (S3). Such constant ratios were obtained by adopting the "double lateral" solution, namely one dripping lateral conveying fresh water and a second one conveying seawater; since every plant was irrigated by a total of five emitters, the desired constant ratios were obtained by arranging them in a way that freshwater was applied by 5, 4, 3, 2 emitters while correspondingly seawater was applied by 0, 1, 2, 3 emitters. Water was applied daily using an automatic system that included a control station, electric valves, totalizing water meters, 16-mm polythene laterals and self-compensating 2 L/h drippers, as described in [5]. Plants were put into the microlysimeters in June and were initially, for 1 month, irrigated exclusively with fresh water. Potential evapotranspiration (PET) was determined as the running average of weekly results given by a class A evaporation pan and those given by the Hargreaves-Samani equation, which have been shown previously to give the best results [6].Water volumes applied to the plants were determined by multiplying PET values by 1.8 to take into account an appropriate leaching fraction: the resulting water amount applied in the period 20 July to 10 November was 223 L, and was the same for all the plants. The NaCl quantities resulting from the various

**Table 2.** Evolution of pH and EC*e* (dS/m) values in the soil.

| | | Control (C) | | 20% seawater (S1) | | 40% seawater (S2) | | 60% seawater (S3) | |
|---|---|---|---|---|---|---|---|---|---|
| | | **Aug** | **Nov** | **Aug** | **Nov** | **Aug** | **Nov** | **Aug** | **Nov** |
| Nicotiana | pH | 8.86 | 9.05 | 8.16 | 9.01 | 8.26 | 9.11 | 8.78 | 9.11 |
| | EC | 0.36 | 0.71 | 5.97 | 12.13 | 12.85 | 26.17 | 16.98 | 37.49 |
| Olive | pH | 8.67 | 8.91 | 8.41 | 9.01 | 8.78 | 9.04 | 9.04 | 9.34 |
| | EC | 0.76 | 0.8 | 8.07 | 10.8 | 19.13 | 21.7 | 29.43 | 32.6 |

treatments are reported in Table 1. About 50% of added NaCl was leached out through drainage.

The soil was a quartz sand poor in carbon (total organic carbon, TOC, 0.44%), total nitrogen (0.04%), total and available phosphorus (1.12 g/kg and 6.02 mg/kg, respectively); it was analyzed in August and November at the depth 0–10 cm.

The saturated soil electrical conductivity (ECe) and the pH are reported in Table 2. Soil pH was determined potentiometrically in an aqueous solution (ratio 1:2.5), total nitrogen by distillation according to Kjeldahl method; and TOC using potassium dichromate as an oxidant for organic matter.

The following determinations were performed on soil samples dried to 10% and sieved to 2 mm:

- Respiration activity, according to [7]
- Soluble C and N by determining microbial biomass according to the method of "fumigation-extraction" [8], by dosing C and N in microbial biomass carbon (MBC) and microbial biomass nitrogen (MBN), according to [9]
- Mineral N in its $NO_3^-$ and $NH_4^+$ forms, according to [10], modified
- Hydrolysis rate of 3,6-diacetyl fluorescein (FDA), according to [11]
- Acid and alkaline phosphatase, according to [7]
- β-glucosidase, according to [12]
- The respiration ratio, calculated as the ratio of MBC to $CO_2$ emitted in 24 h, indicating the activity of microbial biomass: it is an indicator of energy required to maintain the organic C balance.

The results were statistically analyzed using the program Statgraphics for determining two-way ANOVA ($p \leq 0.05$), standard deviations and variation coefficients.

## Results

In August, the MBC values in olives were significantly higher in saline treatments than in the control, with the highest value being with treatment S3

**Table 3.** Evolution of microbial biomass carbon (MBC) and respiration activity.

| **Parameters** | **Month** | **Olive** | | | |
|---|---|---|---|---|---|
| | | **C** | **S1** | **S2** | **S3** |
| MBC (mg/kg soil) | Aug | 54.20c | 74.40b | 92.20a | 94.20a |
| | Nov | 35.70d | 38.60cd | 36.20d | 47.30c |
| Sol C in MBC (mg/kg soil) | Aug | 140.10e | 169.30cd | 176.80c | 268.70a |
| | Nov | 186.00c | 125.90e | 155.60d | 205.50b |
| C–$CO_2$ (mg/kg DM) | Aug | 2.40d | 3.45bc | 2.95c | 3.01c |
| | Nov | 4.58ab | 4.44ab | 5.32a | 4.15b |
| $qCO_2$ (µg C–$CO_2$/ mg MBC) | Aug | 44.40d | 46.60d | 56.10c | 56.60c |
| | Nov | 127.20a | 113.80a | 83.60b | 88.30b |

| **Parameters** | **Month** | **Nicotiana** | | | |
|---|---|---|---|---|---|
| | | **C** | **S1** | **S2** | **S3** |
| MBC (mg/kg soil) | Aug | 32.20b | 22.85c | 17.20d | 21.70c |
| | Nov | 51.70a | 24.90c | 37.70ab | 46.15a |
| Sol C in MBC (mg/kg soil) | Aug | 198.80b | 264.00a | 241.60ab | 262.20a |
| | Nov | 214.60b | 264.70a | 204.40b | 236.30ab |
| C–$CO_2$ (mg/kg DM) | Aug | 2.41d | 2.80c | 3.78bc | 3.72b |
| | Nov | 4.17ab | 4.41a | 4.26a | 4.68a |
| $qCO_2$ (µg C–$CO_2$/ mg MBC) | Aug | 99.40de | 104.80d | 250.60a | 127.30c |
| | Nov | 80.20c | 97.90de | 176.40b | 101.80d |

DM, dry matter

(Tab. 3); all such values had decreased consistently by November, when ECe values became higher. In contrast, in nicotiana, the highest MBC values were recorded for the control in both months (Tab. 3): in fact, at both sampling times, plants in treatments S1, S2 and S3 showed a lower MBC content compared to the control, although values increased from August to November, i.e., a reverse of the trend seen in olives. Thus, there is no consistent response to salinity. Soluble carbon (sol C) was lower in the control for both species in August, whereas in November the responses were not consistent. Respiration activity (C–$CO_2$) was generally, although not always, lower in the control for both species at both sampling months, and consistently increased with time (Tab. 3), but gave no significant response to salinity. The respiration ratio ($qCO_2$) values showed an opposite trend in the soils with olives to those with nicotiana, sharply increasing with time in the former, but decreasing in the latter; once again the response to soil salinity was not consistent.

MBN values decreased steadily with time in both plant species, with somewhat higher values in olive than in *Nicotiana* (Tab. 4) and the same decreasing trend with time was evident in all the N mineral fractions; however, no significant response was found to salinity. The MBC to MBN ratios in all cases were lower in the control in olive plants than in *Nicotiana*; however, for both

**Table 4.** Evolution of microbial biomass nitrogen (MBN) and its mineral fractions in olive and nicotiana.

| **Parameters (mg/kg soil)** | **Month** | **Olive** | | | |
|---|---|---|---|---|---|
| | | **C** | **S1** | **S2** | **S3** |
| MBN | Aug | 12.86a | 9.45bc | 9.30c | 10.96b |
| | Nov | 7.95d | 3.87g | 6.35e | 4.74f |
| Sol N (MBN) | Aug | 9.28ab | 6.98c | 8.11b | 10.81a |
| | Nov | 3.10e | 3.33e | 2.94e | 5.98d |
| $N–NO_3+NO_2$ | Aug | 5.23a | 3.37cd | 3.60c | 3.14d |
| | Nov | 4.48ab | 2.91de | 2.58e | 2.82de |
| $N–NH_4$ | Aug | 4.81a | 4.05b | 4.85a | 4.85a |
| | Nov | 3.83b | 3.83b | 4.05b | 4.05b |
| Mineral N | Aug | 10.04a | 7.42bc | 8.46b | 10.04a |
| | Nov | 8.31b | 6.74c | 6.64c | 7.67bc |

| **Parameters (mg/kg soil)** | **Month** | **Nicotiana** | | | |
|---|---|---|---|---|---|
| | | **C** | **S1** | **S2** | **S3** |
| MBN | Aug | 5.78a | 5.16a | 2.69c | 2.59cd |
| | Nov | 3.16b | 3.34b | 2.27d | 3.66b |
| Sol N (MBN) | Aug | 10.45a | 10.21a | 11.35a | 10.41a |
| | Nov | 3.21d | 3.77cd | 4.22c | 5.50b |
| $N–NO_3+NO_2$ | Aug | 5.97a | 4.82b | 3.94c | 4.43bc |
| | Nov | 2.04d | 2.03d | 1.29e | 1.27e |
| $N–NH_4$ | Aug | 4.01a | 4.07a | 4.09a | 4.05a |
| | Nov | 4.05a | 2.72b | 4.00a | 4.13a |
| Mineral N | Aug | 10.03a | 8.88b | 8.00b | 8.49b |
| | Nov | 6.10d | 6.20cd | 5.35d | 5.33d |

species an increase was evident with time (Tab. 5), as an obvious consequence of MBN decrease reported above. The soluble-to-mineral N ratio (Tab. 5) was relatively constant between the two plant species in all the treatments, although there was a decrease from August to November. The enzyme activities increased with time for both plant species (Tab. 6); more interestingly, their activity in all cases was higher in the control, and evidenced a consistent reduction with salinity both in August and November.

Multiple regressions gave some interesting insights into the plant-soil relationships. Acid phosphatase and FDA on olea and nicotiana showed a stricter correlation to independent variables than simple regressions when simultaneously considering soil EC and pH (Tabs 7 and 8).

Figure 1 shows quite similar trends of FDA reduction in the soil under olive (slope 1.47) and nicotiana (slope 1.34), which suggests, at least in a first approximation, that this parameter may be useful to represent, in an integrated way, the salinity impact on the plant and the soil. Of course, such first results need to be confirmed and extended through further research.

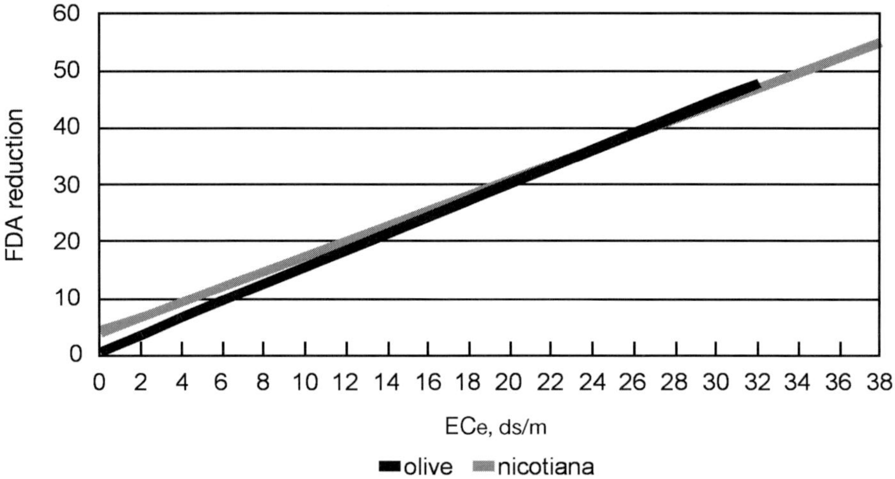


**Figure 1.** The parallel evolution of fluorescein diacetate (FDA) reduction (%) in olive and nicotiana as affected by electrical conductivity (ECe) values.

**Table 5.** Variations in time of MBC/MBN and N sol /N min ratios.

| Parameters | Month | Olive | | | | Nicotiana | | | |
|---|---|---|---|---|---|---|---|---|---|
| | | C | S1 | S2 | S3 | C | S1 | S2 | S3 |
| MBC/MBN | Aug | 0.2 | 7.9 | 9.9 | 3.3 | 5.6 | 4.3 | 6.4 | 8.4 |
| | Nov | 4.5 | 10.0 | 14.8 | 10.0 | 16.4 | 7.4 | 7.6 | 12.6 |
| N sol/N min | Aug | 0.9 | 0.9 | 1.0 | 1.1 | 1.0 | 1.1 | 1.4 | 1.2 |
| | Nov | 0.4 | 0.5 | 0.4 | 0.8 | 0.5 | 0.6 | 0.8 | 1.0 |

## Discussion and conclusions

Microbiological metabolism and enzyme activities are influenced by treatments and plant species. The plant influence on microbial biomass was demonstrated by its different responses, in terms of MBC and MBN and of their ratios, in the rhizosphere of the two plant species.

Respiration increase in saline treatments demonstrates an intense process of mineralization depending both on climatic conditions [13] and the high electric conductivity of the soil. Microbial population used available organic carbon in response to an increased energy requirement, thus shifting the balance towards oxidative rather than growth processes. An increasing soluble microbiological C fraction is related to a decreasing soluble N: prevailing mineralization processes determined an increase in oxidized N forms ($NO_3^-$, $NO_2^-$), compared to the others, while the sandy soil encouraged their leaching. This makes a thorough interpretation of MBN data rather difficult, since the

**Table 6.** Evolution of enzyme activities FDA (% FDA/h per g DM), acid phosphatase, alkaline phosphatase and β-glucosidase (all in mmol *p*-nitrophenol/g DM) in olive and nicotiana – in italics % reduction compared to the control is reported.

| **Parameter** | **Month** | **Olive** | | | |
|---|---|---|---|---|---|
| | | **C** | **S1** | **S2** | **S3** |
| FDA(% reduction) | Aug | 11.57c | 9.97d<br>*13.83* | 7.78e<br>*32.76* | 5.49f<br>*52.55* |
| | Nov | 21.23a | 16.59b<br>*21.86* | 16.56b<br>*22.00* | 11.93d<br>*43.81* |
| Acid phosphatase (% reduction) | Aug | 0.08c | 0.07c<br>*12.50* | 0.04d<br>*50.00* | 0.04d<br>*50.00* |
| | Nov | 0.14a | 0.11b<br>*21.43* | 0.11b<br>*21.43* | 0.10b<br>*28.57* |
| Alkaline phosphatase (% reduction) | Aug | 0.08c | 0.07c<br>*12.5* | 0.02d<br>*75.00* | 0.03d<br>*62.50* |
| | Nov | 0.20a | 0.18a<br>*10.0* | 0.18a<br>*10.0* | 0.15b<br>*25.00* |
| β-glucosidase (% reduction) | Aug | 0.06b | 0.05b<br>*16.67* | 0.03cd<br>*50.00* | 0.02d<br>*66.67* |
| | Nov | 0.11a | 0.06b<br>*45.46* | 0.04c<br>*63.63* | 0.04c<br>*63.63* |

| **Parameter** | **Month** | **Nicotiana** | | | |
|---|---|---|---|---|---|
| | | **C** | **S1** | **S2** | **S3** |
| FDA(% reduction) | Aug | 10.84c | 8.63d<br>*20.39* | 7.51d<br>*30.72* | 7.37d<br>*32.01* |
| | Nov | 21.93a | 19.67a<br>*10.31* | 13.23b<br>*39.67* | 10.85c<br>*50.52* |
| Acid phosphatase (% reduction) | Aug | 0.07bc | 0.08a<br>- | 0.06c<br>*14.29* | 0.03d<br>*57.14* |
| | Nov | 0.16a | 0.08b<br>*50.00* | 0.08b<br>*50.00* | 0.07bc<br>*56.25* |
| Alkaline phosphatase (% reduction) | Aug | 0.16b | 0.12c<br>*25.00* | 0.12c<br>*25.00* | 0.08d<br>*50.00* |
| | Nov | 0.24a | 0.21a<br>*12.50* | 0.15b<br>*37.50* | 0.13c<br>*45.83* |
| β-glucosidase (% reduction) | Aug | 0.06a | 0.05ab<br>*16.67* | 0.05ab<br>*16.67* | 0.04b<br>*33.33* |
| | Nov | 0.05ab | 0.05ab<br>- | 0.06a<br>- | 0.04b<br>*20.00* |

mineral N decrease (particularly at the end of the irrigation season) can be also related to such leaching processes [14].

In the saline treatments the levels of soluble N in microbial biomass (solN) were less than in the control, and the ratio sol N/min N was higher, denouncing a loss of min N as a consequence of the presence of NaCl.

**Table 7.** Correlation coefficients between ECe and reduction in enzyme activities compared to the control (ns, not significant).

| Enzyme activity | Correlation coefficient r (%) | |
|---|---|---|
| | Olive | Nicotiana |
| FDA | 94.73 | 92.52 |
| Acid phosphatase | 79.13 | 81.31 |
| Alkaline phosphatase | ns | 81.74 |
| β-Glucosidase | 93.54 | ns |

**Table 8.** Coefficients of determination in the multiple regression between two enzyme activities (dependent variables) and soil pH + EC (independent variables)

| Enzyme activity | Determination coefficient $R^2$ (%) | |
|---|---|---|
| | Olive | Nicotiana |
| FDA | 90.63 | 96.61 |
| Acid phosphatase | 70.07 | 71.22 |

In conclusion, no consistent significant relationship was evident between microbiological metabolism and salinity in relation to the different impact of the plant species on the soil. In contrast, the variations in enzyme activities among the treatments at the end of the experiment rather reflect consistently the plant response to seawater application, although with some differences between *Olea* and *Nicotiana*. The different responses of acid and alkaline phosphatases can be linked to soil reaction, since reportedly their release and synthesis by microorganisms are dependent on soil pH [12], a fundamental factor in triggering such activities.

Seawater addition, although decreasing the intensity of enzyme activities compared to the control, provided pH gradients more favorable to the activation of alkaline than acid phosphatases. The response to salinity of β-glucosidase was similar to that of acid phosphatase due to its specificity for acid environments. The reaction of FDA to soil salinity levels was the most evident and consistent, with highly significant correlation coefficients, following a practically parallel pathway in the two plant species. Thus, it may be possible to use FDA as an integrated indicator of plant and soil response to salinity. The combined action of EC and pH on FDA and acid phosphatase (Tab. 8) demonstrates the interactions occurring between soil (through EC) and plant (through pH) at the rhizosphere level.

## References

1 Schnurer J, Rosswall T (1990) Hydrolysis fluorescein diacetate as a measure of total microbial activity in soil and litter applied and environmental microbiology. *Soil Biol Biochem* 21: 1256–1261

2 Sparling GP, Ord BG, Vaughan D (1981) Microbial biomass and activity in soil amended with glucose. *Soil Biol Biochem* 13: 99–104

3 Jenkinson DS (1981) The fate of plant and animal residues in soil. In: DJ Greenland, MHB Hayes (eds): *The chemistry of soil processing*. John Wiley & Sons, New York

4 Jenkinson DS, Ladd JN (1981) Microbial biomass in soil: Measurement and turnover. In: *Soil Biochemistry,* vol. 5. Marcel Dekker, New York, 415–471

5 Sardo V (1999) Establishment of an experimental field for saline irrigation in Sicily. In: A Hamdy et al. (eds): *Halophyte uses in different climates*. Backhuys, Leiden

6 La Loggia F, Pennisi M, Sardo V (1996) Hargreaves-Samani method and evaporation pan in the estimation of reference evapotranspiration. In: *Proceedings of the 2nd International Symposium on Irrigation of Agricultural Crops*. Chania, Crete

7 Tabatabai MA (1994) Soil enzymes. Method of soil analysis. II. Microbiological and biochemical properties. SSSA Book Series no. 5: 775–833

8 Sparling GP, West AW (1988) Modification of the fumigation-extraction technique to permit simultaneous extraction and estimation of soil microbial C and N. *Commun Soil Sci Plant Anal* 19: 77–87

9 Vance ED, Brookes PC, Jenkinson DS (1987) An extraction method for measuring microbial biomass C. *Soil Biol Biochem* 19: 703–707

10 Cataldo DA, Harron M, Schrader LF, Youngs VL (1975) Rapid colorimetric determination of nitrate in plant tissue by nitration of salicylic acid. *Commun Soil Sci Plant Anal* 6: 71–80

11 You MP, Sivasithamparam K (1994) Hydrolysis of fluorescein diacetate in an avocado plantation mulch suppressive to *Phytophtora cinnamoni* and its relationship with certain biotic and abiotic factors. *Soil Sci Biochem* 26: 1355–1361

12 Eivazi F, Tabatabai MA (1977) Phosphatases in soil. *Soil Sci Biochem* 9: 167–172

13 Kirchner MJ, Wollum, AG, King LD (1993) Soil microbial populations and activities in reduced chemical input agroecosystem. *Soil Sci Soc Am J* 57: 1289–1295

14 Fauci MF, Dick RP (1994) Microbial biomass as on indicator of soil quality: Effects of long-term management and recent soil amendments. In: *Defining Soil Quality for a Sustainable Environment.* SSSA Special Publication 35–40

Biosaline Agriculture and High Salinity Tolerance
Edited by Chedly Abdelly, Münir Öztürk, Muhammad Ashraf and Claude Grignon


# Halophyte plant diversity in the Irano-Turanian phytogeographical region of Turkey

Münir Öztürk[1], Aykut Güvensen[1], Serdal Sakçalı[2], Güven Görk[3]

[1] *Ege University, Science Faculty, Botany Department, Bornova-Izmir, Turkey*
[2] *Fatih University, Faculty of Science & Arts, Biology Department, Istanbul, Turkey*
[3] *Mugla University, Faculty of Science & Arts, Biology Department, Kötekli-Mugla, Turkey*

## Abstract

Irano-Turanian phytogeographical region includes the central, east and southeast Anatolian geographical divisions of Turkey. The region is characterized by a continental type of climate with very cold winters and hot summers. Precipitation in the form of a heavy snowfall is seen usually in winter and rains are common in spring and autumn. However, the southeast experiences a steppe climate with very hot summers, severe aridity and evaporation. The area of saline and degraded soils in the region is increasing. The factors responsible for this are over-irrigation practices, and salt accumulation due to evaporation. In the region, 137 halophytic taxa belonging to 88 genera and 34 families are found. The genera *Salsola*, *Chenopodium*, *Limonium*, *Alhagi* and *Allium* include the highest number of taxa. Phytogeographically 80.29% of these taxa are Irano-Turanian and 13.14% are common. The number of endemic taxa is around 42. For a management of the saline-alkaline habitats sound information on plant diversity, vegetation cover, habitat types, and locations of species communities is needed, together with land cover maps, to follow the habitat deterioration and its causes. The present study will thus help in the determination of site quality.

## Introduction

The Irano-Turanian phytogeographical region includes the central, east and southeast Anatolian geographical divisions of Turkey with a total area of approximately 38 900 000 ha [1]. A major part of this region is covered by grass, some of which are harvested for winter fodder and some used for grazing in summer. The area embodies the richest plant diversity in Turkey and is dominated by a typical steppe vegetation. These resemble the steppes of Iran and Central Asia [2, 3]. The natural *Artemisia* steppes are distributed around the uncultivated areas far from the villages near the Yesilhisar plains between Kayseri-Nigde and Ulukisla-Karaman-Eregli. These have been designated as natural protection sites. Similarly, the Urfa Ceylan Pinar Government Production Ranch and some sites around the army headquarters represent protected steppes with a natural species cover. In the cultivated areas unploughed sites between the ploughed fields are forced growth habitats of natural steppe plants. In places there is an intermixing with the Mediterranean and Euro-Siberian floral divisions. This region is home for more than 800 plant taxa of endemics [4, 5]. This is nearly 49.18% of the total endemics distributed in Turkey. Many papers have been published on its general flora but studies on the halophytic plant cover are rare [6–17]. Here an attempt is made to present an overview of the halophytic plant diversity of this region.

## Climate and soils

The Irano-Turanian phytogeographical region exists at a place where different types of climates inter-cross. It is rough topographically, plateaus being very high around Erzurum, Agri, Van and Hakkari, where high altitude mountains are gathered together [18, 19]. The climate is continental type, with cold winters and hot summers, and includes the coldest parts of Turkey. The southeast part experiences a steppe climate with very hot summers and severe aridity, with evaporation reaching up to 1000–2000 mm [1].

Temperatures increase gradually from the north to the south and from the east to from the west. The annual mean temperatures in Central Anatolia vary between 10° and 12°C, but in the winter values are 1–2°C and in summers 22–23°C. In East Anatolia, annual mean temperature in the north lies between 4° and 6°C, in winter between –7° and –10°C and in summer between 17° and 19°C. The hottest month in the East Anatolia is August, with a mean maximum temperature of 24–28°C, but in the coolest month mean minimum varies between –16° and –17°C. Extremely cold temperatures of –45.6°C have been recorded around Agri, whereas around Malatya and Elazig summer temperatures reach 42°C. In Southeast Anatolia winters hardly experience frost but severe dry winds dominate the summer season. Mean annual temperature varies between 15° and 16°C, winter temperatures are 3–4°C, summer temperatures 30–35°C [20].

The precipitation in the form of snow is a regular feature in winter, and rains are common in spring and autumn. Mean annual precipitation is 430–790 mm. Summer rains are more extreme than those in winter, particularly in the northeast around the states of Erzurum and Kars. Highest rainfall is observed in Tunceli (1003.8 mm) and lowest in Erzincan (359.6 mm). It ranges between 300 and 400 mm in Central Anatolia. In Northeast Anatolia, values lie between 500 and 600 mm, and between 400 and 700 mm in Southeast Anatolia, but the Harran plain gets only 330 mm of annual average rainfall, most of which falls in the winter [16, 19, 20]. The climatic diagrams of the representative States are given in Figure 1.

The harsh climatic conditions in the region result in higher mechanical than chemical weathering of parent material, as such soils are pebbly in nature. Main soil types met within the area are: alluvial, colluvial, chestnut,

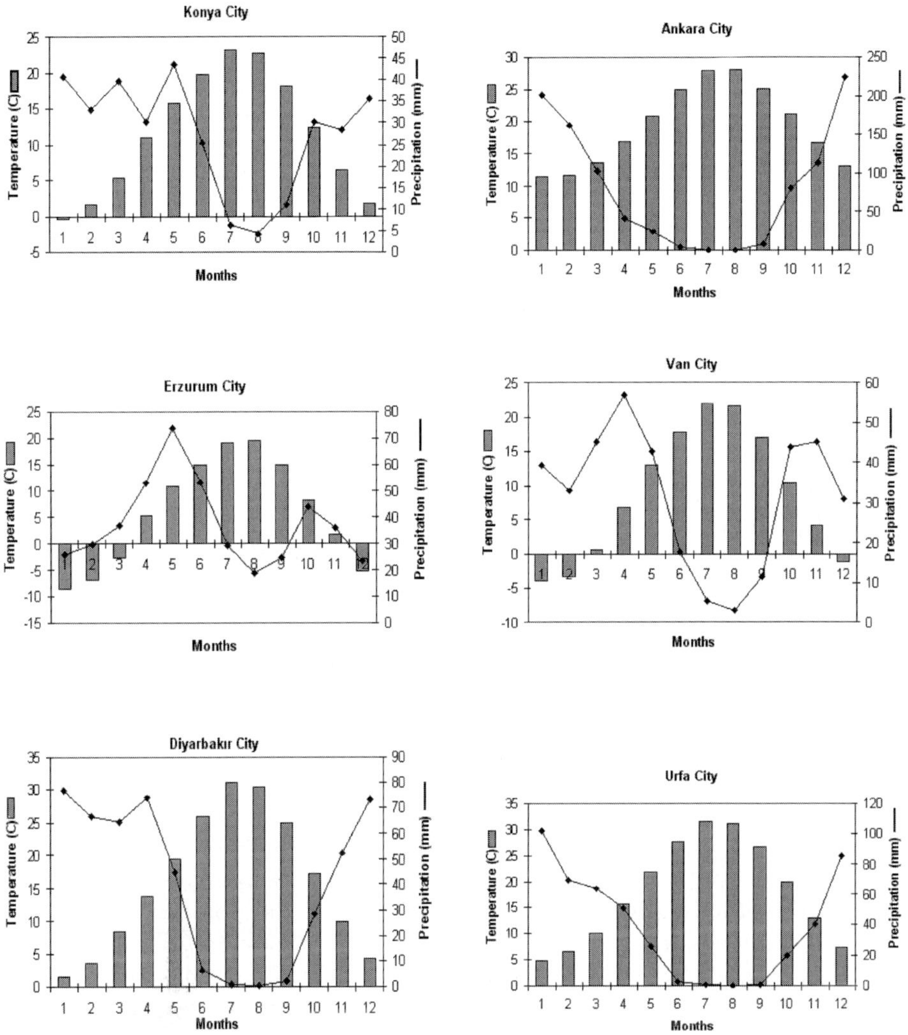


**Figure 1.** Climatic diagrams of the representative States.

brown, regosols, basaltic, organic, and arid soils [18–20]. Soil characteristics of representative samples from different basins reveals that pH varies between 7.5 and 7.9, and the ranges of electrical conductivity (EC), $Na^+$, $K^+$, $Ca^{2+}$, $Mg^{2+}$, $Cl^-$, $SO_4^-$ and ESP values are 9.65–60.7 (mmhos/cm), 2.4–20.50 (meq/100 g), 0.01–0.86 (meq/100 g), 1.5–31.97 (meq/100 g), 10.82–84.75 (meq/100 g), 0.58–29.41 (meq/100 g), 1.68–21.55 (meq/100 g), and 10.82–84.75, respectively [18]. The soils of the Harran plain are alluvial, dark-red or reddish brown clay, humid and calcareous, with a pH lying between 7.3 and 7.6 and total salt level varies from 0.075% to 1.450% [21].

**Table 1.** Area of degraded and saline soils Irano-Turanian geographical divisions of Turkey.

| Geographical divisions | Area (ha) | Saline soils (ha) | Saline soils (% of total area) | Total degraded soils (ha) | Total degraded soils (% of total area) |
|---|---|---|---|---|---|
| Central Anatolia | 15100000 | 681147 | 4.51 | 1614.181 | 10.68 |
| Southeast Anatolia | 7500000 | 236 | 0.003 | 6.336 | 0.08 |
| East Anatolia | 16300000 | 168270 | 1.03 | 449.884 | 2.76 |
| Total | 38900000 | 849653 | 2.18 | 2070.401 | 5.32 |

## Salinity and alkalinity problems

The Irano-Turanian phytogeographical region of Turkey has been a birthplace for many civilizations. Irrigation and agriculture have played an important role in the area around water basins like the Euphrates, Aras, Van, and Tigris. However, as the maintenance of irrigation and drainage networks was neglected, the salinization of soils made the land unsuitable for agriculture. In the fertile crescent, the Tigris-Euphrates rivers salted up 5–2 millennia ago, resulting in the collapse of civilizations. In fact, salt accumulation in the irrigated lands led to desertification in Mesopotamia, and the lands bordering the Mediterranean more than 2000 years ago [14, 18, 19].

Some work has been published on the salinity-alkalinity problems of this region [9, 18, 19, 21, 22]. The total area of saline and degraded soils is given in Table 1. The factors responsible for this can be summarized as: the accumulation of salts in the plains due to heavy rains, a long-standing high water table and geological features, the existence of saline areas as internal seas or soda lakes, and over-irrigation practices together with arid and semi-arid climatic conditions. The area of saline sodic soils in the Southeast Anatolian States of Adiyaman, Urfa, Gaziantep, Siirt, Diyarbakir, Mardin and in the East Anatolian States of Malatya, Elazigi, Tunceli and Bingöl is less than 500 ha, but 500–5000 ha in the States of Erzurum, Bitlis and Agri. In the States of Eskisehir, Ankara, Kirsehir, Nevsehir, Sivas, Çorum and Çankırı in Central Anatolia and Erzincan, Kars, Mus, Van and Hakkari in East and Southeast Anatolia areas varies between 5000 and 50000 ha. In the State of Kayseri (Central Anatolia) the area varies between 50000 and 100000 ha. However, in the States of Konya and Nigde in same region, it is more than 100000 ha [14, 18]. The salinity levels of the alluvial and hydromorphic soils in different basins are given in Figures 2 and 3. Recent studies in Southeast Anatolia, in particular the Harran plain, have shown that halophytic areas have increased considerably [16]. The Euphrates, Tigris and Van basins are presenting an alarming situation with over 75000 ha facing salinity-alkalinity problems (Tab. 2). After the completion of the Southeast Anatolian Irrigation Project, over-watering has lead to an increase in the salinity levels in the soil [16]. When these cultivated areas were abandoned due to high salt levels, these

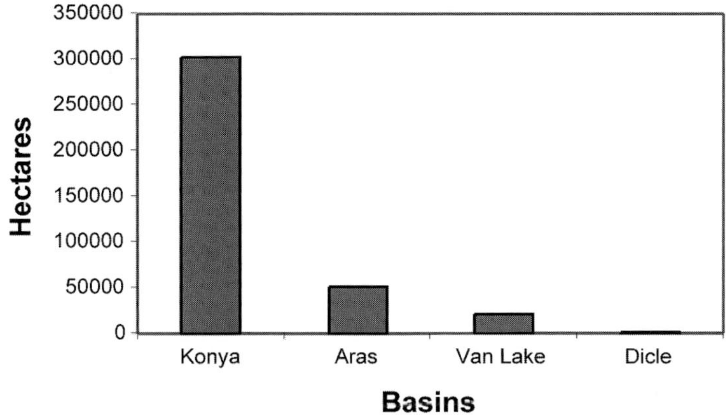


**Figure 2.** Alluvial soils affected by salinity (total soluble salts 0.15–0.35%; ESP around 15%) in different basins (modified from [17]).

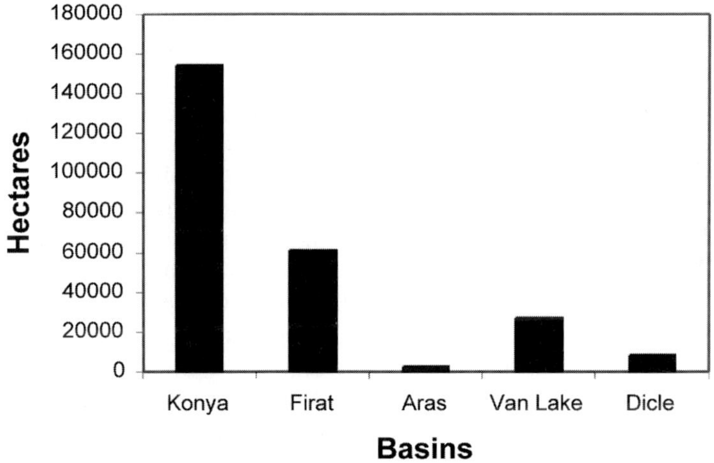


**Figure 3.** Hydromorphic alluvial soils affected by salinity (total soluble salts 0.15–0.35%; ESP around 15%) in different basins (modified from [17]).

were occupied by the secondary halophytes. Out of 225 000 ha in the Harran plain 8513 ha are classified as very highly saline [23]. These areas are especially found around Akcakale and their total salinity level varies between 0.075 and 1.450% [21]. In 1987, out of a total of 36 166 ha in the Harran plain, approximately 30 617 ha had no salinity problems, but in 2000 this area was reduced to 24 674 ha. This means that within a period of 13 years 5549 ha became salinized [16].

**Table 2.** Salinity and alkalinity problems in different basins.

| | Euphrates (ha) | Aras (ha) | Van (ha) | Tigris (ha) |
|---|---|---|---|---|
| Slightly saline | 46500 | 28145 | 2080 | 8250 |
| Saline | 3000 | 8938 | 2955 | 500 |
| Slightly saline alkaline | 6000 | 4922 | 15 903 | 1150 |
| Saline alkaline | 6000 | 9151 | 6424 | |

## Halophyte diversity

Present investigations revealed that 137 halophytic taxa, belonging to 88 genera and 34 families constitute the halophytic plant diversity of Irano-Turanian floral division (Tab. 3). The dominant families are Asteraceae, Chenopodiaceae, Fabaceae, and Liliaceae, with the dominant genera with highest number of taxa are *Salsola*, *Chenopodium*, *Limonium*, *Alhagi* and *Allium*. Phytogeographically, 80.29% of these taxa are Irano-Turanian, 13.14% are common, 5.84% imperfectly known and 0.73% cosmopolitan. The life form spectrum reveals that 43.07% of the halophytes are hemicryptophytes, 28.47% therophytes, 21.16% chamaephytes, 6.57% cryptophytes and 0.73% phanerophytes, with the majority being hygrohalophytes (47 taxa) followed by xerohalophytes (33 taxa) and xerophytes (20 taxa). The number of endemic taxa is about 42, the highest in Asteraceae (9 taxa), Plumbaginaceae (5 taxa), and Liliaceae (5 taxa) (Tabs 4, 5). The red data book of Turkish plants [24] shows that 39 endemic and 1 non-endemic halophytic taxa are in danger of extinction (Tab. 6). Recent studies by Adigüzel et al. [25] have shown that around Tuz gölü–Konya species like *Kalidiopsis wagenitzii* and *Salvia halophila* distributed among the natural steppes formation are accepted as globally endangered species.

In most places in Southeast Anatolia, halophytic taxa like *Salsola dendroides*, *Cressa cretica*, and *Tamarix smyrnensis* have migrated into the abandoned fields created due to the uncontrolled and over watering, particularly in the cotton growing areas. *Amaranthus retroflexus*, *A. albus*, *Xanthium strumarium*, *X. spinosum*, *Chenopodium album* subsp. *album* var. *album*, *Polygonum arenastrum*, *P. aviculare*, *Rumex pulcher*, *Hyoscyamus reticulatus*, *Tribulus terrestris*, *Urtica dioica*, *U. pulifera*, *Parietaria judaica*, *Physalis alkekengi*, *Salsola tragus*, *Portulaca oleracea*, *Heliotropium circinatum*, *H. europaeum* are the adventive taxa that occupy saline habitats as a result of succession originating from anthropogenic factors. In this area, particularly the Harran plain, when species like *Salsola dendroides*, *Cressa cretica*, and *Phragmites communis* migrate into the salinized areas, the species like *Stipa holosericea*, *Artemisia herba-alba*, *A. absinthium*, *Festuca callieri*. ssp. *callieri*, *Poa bulbosa*, *Teucrium polium*, *Astragalus microcephalus*, *Phlomis pungens* var. *pungens*, *Bromus japonicus* ssp. *japonicus*, *Thymbra spicata* var. *spicata*, *Eryngium campestre* var. *virens* abandon these habitats [16].

**Table 3.** List of halophytic taxa together with their ecological features.

| Families | Taxa | Life form | Choro-type | Eco-type | Altitude (m above sea level) | Flower-ing time (months) |
|---|---|---|---|---|---|---|
| Amaranthaceae | *Amaranthus retroflexus* | T | IN | HG | 1900 | 5–7 |
| Apiaceae | **Bupleurum heldreichii* | T | IT | H | 800–1200 | 5–7 |
| Apiaceae | **B. turcicum* | HE | IT | H | 900–1100 | 6–7 |
| Apiaceae | *Echinophora orientalis* | CH | IT | PH | 1300–2000 | 7–8 |
| Apiaceae | *Turgenia latifolia* | HE | CM | X | 0–3000 | 4–6 |
| Asteraceae | **Achillea goniocephala* | HE | IT | X | 1300–1900 | 6–8 |
| Asteraceae | **A. schischkinii* | HE | IT | XH | 700–2200 | 5–7 |
| Asteraceae | **A. sieheana* | HE | IT | X | 950–1200 | 6–9 |
| Asteraceae | *Artemisia herba-alba* | CH | IT | XH | 320 | 2–4 |
| Asteraceae | *Carthamus persicus* | T | IT | X | 340–1750 | 6–9 |
| Asteraceae | *Centaurea balsamita* | T | IT | XH | 650–1900 | 7 |
| Asteraceae | *C. pterocaula* | T | IT | X | 900–1200 | 7–8 |
| Asteraceae | *C. virgata* | CH | IT | XH | 100–2000 | 6–9 |
| Asteraceae | *Cirsium alatum* subsp. *alatum* | HE | IT | HG | 900–1800 | 6–9 |
| Asteraceae | **Cousinia birandiana* | HE | IT | HG | 850–1200 | 6–8 |
| Asteraceae | **C. Humilis* | HE | IT | HG | – | – |
| Asteraceae | **C. Iconica* | HE | IT | X | 850–2000 | 5–9 |
| Asteraceae | *Crepis foetida* subsp. *rhoeadifolia* | T | CM | PH | 0–2000 | 5–10 |
| Asteraceae | *Crepis sancta* | T | CM | PH | 0–2450 | 3–7 |
| Asteraceae | *Inula aucherana* | C | IT | XH | 550–1900 | 7–9 |
| Asteraceae | **Onopordum poly-cephalum* | HE | IT | R | 1400–2130 | 6–8 |
| Asteraceae | *O. turcicum* | HE | IT | R | 800–1250 | 7–8 |
| Asteraceae | *Scariola orientalis* | CH | IT | X | 800–2700 | 7–9 |
| Asteraceae | *Senecio doriiformis* subsp. *orientalis* | HE | IT | HG | 1800–2700 | 7–9 |
| Asteraceae | **Taraxacum farinosum* | HE | IT | XH | 800–1200 | 6–9 |
| Asteraceae | **Tripleurospermum callosum* | HE | IT | HG | 1100–2730 | 6–8 |
| Boraginaceae | *Coccinia macranthera* | CH | IT | PH | | |
| Boraginaceae | *Heliotropium lasio-carpum* | T | IT | PH | 50–1050 | 6–8 |
| Boraginaceae | *Moltkia coerulea* | HE | IT | HG | 700–1850 | 4–6 |
| Brassicaceae | **Alyssum pateri* subsp. *pateri* | CH | IT | X | 500–3000 | 5–6 |
| Brassicaceae | *Isatis kotschyana* | CH | IT | HG | 1800 | 6–7 |
| Brassicaceae | **Lepidium caespitosum* | HE | IT | XH | 800–1300 | 5–6 |
| Brassicaceae | *Sinapis arvensis* | T | CM | XH | 0–1800 | 4–6 |
| Brassicaceae | *Sisybrium loeselii* | CH | CM | H | 0–1800 | 3–6 |
| Caryophyllaceae | **Gypsophila bitlisensis* | T | IT | PH | 1650–1800 | 6–7 |

**Table 3 (continued).** List of halophytic taxa together with their ecological features

| Families | Taxa | Life form | Choro-type | Eco-type | Altitude (m above sea level) | Flower-ing time (months) |
|---|---|---|---|---|---|---|
| Caryophyllaceae | **G. oblanceolata* | C | IT | XH | 950–1050 | 6–8 |
| Caryophyllaceae | *Minuartia urimiensis* | T | IT | XH | 1000 | 5–6 |
| Caryophyllaceae | **Silene salsuginea* | HE | IT | XH | 900 | 7 |
| Chenopodiaceae | *Atriplex davisii* | T | IT | PH | 900–1000 | 5–7 |
| Chenopodiaceae | *A. hortensis* | T | IN | H | 0–1250 | 5–7 |
| Chenopodiaceae | *A. nitens* | T | IN | H | 800–1900 | 5–7 |
| Chenopodiaceae | *Chenopodium album* subsp. *iranicum* | T | IN | XH | 1200–1900 | 5–8 |
| Chenopodiaceae | *C. botrys* | T | CM | PH | 0–1900 | 5–7 |
| Chenopodiaceae | *C. foliosum* | CH | CM | XH | 1200–2800 | 5–7 |
| Chenopodiaceae | *C. murale* | T | Cosm. | XH | 0–1400 | 5–7 |
| Chenopodiaceae | *C. sosnowskyi* | T | IT | XH | 800–1500 | 5–8 |
| Chenopodiaceae | **Kalidiopsis wagenitzii* | CH | IN | H | 940 | 5–6 |
| Chenopodiaceae | *Kochia prostrata* | T | CM | XH | 0–1900 | 6–8 |
| Chenopodiaceae | **Salsola anatolica* | T | IT | H | 1000 | 6–8 |
| Chenopodiaceae | *S. inermis* | T | IT | H | 900–950 | 5–7 |
| Chenopodiaceae | *S. laricina* | CH | IT | HG | 950 | 5–7 |
| Chenopodiaceae | *S. macera* | T | IT | X | 850–950 | 5–7 |
| Chenopodiaceae | *S. nitrarira* | T | IT | H | 900–1200 | 6–8 |
| Chenopodiaceae | *S. ruthenica* | T | CM | H | 0–1750 | 5–7 |
| Chenopodiaceae | *Suaeda confusa* | T | IN | H | 800 | 6–8 |
| Cistaceae | *Helianthemum ledifolium* var. *mucrocarpum* | T | CM | HG | 500–1400 | 5–6 |
| Cyperaceae | *Carex diluta* | HE | IT | HG | 20–2000 | – |
| Cyperaceae | *C. kukkonenii* | HE | IT | HG | 2400–3140 | – |
| Cyperaceae | *Schoenoplectus lacustris* subsp. *tabernamontani* | HE | IT | HG | 0–3050 | 4–8 |
| Euphorbiaceae | *Euphorbia macroclada* | HE | IT | HG | 250–2500 | 5–9 |
| Euphorbiaceae | *E. falcata* | T | CM | XH | 0–1900 | 4–8 |
| Fabaceae | *Alhagi pseudalhagi* | CH | IT | XH | 1–1200 | 6–8 |
| Fabaceae | **A. karamasicus* | HE | IT | R | 450–2060 | 6–7 |
| Fabaceae | **A. micropterus* | CH | IT | X | 850–1860 | 6–7 |
| Fabaceae | *A. tribuloides* | T | IT | PH | 920–1050 | 5–6 |
| Fabaceae | *Astragalus microcephalus* | CH | IT | X | 850–2700 | 6–8 |
| Fabaceae | *A. onobrychis* | CH | CM | X | 975–2400 | 5–7 |
| Fabaceae | *A. shelkovnikovii* | CH | IT | PH | 800 | 5–6 |
| Fabaceae | *Medicago lupulina* | HE | CM | XH | 0–200 | 5–7 |
| Fabaceae | **Sphaerophysa kotschyana* | HE | IT | HG | 1000 | 6 |
| Fabaceae | *Trigonella coerulescens* | HE | IT | HG | 300–1300 | 4–7 |
| Fabaceae | *T. orthoceras* | T | IT | X | 1200–1900 | 6 |

**Table 3 (continued).** List of halophytic taxa together with their ecological features

| Families | Taxa | Life form | Choro-type | Eco-type | Altitude (m above sea level) | Flower-ing time (months) |
|---|---|---|---|---|---|---|
| Globulariaceae | *Globularia trichosantha* | HE | IT | HG | 200–2470 | 4–7 |
| Guttiferae | **Hypericum salsugineum* | HE | IT | XH | 900 | 8 |
| Iridaceae | **Gladiolus halophilus* | C | IT | HG | 900–1200 | 6–7 |
| Iridaceae | *G. atroviolaceus* | HE | IT | X | | |
| Iridaceae | *Iris paradoxa* | C | IT | HG | 1750–2800 | 4–6 |
| Iridaceae | *I. spuria* subsp. *musulmanica* | HE | IT | HG | 800–1900 | 5–7 |
| Juncaceae | *Juncus gerardi* subsp. *libanoticus* | C | IT | HG | 50–2500 | 5–8 |
| Juncaceae | *J. heldreichianus* subsp. *orientalis* | HE | IT | HG | | |
| Lamiaceae | **Phlomis armeniaca* | HE | IT | HG | 800–2350 | 6–8 |
| Lamiaceae | **Salvia halophila* | HE | IT | XH | 950–1000 | 8–10 |
| Lamiaceae | *S. staminea* | CH | IT | HG | 1700–3150 | 5–8 |
| Lamiaceae | *S. syriaca* | HE | IT | XH | 450–1850 | 4–7 |
| Lamiaceae | *Scutellaria orientalis* subsp. *orientalis* | HE | IT | XH | 450–1500 | 6–8 |
| Liliaceae | **Allium cupani* subsp. *hirtovaginatum* | C | IT | HG | 950–1650 | 6–8 |
| Liliaceae | *A. macrochaetum* subsp. *macrochaetum* | C | IT | HG | 950–1650 | 6–8 |
| Liliaceae | **A. nevsehirense* | C | IT | R | 800–1900 | 7–8 |
| Liliaceae | **A. sieheanum* | C | IT | HG | 900–1200 | 7–8 |
| Liliaceae | **Asparagus lycaonicus* | HE | IT | HG | 1000 | 8 |
| Liliaceae | *A. persicus* | HE | IT | XH | 800–1700 | 5–7 |
| Liliaceae | **Bellevalia forniculata* | T | IT | HG | 1800–2400 | 5 |
| Liliaceae | *Gagea gageoides* | HE | IT | HG | 1300–3000 | 4–5 |
| Liliaceae | *Puschkinia scilloides* var. *libanotica* | HE | IT | HG | 1700–3500 | 4–8 |
| Liliaceae | *Merendera sobolifera* | HE | IT | HG | 1000–2400 | 3–6 |
| Linaceae | *Linum seljukorum* | T | IT | HG | 1100 | 6–8 |
| Linaceae | *L. mucronatum* | HE | IT | XH | 450–1200 | 4–6 |
| Papaveraceae | *Glaucium leiocarpum* | CH | CM | HG | 15–1600 | 6–7 |
| Plantaginaceae | **Plantago anatolica* | HE | IT | HG | 1750 | 5 |
| Plantaginaceae | *P. maritima* | HE | IT | H | 0–2400 | 5–8 |
| Malvaceae | **Alcea calvertii* | CH | IT | XH | 600–1500 | 6–8 |
| Plumbaginaceae | **Acanthalimon halophilum* | CH | IT | X | 900–110 | 6 |
| Plumbaginaceae | **Limonium anatolicum* | CH | IT | H | 900–1000 | 6–9 |
| Plumbaginaceae | *L. globuliferum* | HE | IT | H | 900–1100 | 6–9 |
| Plumbaginaceae | **L. iconicum* | HE | IT | H | 900–1040 | 6–9 |

**Table 3 (continued).** List of halophytic taxa together with their ecological features

| Families | Taxa | Life form | Choro-type | Eco-type | Altitude (m above sea level) | Flowering time (months) |
|---|---|---|---|---|---|---|
| Plumbaginaceae | **L. lilacinum* | HE | IT | H | 900–1200 | 6–9 |
| Plumbaginaceae | **L. tamaricoides* | HE | IT | H | – | – |
| Poaceae | *Agropyron repens* | HE | IT | HG | – | – |
| Poaceae | *Apera intermedia* | T | IT | PH | 0–2000 | 5–7 |
| Poaceae | *Crypsis aculeata* | T | CM | PH | 0–1510 | 6–10 |
| Poaceae | *Hordeum violaceum* | HE | IT | XH | 1600–3000 | 6–8 |
| Poaceae | *Phragmites australis* | HE | CM | HG | 0–2400 | 8–10 |
| Poaceae | *Puccinellia distans* subsp. *sevagensis* | HE | IT | H | 300–1920 | 7–8 |
| Poaceae | *Sphenopus divaricatus* | T | IT | HG | 1–1000 | – |
| Primulaceae | *Glaux maritima* | HE | IT | PH | 0–1720 | 5–8 |
| Primulaceae | *Primula algida* | HE | IT | HG | 2000–3600 | 5–8 |
| Polygonaceae | *Atraphaxis spinosa* | CH | IT | XH | 900–1600 | 6–8 |
| Polygonaceae | *Polygonum arenastrum* | T | IN | HG | 0–2300 | 6–11 |
| Ranunculaceae | *Consolida anthoroidea* | CH | IT | XH | 2000 | 7–9 |
| Ranunculaceae | **C. glandulosa* | T | IT | X | 800–1600 | 6–9 |
| Ranunculaceae | *C. orientalis* | CH | IT | X | 0–1900 | 5–8 |
| Ranunculaceae | **Delphinium verulosum* | T | IT | X | 200–1200 | 7–8 |
| Ranunculaceae | *Ranunculus kotschyi* | CH | IN | HG | 1300–2400 | 5–6 |
| Rubiaceae | *Cruciata taurica* | CH | IT | XII | 300–3300 | 3–7 |
| Santalaceae | *Thesium compressum* | T | IT | XH | 1000–2200 | 6–8 |
| Scrophulariaceae | *Lagotis stolonifera* | HE | IT | HG | 1300–2300 | 4–6 |
| Scrophulariaceae | **Verbascum campestre* | HE | IT | X | 860–1370 | 5–6 |
| Scrophulariaceae | **V. helianthemoides* | HE | IT | HG | 960–1200 | 6–8 |
| Scrophulariaceae | **V. pyroliforme* | HE | IT | HG | 1000 | 6–9 |
| Scrophulariaceae | *Veronica viscosa* | T | IT | HG | 500–1600 | 3–5 |
| Solanaceae | **Lycium anatolicum* | CH | IT | X | 450–1750 | 4–11 |
| Solanaceae | *L. ruthenicum* | CH | IT | X | 915–1750 | 6–7 |
| Tamaricaceae | *Tamarix parviflora* | P | IN | H | 0–300 | 3–6 |
| Typhaceae | *Typha latifolia* | HE | CM | HG | 0–1850 | 6–10 |
| Valerianaceae | *Centranthus longiflorus* subsp. *longiflorus* | HE | IT | HG | 0–2300 | 4–9 |
| Valerianaceae | *Valerianella vesicaria* | HE | CM | XH | 0–2000 | 4–6 |
| Zygophyllaceae | *Nitraria schoberi* | CH | IT | PH | 200–1000 | – |

Life forms: HE, hemicryptophytes; CH, chamaephytes; T, therophytes; C, cryptophytes; P, phanerophytes.
Chorotypes: IT, Irano-Turanian; CM, common; Cosm., Cosmopolitan; IN, imperfectly known.
Ecological types: H, halophyte; HG, hygrohalophytes; X, xerophyte; R, ruderal; XH, xerohalophytes; PH, psammohalophytes.
FL, flowering time.
*, endemic.

**Table 4.** Chrotypes, life forms, halophyte types, major genera and families.

| Choro-types | % | Life forms | % | Eco-type | Number of plant taxa | % of the halo-phyte type | Major genera | Major families |
|---|---|---|---|---|---|---|---|---|
| IT | 80.29 | HE | 43.07 | H | 19 | 13.86 | *Salsola* | Asteraceae |
| CM | 13.14 | T | 28.47 | HG | 47 | 34.31 | *Chenopodium* | Chenopodiaceae |
| Cosm. | 0.73 | C | 6.57 | X | 20 | 14.6 | *Limonium* | Fabaceae |
| IN | 5.84 | CH | 21.16 | R | 4 | 2.92 | *Alhagi* | Liliaceae |
| | | P | 0.73 | XH | 33 | 24.09 | *Allium* | |
| | | | | PH | 14 | 10.22 | | |

Life forms: HE, hemicryptophytes; CH, chamaephytes; T, therophytes; C, cryptophytes; P, phanerophytes.
Chorotypes: IT, Irano-Turanian; CM, common; Cosm., Cosmopolitan; IN, imperfectly known.
Ecological types: H, halophyte; HG, hygrohalophytes; X, xerophyte; R, ruderal; XH, xerohalophytes; PH, psammohalophytes.

## Conclusions

There is an urgent need to increase crop production under saline conditions to meet the greater demand for food at a time when the size of agricultural lands is decreasing due to soil salinization. Halophytes could be evaluated as potential agricultural crops by growing them on saline soils [26, 27]. During the last two decades much work has been published on halophytes in particular on biosaline agriculture [28–30]. These studies will probably lead to new approaches for more effective solutions to problems related to salinity. Halophytes can serve as indicators of salinity-sodicity, and areas with ground waters possessing high salinity could be used for halophytic forage crop production [31–34]. There is a great potential in halophytic plant cover for consumption as well as amelioration of degraded lands. In Turkey some work has already begun in this connection [14, 35], but many more studies and experiments need to be carried out to develop halophytes successfully as agricultural cash crops.

**Table 5.** Number of families, genera , species and endemic taxa.

| Families found in halophytic habitats | Genera | Species | Endemic taxa |
|---|---|---|---|
| Amaranthaceae | 1 | 1 | |
| Apiaceae | 3 | 4 | 2 |
| Asteraceae | 13 | 21 | 9 |
| Boraginaceae | 3 | 3 | |
| Brassicaceae | 5 | 5 | 2 |
| Caryophyllaceae | 3 | 4 | 3 |
| Chenopodiaceae | 6 | 17 | 2 |
| Cistaceae | 1 | 1 | |
| Cyperaceae | 2 | 3 | |
| Euphorbiaceae | 1 | 2 | |
| Fabaceae | 5 | 11 | 3 |
| Globulariaceae | 1 | 1 | |
| Guttiferae | 1 | 1 | |
| Iridaceae | 2 | 4 | 1 |
| Juncaceae | 1 | 2 | |
| Lamiaceae | 3 | 5 | 2 |
| Liliaceae | 6 | 10 | 5 |
| Linaceae | 1 | 2 | |
| Papaveraceae | 1 | 1 | |
| Plantaginaceae | 1 | 2 | 1 |
| Malvaceae | 1 | 1 | 1 |
| Plumbaginaceae | 2 | 6 | 5 |
| Poaceae | 7 | 7 | |
| Primulaceae | 2 | 2 | |
| Polygonaceae | 2 | 2 | |
| Ranunculaceae | 3 | 5 | 2 |
| Rubiaceae | 1 | 1 | |
| Santalaceae | 1 | 1 | |
| Scrophulariaceae | 3 | 5 | 3 |
| Solanaceae | 1 | 2 | 1 |
| Tamaricaceae | 1 | 1 | |
| Typhacaea | 1 | 1 | |
| Valerianaceae | 2 | 2 | |
| Zygophyllaceae | 1 | 1 | |
| **Total:** | **88** | **137** | **42** |

**Table 6.** Endangered endemic halophytic taxa.

| **Families** | **Endemic taxa** | **Threat categories** |
|---|---|---|
| Chenopodiaceae | *Kalidiopsis wagenitzii* | EN |
| Asteraceae | *Achillea goniocephala* | VU |
| | *A. sieheana* | VU |
| | *Cousinia birandiana* | LR (lc) |
| | *C. humilis* | CR |
| | *C. iconica* | LR (cd) |
| | *Onopordum polycephalum* | LR (lc) |
| | *Taraxacum farinosum* | LR (lc) |
| | *Tripleurospermum callosum* | LR (lc) |
| Plumbaginaceae | *Acanthalimon halophilum* | LR (lc) |
| | *Limonium anatolicum* | VU |
| | *Limonium iconicum* | LR (lc) |
| | *L. lilacinum* | LR (lc) |
| | *L. tamaricoides* | EN |
| Liliaceae | *Asparagus lycaonicus* | EN |
| | *A. schischkinii* | LR (lc) |
| | *Allium nevsehirense* | LR (lc) |
| | *A. sieheanum* | LR (lc) |
| | *Bellevalia forniculata* | LR (lc) |
| Scrophulariaceae | *Verbascum campestre* | LR (cd) |
| | *V. helianthemoides* | VU |
| | *V. pyroliforme* | VU |
| Fabaceae | *Astragalus shelkovnikovii* | VU |
| | *Alhagi micropterus* | LR (lc) |
| | *Sphaerophysa kotschyana* | LR (cd) |
| Malvaceae | *Alcea calvertii* | LR (lc) |
| Apiaceae | *Bupleurum heldreichii* | LR (cd) |
| | *B. turcicum* | LR (nt) |
| Ranunculaceae | *Consolida glandulosa* | LR (lc) |
| | *Delphinium verulosum* | LR (lc) |
| Caryophyllaceae | *Gypsophila bitlisensis* | LR (cd) |
| | *Silene salsuginea* | EN |
| Iridaceae | *Gladiolus halophilus* | VU |
| Guttiferae | *Hypericum salsugineum* | VU |
| Solanaceae | *Lycium anatolicum* | LR (lc) |
| Brassicaceae | *Isatis kotschyana* | VU |
| Lamiaceae | *Phlomis armeniaca* | VU |
| | *Salvia halophila* | VU |
| Plantaginaceae | *Plantago anatolica* | LR (cd) |

CR, critically endangered; VU, vulnerable; EN, endangered; LR, lower risk; (cd) conservation dependent; (lc) least concern; (nt) near threatened.
The non-endemic taxon facing threat of extinction is *Salsola inermis* (vulnerable category).

## References


1 DIE (2000) Statistical Yearbook of Turkey. State Institute of Statistics of Prime Ministry, Publ. no. 1720, Ankara

2 Breckle SW (1986) Studies on halophytes from Iran and Afghanistan. Ecology of halophytes along salt gradient. Proc R Soc Edinburgh 89 B, 203–215

3 Akhani H (2006) Biodiversity of halophytic and sabkha ecosystems in Iran. In: MA Khan, B Böer, GS Kust, HJ Barth (eds): *Sabkha Ecosystems*, vol. II. Springer, Dordrecht, 71–88

4 Davis PH (1965–1988) *Flora of Turkey and East Aegean Island*, vol. 1–10. Edinburgh University Press, UK

5 Guner A, Ozhatay N, Baser KHC (2000) *Flora of Turkey and The East Aegean Islands*, vol. 11. Edinburgh University Press, UK

6 Birand H (1960) Erste Ergebnisse der Vegetations-Untersuchungen in der zentralanatolischen Steppe. I. Halophytengesellschaften des Tuz Gölü. *Bot Jahrb* 79: 255–296

7 Yurdakulol E (1974) Studies on the plant sociology of the vegetation of arid marshes in the Konya plain. *Bitki* 1: 257–277

8 Yurdakulol E, Ercoskun T (1990) Ecological and syntaxonomical studies on the vegetation of arid lands in Central Anatolia. *Turk J Bot* 14: 109–123

9 Ozturk M, Ozcelik H, Behcet L, Guvensen A, Ozdemır F (1995) Halophytic flora of Van Lake basin – Turkey. In: MA Khan, IA Ungar (eds): *Biology of Salt Tolerant Plants Book*. Crafters, Michigan, 306–315

10 Aydogdu M, Hamzaoglu E, Kurt L (2002) New halophytic syntaxa from Central Anatolia (Turkey). *Isr J Plant Sci* 50: 313–323

11 Avci K, Kutbay G, Gencoglu S (2004) Halophytic vegetation of Bafra plain. *Annual Report*, Directorate of Rural Services, Ankara, 56

12 Dogan O, Ustun H, Gurlesen N, Tomul F, Tıpırdamaz R, Sozeri S (2004) Ecology of halophytic vegetation in the Seyfe lake basin and determination of grassland and agricultural plants suitable for growth on arid soils. *Annual Report*, Directorate of Rural Services, Ankara, 67

13 Gidirisli A, Cakır R, Yarci C (2004) Halophytic vegetation of saline soils in Gelibolu-Gokbuet and Ipsala region. *Annual Report*, Directorate of Rural Services-Ankara, 53

14 Guven E, Vural M, Bahceci I, Sonmez B (2001) Ecology of Central Anatolian Region Halophyte Vegetation. *Soil Plant Sources Res Yearb* 117: 265–280

15 Hamzaoglu E, Aksoy A (2006) Studies on the phytosociology of halophytic communities of sultansazligi marsh (Central Anatolia-Kayseri). *Ekoloji* 15: 8–15

16 Atamov V, Yavuz M, Cetin E, Cevheri AC, Aslan M (2005) Halophytes and Salinization in Harran Plain Şanlıurfa-Turkiye. Mugla Proceedings

17 Guvensen A, Gork G, Ozturk M (2006) An overview of the halophytes in Turkey. In: MA Khan, B Böer, GS Kust, H-J Barth (eds): *Sabkha Ecosystems*, vol. II. Springer, Dordrecht, 9–30

18 Ozturk M, Guvensen A, Aksoy A, Beyazgul M (2006) An overview of the soils and sustainable land use in Turkey. In: R Gerger, K Yenigün, M Kömür, H Bulut, MI Yeşilçınar (eds): *GAP V. Muhendislik Kongresi*. Harran Universitesi, Urfa, 1548–1555

19 Ozturk M, Guvensen A, Sakcali MS, Bahadir H (2006) An overview of the land degradation problems in East Anatolia. In: R Gerger, K Yenigün, M Kömür, H Bulut, MI Yeşilçınar (eds): *GAP V. Muhendislik Kongresi*. Harran Universitesi, Urfa, 1548–1561

20 Ozturk M, Ozcelik H, Sakcali MS, Guvensen A (2004) Land degradation problems in the Euphrates basin, Turkey. *Environews* 10: 7–9

21 Dinc U, Senol S (1988) Güneydogu Anadolu Bölgesi Toprakları (GAT). I. Harran Ovası, TÜBITAK Tarım ve Ormancılık Araştırma Grubu Güdümlü Araştırma Kesin Raporu, Proje no: TOAG-534
22 Kaynak L, Imamgiller B, Ersoy N, Yazici K (2000) *Salinity problems in agricultural soils.* GAP Environmental Congress, Sanliurfa, 555–561
23 Ozkutlu F, Ince F (1999) *Present state of salinity in the Harran plain and potentials for its expansion. GAP 1*, vol. 2. Agricultural Congress, Şanlıurfa, 909–914
24 Ekim T, Koyuncu M, Vural M, Duman H, Aytac Z, Adıguzel N (2000) *Red data book of Turkish plants*, Ankara
25 Adıgüzel N, Byfield A, Duman H, Vural M (2005) Tuz Gölü ve Stepleri. Turkiye' nin 122 önemli Bitki Alanı. In: A Byfield, N Özhatay, S Atay (eds): *WWF Türkiye Dogal Hayatı Koruma Vakfı Yayını,* Istanbul, 289–292
26 Khan MA, Duke NC (2001) Halophytes – A resource for the future. *Wetlands Ecol Manag* 6: 455–456
27 Lieth H, Mochtchenko M (2003) *Cash Crop Halophytes*. Kluwer Academic Publishers, Dordrecht
28 Ahmed R, Pietro AS (1986) *Prospects for Biosaline Research.* Shamim Printing Press, Karachi
29 Ahmed R, Malik KA (2002) *Prospects of Saline Agriculture.* Kluwer Academic Press, Dordrecht
30 Ozturk M, Waisel Y, Khan MA, Gork G (2006) *Biosaline Agriculture and Salinity Tolerance in Plants*. Birkhäuser, Basel
31 Ashraf M (1999) Breeding for salinity tolerance proteins in plants. *Crit Rev Plant Sci* 13: 17–42
32 Yensen NP (2006) Halophyte uses for the twenty-first century and a new hypothesis the role of sodium in $C_4$ physiology. In: MA Khan, DJ Weber (eds): *Ecophysiology of High Salinity Tolerant Plants*. Springer, Dordrecht, 367–396
33 Khan MA, Qaiser M (2006) Halophytes of Pakistan: Distribution, Ecology, and Economic Importance In: MA Khan, H-J Barth, GC Kust, B Boer (eds): *Sabkha Ecosystems: vol. II: The South and Central Asian Countries*. Springer, Dordrecht, 135–160
34 Weber DJ, Ansari R, Gul B, Khan MA (2007) Potential halophytes as source of edible oil. *J Arid Environ* 68: 315–321
35 Cullu MA (2002) *Application of GIS monitoring soil salinization in the Harran plain.* International Conference on Sustainable Land use and Management, Canakkale, Turkey, 326–333

Biosaline Agriculture and High Salinity Tolerance
Edited by Chedly Abdelly, Münir Öztürk, Muhammad Ashraf and Claude Grignon


# Potential use of halophytes with emphasis on fodder production in coastal areas of Pakistan

M. Ajmal Khan and Raziuddin Ansari

*Institute of Sustainable Halophyte Utilization, University of Karachi, Karachi-75270, Pakistan*

## Abstract

Fresh water resources both for domestic and agricultural use are constantly depleting worldwide and crop yields suffer from a steady increase in soil salinity, especially in the arid and semi-arid areas. A burgeoning world population is a further threat to sustained food supply. Equally or even more affected in some cases, are other resources like fodder for animals and fuel wood for the rural poor. Efforts are hence needed to find an alternate source of water and utilization of saline lands for economic benefit. The varied climatic conditions of Pakistan offer opportunities for selecting suitable halophytes for specific purposes. Whereas halophytes may be used for a variety of purposes like food, fiber, fuel wood, medicines, source of chemicals, landscaping, ornamental, carbon sequestration, etc., one of the very important utilities lies in their use as fodder. An animal feeding trial showed that traditional green fodder (maize) and a halophytic grass (*Panicum*) were equally good for growth and development of 1-year-old cow calves. Meat from animals fed 100% *Panicum* was leaner and hence better for human consumption from health point of view.

## Introduction

Soil factors like availability of nutrients and moisture, soil reaction, microbes, pollutants and factors of the environment like light, temperature, humidity, wind, fire, predators, etc., all affect plant growth. Even near optimal conditions are rare in nature and the plants inevitably suffer from one or the other stress during their life cycle. One such stress, commonly encountered by plants in arid and semi-arid regions is salinity of the soil and water. The salinity hazard of today is mostly man-made, and is largely brought about by the lack of appropriate irrigation management [1]. About a billion hectares of lands have become salt-affected as a consequence, and an additional two million hectares go out of productivity worldwide each year due to secondary salinization [2, 3]. The scarcity of good quality irrigation water coupled with the brackish nature of the underground water and insufficient rainfall causes further soil deterioration.

The current state of affairs suggests that the trend of ever-increasing deterioration of soil and water resources due to salinity is likely to persist at least in the near future, whereas most of our conventional crops have marginal salt tolerance. This necessitates the need to explore alternate options for meeting the challenge. A number of halophytes with potential for use as cash crops could come to our rescue in such a situation. Research has indicated the potential of many halophytes to withstand high soil salinity and saline water irrigation, some even with seawater [4]. This tolerance is achieved through a number of adaptations such as selective uptake and transport of ions, localization of ions in vacuoles, synthesis of compatible organic solutes for cytoplasm balancing and protection of enzyme systems, developing succulence to dilute salt injury, reducing salt load by excretion through specialized salt glands or shedding plant parts, etc. [5]. These plants are hence able to exhibit sustained growth by establishing homeostatic conditions.

Pakistan, with an area of about 800 000 $km^2$ spans a distance of 1600 km from the Arabian Sea to plains, prairies and playas of temperate northern mountains across deserts. The varied climatic conditions of the country ranging from mild coastal to extremes of hot and cold have created a richly diversified flora of halophytes. A recent survey reports that out of a total of about 2500 halophytic species found worldwide, some 410 are found in Pakistan out of which about 100 are present on the coastal belt of Karachi and the adjoining Balochistan province [6].

## Utilization potential of halophytes

Utilization of *Alhaji maurorum* by the Sumerians as a soil ameliorant is the oldest report on the use of halophytes. However, in recent times the credit goes to the Israelis for rediscovering the halophytes for utilization of saline water and salinized lands [7]. Subsequently, systematic research on the subject gathered momentum during the later half of the last century not only in Israel [8] but also in other countries [1, 3, 5, 9–13] because of the realization that our field crops are generally salt sensitive with a limited potential for improvement of this trait [14]. However, there is increasing evidence in support of utilization of halophytes for various purposes [1, 15]. The problem of salinization has global dimensions and most regions are affected by it to various degrees, although arid and semi-arid areas are the worst affected. Agriculture has so far managed to meet the food demand; however, if the present trends of an increasing population and decreasing crop yields due to salinization continues, it would necessitate remedial measures. It may be too early to predict when we shall start consuming halophytic food, due to our long-established eating preferences, but one thing is pretty certain that the salt sensitivity of our staple food crops will ultimately make them redundant in the future.

Halophytes are non-conventional crops and it would take a while to make people believe that they represent good food, although some are already used by certain communities for this purpose [2]. However, there are potentials to extract good quality oil from halophytes [16], and they could also serve as a source of feed, fiber and forage, and this would not have any problem of acceptability. Their potential as a land cover cannot be denied, which is not only aesthetically pleasing but also checks land erosion and degradation. The more important opportunities relate to reforestation or replanting and ecological recovery of saline areas that have fallen into disuse [17, 18], coastal development and protection [19], production of cheap biomass for renewable energy [20], and environment conservation through carbon sequestration [21–23]. Mangroves, a dominant vegetation of many tropical and subtropical seacoasts, play an important role in stabilization of coasts and beaches, food chain and life support systems, aquaculture, agriculture, and support in development of wild-life sanctuary and recreation areas [24]. They also provide tannin, thatching material, fodder, fish poison, food products, medicine and wood for building purposes, fuel, and boat and canoe making for the residents of coastal areas [25]. For a more extensive coverage on the potential uses of halophytes, Aronson's HALOPH is recommended [4]. There is no doubt that saline agriculture is here to stay but it will need fine tuning in different ecological situations.

## Forages and fodders

Capitalizing on the presence of a rich halophytic flora, especially in the coastal areas of Pakistan, the eco-physiology group at the Institute of Sustainable Halophyte Utilization, University of Karachi, has embarked upon an elaborate program of exploring the potential of these plants for use as cash crops. It has been realized that one of the greatest potential of halophytes probably rests in their utilization as forage and fodder [26]. This is particularly relevant for Pakistan, a predominantly agricultural country with serious soil salinity problems, where growing halophytic forages offers the opportunity to reduce the national deficit substantially in forage for livestock. Examples of use of halophytes as forage are found in many countries, e.g., *Haloxylon* and *Kochia* sp. in Iran, *Atriplex* sp. in Argentina and Australia, *Spartina* on the east coast and *Salicornia bigelovii* at the western coast of USA, and planting of halophytes is now a widely accepted option in low rainfall saline regions of Australia and many countries of north Africa to provide valuable fodder reserves when other supplies are exhausted [9–13].

There are about 95 halophytic species that could be used as either forage or fodder. Animals browse most of the halophytic grasses found in Pakistan, which contain variable quantities of fiber and proteins [17]. *Leptochloa fusca*, *Lasarius scindicus*, *Panicum turgidum*, *Dactyloctenium sindicum*, *Cynodon dactylon*, *Paspalum vaginatum*, *Sporobolus marginatus*, *S. ioclados*, *Chloris gayana*, *Chloris virgata*, *Echinochloa turnerana*, *Echinochloa colonum* and *Puccinellia distans* are some common grass species found in saline and alka-

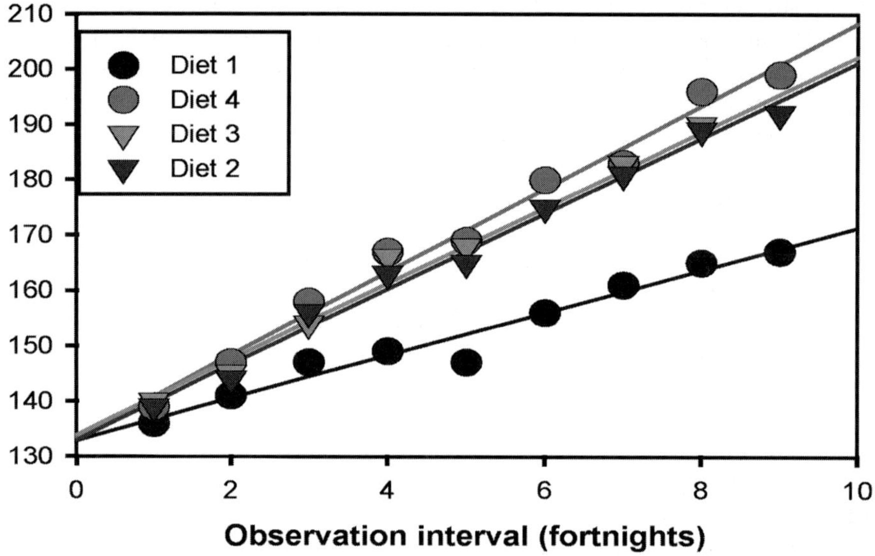


**Figure 1.** Effect of different diets on the weight of calfs.

line areas and used as forages [6]. Many species of *Alhagi*, *Salicornia*, *Chenopodium*, *Atriplex*, *Salsola*, *Suaeda* and *Kochia* are common fodder shrubs [2]. Among trees, species of *Acacia*, *Prosopis*, *Salvadora* and *Zizyphus* are traditional fodder of arid regions, while on the sea coast, species of mangroves (*Avicennia marina*, *Aegiceras corniculata*, *Ceriops tagal*, *Rhizophora mucronata*) are consumed extensively as forage by camels [20, 24].

A search of local halophytic flora of Sindh/Balochistan seacoast of Pakistan and adjoining areas resulted in the identification of a number of grasses with potential for use as fodder/forage; these were brought to Zia Model Farm, Hub Kund, Balochistan for field testing and further scientific assessment. This farm, situated at 24°57'07.99" N, 66°45'23.74" E has saline soil (EC 10–15 mS/cm) and brackish underground water (EC 15 mS/cm) for irrigation. Based on growth performance under the above-mentioned conditions of the Farm, *Panicum turgidum*, a perennial grass was selected for further testing. It grows well by transplanting root stocks and attains a height of about 1 m in 25–30 days in summer (maximum temperatures generally between 30 and 35°C). However, during the winter months of December to February (minimum 15–18°C) it requires 35–40 days for comparable growth.

An animal feeding trial was conducted at the Farm on 16 young (1 year old) cow calves. The animals were divided into four groups of four and fed a diet containing combinations of traditional (maize) or the halophytic fodder (*Panicum*), while 'concentrates' were omitted from the control treatment. It was observed (Fig. 1) that diets containing maize or *Panicum* were both equally better than control. A progressive increase in the live weight of all animals was observed during the course of experiment except in the month of August. This coincides with the monsoon rains, which sometime developed muddy conditions in the animal house and, although efforts were made to drain out water as quickly as possible and dry the floor, it seems that the animals were stressed and their growth was hindered during this period. Normal weight gain was, however, subsequently attained. Abdominal fat indicated that meat from animals fed 100% *Panicum* was leaner and hence better for human consumption from health point of view. For confirmation of the results, the experiment will be repeated on a fresh batch of animals.

Considering the acute shortage of fodder in the country, it is encouraging that we have found a suitable grass in the local flora, which can completely replace maize (conventional green fodder) and thus result in considerable savings on its purchase. The halophytic grass is a non-accumulator of salts, which not only improves its acceptability by the animals but it also does away with the additional water requirement caused by the increased thirst of the animals fed high-salt fodder. To the best of our knowledge, few previous studies have looked at big ruminants, especially cow calves, that have been raised entirely on halophytic green fodder, although a number of grazing trials have been conducted on goats and sheep. This work is hence unique in this respect. Efforts are also being made to find a suitable replacement for the dry fodder component of the animal feed to also reduce dependence on this expensive item.


## *Acknowledgements*

We are grateful to M/S S. Zia-al-Haq & Sons, especially their manager, General Services, Mr. Zaleir Abuwath, for facilities to conduct animal trials at their farm.


## References


1 Khan MA, Weber DJ (eds) (2006) *Ecophysiology of high salinity tolerant plants*. Springer, Dordrecht

2 Menzel U, Lieth H (1999) Halophyte Database Vers. 2.0. *Prog Biometeriol* 13: 77–88

3 Yensen NP (2006) Halophyte uses for the twenty-first century and a new hypothesis: The role of sodium in $C_4$ physiology. In: MA Khan, DJ Weber (eds): *Ecophysiology of high salinity tolerant plants*. Springer, Dordrecht, 367–396

4 Aronson J (1989) *HALOPH: Salt tolerant plants for the world – A computerized global data base of halophytes with emphasis on their economic uses*. University of Arizona Press, Tucson

5 Khan MA, Ansari R, Gul B, Qadir M (2006) Crop diversification through halophyte production on salt-prone land resources. *CAB Reviews: Perspective in Agriculture, Veterinary Science, Nutrition and Natural Resources*. 1, No. 048, 8

6 Khan MA, Qaiser M (2006) Halophytes of Pakistan: Distribution, ecology, and economic importance In: MA Khan, H-J Barth, GC Kust, B Boer (eds): *Sabkha Ecosystems: Vol. II. The South and Central Asian Countries*. Springer, Dordrecht, 135–160
7 Boyko H (1964) Principles and experiments regarding irrigation with highly saline and seawater without desalinization. *Trans NY Acad Sci Series* 2: 1087–1102
8 Pasternak D, Nerd A (1996) Research and utilization of halophytes in Israel. In: R Choukr-Allah, CV Malcolm, A Hamdy (eds): *Halophytes and biosaline agriculture*. Marcel Dekker, New York, 325–348
9 Ahmed, R, Malik KA (2002) *Prospects of saline agriculture.* Kluwer Academic Press, Dordrecht
10 El-Shaer HM, Kandil (1999) Potential of *Atriplex* species as fodder shrubs under the arid conditions of Egypt. *Prog Biometeriol* 13: 87–94
11 Glenn E, Hicks N, Riley J, Swingle S (1996) Sea water irrigation of halophytes for animal feed. In: R Choukr-Allah, CV Malcolm, A Hamdy (eds): *Halophytes and biosaline agriculture.* Marcel Dekker, New York, 221–236
12 Malcolm CV (1996) Characteristics and methods for determining the best forage species for particular sites. In: R Choukr-Allah, CV Malcolm, A Hamdy (eds): *Halophytes and biosaline agriculture*. Marcel Dekker, New York, 97–114
13 O'Leary JW (1984) The role of halophytes in irrigated agriculture. In: RC Staples, GH Toenniessen (eds): *Salinity tolerance in plants: Strategies for crop improvement*. John Wiley and Sons, New York, 285–300
14 Ashraf M (1999) Breeding for salinity tolerance proteins in plants. *Crit Rev Plant Sci* 13: 17–42
15 Flowers TJ, Garcia A, Koyama M, Yeo AR (1997) Breeding for salt tolerance in crop plants – The role of molecular biology. *Acta Physiol Plant* 19: 427–433
16 Weber DJ, Ansari R, Gul B, Khan MA (2007) Potential halophytes as source of edible oil. *J Arid Environ* 68: 315–321
17 Aronson JA (1985) Economic halophytes – A global review. In: GE Wickens, JR Goodin, DV Field (eds): *Plants for arid lands*. George Allen and Unwin, London, 177–188
18 Garg VK (1998) Interaction of tree crops with a sodic soil environment: Potential for rehabilitation of degraded environments. *Land Degrad Dev* 9: 81–93
19 Khan MA, Ungar IA (eds) (1995) *Biology of Salt Tolerant Plants*. University of Karachi, Pakistan
20 Dagar JC (1995) Characteristics of halophytic vegetation in India. In: MA Khan, IA Ungar (eds): *Biology of Salt Tolerant Plants*. University of Karachi, Pakistan, 255–276
21 Glenn EP, Squires V, Olsen M, Frye R (1993) Potential for carbon sequestration in the dry lands. *Water Air Soil Pollut* 70: 341–356
22 Geesing D, Felker P, Bingham RL (2000) Influence of mesquite (*Prosopis glandulosa*) on soil nitrogen and carbon development: Implications for global carbon sequestration. *J Arid Environ* 46: 157–180
23 Scurlock JMO, Hall DO (1998) The global carbon sink: A grassland perspective. *Global Change Biol* 4: 229–233
24 Khan MA, Aziz I (2001) Salinity tolerance of some mangroves from Pakistan. *Wetland Ecol Manag* 9: 228–232
25 Cintrón G, Lugo AE, Pool DJ, Morris G (1978) Mangroves of arid environments in Puerto Rico and adjacent islands. *Biotropica* 10: 110–121
26 Pasternak D (1990) Fodder production with saline water. *Project Report, The Institute for Applied Research*, Ben-Gurion University of the Negev, Israel

Biosaline Agriculture and High Salinity Tolerance
Edited by Chedly Abdelly, Münir Öztürk, Muhammad Ashraf and Claude Grignon


# Role of seed banks in the dynamics of desert communities

M. Ajmal Khan, Farhat Agha and Bilquees Gul

*Department of Botany, University of Karachi, Karachi-75270, Pakistan*

## Abstract

Sub-tropical deserts of Karachi, Pakistan are dominated by perennial shrubs and grasses, usually in a mono-specific stands. Seed bank dynamics of three stands about a kilometer away from the coast were selected and monitored for 2 years. These stands were dominated by *Atriplex stocksii*, *Cyperus arenarius* and *Cressa cretica*, respectively, with few individuals of other species in each stand. Soil analysis of the *C. arenarius* community showed seeds of *A. stocksii* and *C. arenarius* were present throughout both of the seasons; however, only *A. stocksii* maintained a persistent seed bank that had some relationship with the aboveground vegetation. The *C. cretica* dominated community had a substantial presence of *C. arenarius* seed in the soil with some seeds of *C. cretica*, *Salsola imbricata* and *A. stocksii*. However, *C. cretica* maintained a small persistent seed bank showing a close relationship between seed bank and vegetation. *A. stocksii* maintained a large persistent seed bank during both seasons indicating a good seed bank-vegetation relationship.

## Introduction

Seed banks may play a significant role in determining the spatial and temporal distribution of halophytes [1]. Salt marsh environments are unfavorable for most plants because of high salinity, drought and flooding stress [2–7]. When conditions for seed germination are not favorable and beyond the tolerance limits for species, un-germinated seeds of halophytes often maintain enforced dormancy in the soil and serve as a transient or persistent seed bank [8–10]. Seed bank size may vary considerably in coastal salt marshes and some community types have large persistent seed banks [11] and others contain few or no seeds in the seed bank [9, 10, 12, 13]. Perennial species usually do not form a persistent seed bank, e.g., east and west coast of North America [14, 15], and in seed banks of sub-tropical coast of Karachi [3, 4]. However, since annual species of halophytes have only a single opportunity to reproduce in their life history, seed banks play a more significant role in determining their spatial and temporal distribution than that of perennials, which generally reproduce through ramets [16–18].

The relationship between seed banks and aboveground vegetation in marsh communities is apparently quite variable [1]. In some systems there is a high similarity between the aboveground vegetation of salt marsh zonal communities and the seed bank [7, 19], whereas in other marsh communities fewer similarities are reported [9, 10, 20, 21]. Zaman and Khan [22] determined that the seed bank and aboveground vegetation of salt deserts were similar in species composition in communities around Karachi. In contrast, Khan [23] reported that the seed bank of a salt desert inland community had a high diversity in comparison to those occurring in aboveground plant communities. *Suaeda fruticosa* seeds formed a persistent seed bank, while other species such as *Haloxylon stocksii* and *C. cretica* had a transient seed bank. Coastal communities were found to have only a small seed bank with 113 seeds $m^{-2}$ or less [3]. Communities dominated by *Arthrocnemum indicum*, *A. macrostachyum*, *S. fruticosa*, *Halopyrum mucronatum*, *Cenchrus pennisetiformis* and *A. stocksii* have an aboveground vegetation similar in species composition to that of the seed bank.

Salt deserts tend to have smaller persistent seed banks than do salt marshes, and only a small number of salt desert perennial species are apparently able to produce a persistent seed bank [23]. Seeds of salt desert species are exposed to stresses like high temperature, drought, and salinity and their interactions. Khan [24] and Zaman and Khan [22] showed that salt desert species in the Karachi region, including *C. cretica*, *H. stocksii*, *S. imbricata* and *Sporobolus ioclados* had a transient seed bank that was depleted after monsoon rains. *S. fruticosa*, however, maintained a persistent seed bank of 233–981 seeds $m^{-2}$ [24]. The seed numbers of *C. cretica*, *S. imbricata*, and *S. ioclados* in the seed bank varied seasonally, decreasing from February until May, and by June there were few seeds in the seed bank [22]. The seed bank of *S. fruticosa* varied in size during the growing season, and ranged from a low of 317 seeds $m^{-2}$ in October to a high of 917 seeds $m^{-2}$ in February [23]. *H. stocksii* and *C. cretica* followed a similar pattern, with low values of 2 and 2.5 seeds $m^{-2}$ and high values of 37 and 26 seeds $m^{-2}$ for the two species, respectively. Aziz and Khan [25] reported that *C. cretica* had a seed bank of 2800 seeds $m^{-2}$ in May in a sand dune near the coast of Karachi. From January to December the seed bank decreased from 1600 to 300 seeds $m^{-2}$ and with the dispersal of new seeds the number of seeds in the seed bank increased to 2800 seeds $m^{-2}$. There was very low recruitment by seeds in these salt desert habitats because of the intense environmental stress caused by high temperatures, low soil moisture, and high salinity to which seeds are exposed [25]. Gul and Khan [26] reported a larger seed bank in the inter-tidal community of *A. macrostachyum* approaching a maximum of 917 135 seeds $m^{-2}$. The number decreased after a few months but the community still maintained a substantial size of the seed bank.

It is evident from the above that the role of the seed bank in proximate and ultimate success of various plant populations is still far from clear. The present research was, therefore, carried out in the coastal region of Pakistan to test the hypothesis that the seed bank has a prominent role in the success of saline desert communities.

## Study site

The studies were conducted from 1999 to 2002 in communities located about a kilometer away from the Gizri Creek, Karachi, Pakistan (24°57'07.99" N, 66°45'23.74" E). The soil was dry, loamy sand and the vegetation was dominated by salt tolerant grasses and shrubs. In this geographical region there are only two seasons: a long summer and a short winter. Monthly mean ambient summer and winter temperatures are 36°C and 25°C, respectively. The monsoon season extends from June to September. The study location is just beyond the area that receives direct inundation. Various pure communities dominated by *A. stocksii* (Chenopodiaceae), *S. fruticosa* (Chenopodiaceae), *Aeluropus lagopoides* (Poaceae), *H. mucronatum* (Poaceae), and *C. cretica* (Convolvulaceae) were found. The present study was conducted on the communities dominated by *A. stocksii*, *C. cretica* and *C. arenarius* because they are among the few most frequently found communities around supra-littoral areas of Karachi, Pakistan.

## Vegetation and seed bank sampling

The vegetation in each community was analyzed by the point-centered quarter method [27]. Density, frequency and cover were measured and an importance value index calculated using data from 20 sample points. To assess the size and species composition of the seed bank, 20 randomly selected soil cores were collected using a 1.5-cm diameter corer to a depth of 15 cm at monthly intervals from November 1999 to October 2002. Seeds were manually sorted immediately after collection with the help of a binocular microscope, identified using a reference collection and counted. Seeds that appeared intact and viable were included in the count. A seed viability test was carried out using Tetrazolium chloride test.

## Salient findings of the study

The three communities studied showed a low species diversity (Tab. 1). Few individual species of other than the dominant species were found. Seasonal profile of the *C. arenarius* seed bank indicated the presence of seeds of *A. stocksii* and *C. arenarius* seeds throughout both seasons (Tab. 2). Most *C. arenarius* seeds were dead, but *A. stocksii* maintained a persistent seed bank showing little relationship with vegetation (Tab. 3). Seeds of other species disappeared soon after dispersal.

**Table 1.** Species composition of three coastal communities showing relative density (RD), relative frequency (RF) and relative cover (RC) for each species.

| **Species** | **Plant community** | | | | | | | | |
|---|---|---|---|---|---|---|---|---|---|
| | ***Atriplex stocksii*** | | | ***Cressa cretica*** | | | ***Cyperus arenarius*** | | |
| | RD | RF | RC | RD | RF | RC | RD | RF | RC |
| *Atriplex stocksii* | 70 | 80 | 92 | 0 | 0 | 0 | 1 | 1 | 7 |
| *Cressa cretica* | 3 | 3 | 1 | 96 | 76 | 96 | 0 | 0 | 0 |
| *Cyperus arenarius* | 15 | 7 | 2 | 0 | 0 | 0 | 97 | 97 | 84 |
| *Heliotropium subulatum* | 1 | 2 | 1 | 0 | 0 | 0 | 1 | 1 | 5 |
| *Suaeda fruticosa* | 10 | 8 | 4 | 4 | 24 | 4 | 1 | 1 | 4 |

The *C. cretica*-dominated community had a substantial presence of *C. arenarius* seeds in the soil (Tab. 4) with a small presence of *C. cretica*, *S. imbricata* and *A. stocksii*. However, *C. cretica* maintained a small persistent seed bank (Tab. 5) with some seeds of *A. stocksii*, *C. arenarius* and *S. imbricata*. Here a close relationship between seed bank and vegetation was witnessed.

The *A. stocksii* community showed relatively higher number of its seeds throughout both the seasons. Seeds of *C. cretica*, *C. arenarius* were also present (Tab. 6). However, a persistent seed bank was maintained by *A. stocksii* during both seasons (Tab. 7), indicating a good seed bank-vegetation relationship.

The role of the seed bank in regulating dynamics of perennial shrub communities is far from clear [1, 26, 28]. Perennial shrubs and grasses produce a large number of seeds but the recruitment from seeds is either rare or infrequent [3, 4]. The size of the seed bank also varies considerably in salt marsh and salt desert habitats [1, 6, 7, 20, 23, 29, 30]. Investigations in coastal salt marshes indicate that some community types have large persistent seed bank [11, 26], while others contains few or no seeds in the seed bank [3, 4, 12, 31]. The present study showed a small sized seed bank in the near coastal communities dominated by perennial shrubs. The largest number of seeds found here was 53078 seed $m^{-2}$; however, the number of viable seeds among them was reduced to 4423 seed $m^{-2}$. It appears that the actual seed bank may be much smaller than previously reported. For example, Gul and Khan [26] reported 917135 seeds $m^{-2}$, probably one of the largest seed bank in the inter-tidal coastal marsh near Karachi. However, little recruitment was reported from the seed bank. When the viability of seeds from the seed bank was investigated most of them were found to be dead (Khan, unpublished data).

The unpredictability or harshness of the environment may be a significant factor in determining the nature of a persistent seed bank [1]. Environmental conditions may play a significant role in determining the longevity of seed banks. Tolerance at the germination stage can be determined by measuring two factors: the capacity of seeds to remain viable under hyper-saline conditions and the maximal salinity limit at which germination occurs [1]. There are

**Table 2.** Number of seeds ($m^{-2}$) of various species in the seed bank of *Cyperus arenarius* community.

| Year | Species | Nov | Dec | Jan | Feb | Mar | Apr | May | Jun | Jul | Aug | Sep | Oct |
|---|---|---|---|---|---|---|---|---|---|---|---|---|---|
| 1999–2000 | *Atriplex stocksii* | 0 ± 0 | 0 ± 0 | 0 ± 0 | 0 ± 0 | 885 ± 88 | 885 ± 88 | 2654 ± 188 | 7077 ± 368 | 3539 ± 235 | 1769 ± 118 | 3539 ± 65 | 0 ± 0 |
| | *Cressa cretica* | 1769 ±117 | 0 ±0 | 0 ± 0 | 2654 ±188 | 885 ± 87 | 2654 ±188 | 885 ± 85 | 0 ± 0 | 0 ± 0 | 0 ± 0 | 0 ± 0 | 0 ± 0 |
| | *Cyperus arenarius* | 1769 ± 118 | 3538 ± 236 | 6192 ± 190 | 2654 ± 184 | 3538 ± 236 | 6192 ± 190 | 13270 ± 322 | 6192 ± 190 | 7962 ± 360 | 12385 ± 400 | 1868 ± 260 | 7962 ± 336 |
| | *Salsola imbricata* | 0 ± 0 | 0 ± 0 | 0 ± 0 | 0 ± 0 | 0 ± 0 | 0 ± 0 | 884 ± 75 | 0 ± 0 | 0 ± 0 | 2654 ± 190 | 885 ± 75 | 885 ± 84 |
| | *Tribulus longipetalus* | 0 ± 0 | 0 ± 0 | 0 ± 0 | 0 ± 0 | 0 ± 0 | 0 ± 0 | 0 ± 0 | 0 ± 0 | 0 ± 0 | 0 ± 0 | 0 ± 0 | 0 ± 0 |
| 2001–2002 | *Atriplex stocksii* | 0 ± 0 | 885 ± 65 | 0 ± 0 | 0 ± 0 | 885 ± 77 | 885 ± 68 | 0 ± 0 | 1770 ± 120 | 0 ± 0 | 5308 ± 350 | 885 ± 69 | 0 ± 0 |
| | *Cressa cretica* | 0 ± 0 | 0 ± 0 | 0 ± 0 | 0 ± 0 | 0 ± 0 | 0 ± 0 | 0 ± 0 | 0 ± 0 | 0 ± 0 | 885 ± 68 | 885 ± 57 | 0 ± 0 |
| | *Cyperus arenarius* | 0 ± 0 | 4423 ± 350 | 4423 ± 147 | 15924 ± 4128 | 28309 ± 1607 | 1769 ± 117 | 6192 ± 269 | 884 ± 69 | 7677 ± 256 | 5308 ± 195 | 12385 ± 399 | 5308 ± 378 |
| | *Salsola imbricata* | 0 ± 0 | 0 ± 0 | 0 ± 0 | 0 ± 0 | 0 ± 0 | 0 ± 0 | 0 ± 0 | 0 ± 0 | 0 ± 0 | 0 ± 0 | 0 ± 0 | 0 ± 0 |
| | *Tribulus longipetalus* | 0 ± 0 | 2653 ± 198 | 0 ± 0 | 0 ± 0 | 3539 ± 271 | 0 ± 0 | 0 ± 0 | 0 ± 0 | 0 ± 0 | 0 ± 0 | 0 ± 0 | 2654 ± 179 |

**Table 3.** Number ($m^{-2}$) of viable seeds in the seed bank of *Cyperus arenarius* community.

| **Month** | ***Atriplex stocksii*** | | ***Cressa cretica*** | | ***Cyperus arenarius*** | | ***Tribulus longipetalus*** | |
|---|---|---|---|---|---|---|---|---|
| | 1999–2000 | 2001–2002 | 1999–2000 | 2001–2002 | 1999–2000 | 2001–2002 | 1999–2000 | 2001–2002 |
| Nov | 0 ± 0 | 0 ± 0 | 1769 + 133 | 0 ± 0 | 0 ± 0 | 0 ± 0 | 0 ± 0 | 0 ± 0 |
| Dec | 0 ± 0 | 0 ± 0 | 0 ± 0 | 0 ± 0 | 0 ± 0 | 0 ± 0 | 0 ± 0 | 0 ± 0 |
| Jan | 0 ± 0 | 0 ± 0 | 0 ± 0 | 0 ± 0 | 0 ± 0 | 0 ± 0 | 0 ± 0 | 1769 + 134 |
| Feb | 0 ± 0 | 0 ± 0 | 0 ± 0 | 0 ± 0 | 0 ± 0 | 885 ± 88 | 0 ± 0 | 0 ± 0 |
| Mar | 885 ± 84 | 885 ± 84 | 885 ± 84 | 0 ± 0 | 0 ± 0 | 0 ± 0 | 0 ± 0 | 0 ± 0 |
| Apr | 885 ± 84 | 0 ± 0 | 2654 + 186 | 0 ± 0 | 0 ± 0 | 0 ± 0 | 0 ± 0 | 3539 + 144 |
| May | 885 ± 84 | 0 ± 0 | 2654 + 188 | 0 ± 0 | 0 ± 0 | 885 ± 84 | 0 ± 0 | 0 ± 0 |
| Jun | 0 ± 0 | 885 ± 84 | 0 ± 0 | 0 ± 0 | 885 ± 84 | 0 ± 0 | 0 ± 0 | 0 ± 0 |
| Jul | 1769 ±133 | 0 ± 0 | 0 ± 0 | 0 ± 0 | 0 ± 0 | 885 ± 84 | 0 ± 0 | 0 ± 0 |
| Aug | 0 ± 0 | 885 ± 84 | 0 ± 0 | 0 ± 0 | 0 ± 0 | 0 ± 0 | 0 ± 0 | 0 ± 0 |
| Sep | 885 ± 84 | 0 ± 0 | 0 ± 0 | 1769 + 133 | 0 ± 0 | 885 ± 84 | 0 ± 0 | 0 ± 0 |
| Oct | 0 ± 0 | 0 ± 0 | 0 ± 0 | 0 ± 0 | 0 ± 0 | 885 ± 84 | 0 ± 0 | 885 ± 84 |

several reports that indicate the differential responses of halophytic seeds to harsh environmental conditions. In cool moist habitats, usually dominated by annuals, when seeds were exposed to hyper-saline conditions, they remained in an enforced dormancy state that was released by exposing seeds to fresh water [8–10, 32–35], while increase in salinity caused a substantial reduction in seed germination of cold desert species like *Chrysothamnus nauseosus* and *Sarcobatus vermiculatus* [36]. The present study showed that seeds from perennial shrubs growing in hot and hyper-saline conditions of subtropical desert are dead except for some species like *A. stocksii*, which maintain a very small but viable bank.

The relationship between seed bank and aboveground vegetation in saline communities is apparently quite variable [1, 19, 32, 37]. In some systems there is a high similarity between the aboveground vegetation of salt marsh zonal communities and their seed banks [19, 38], whereas in other marsh communities low similarities are reported [9, 10, 20, 21, 39]. The Arabian Sea coast communities showed a close relationship between vegetation and seeds of flora found in the soil [3, 4, 28]. The present study showed that more seeds of the dominant species occur in the seed bank. However, except for *A. stocksii* and *C. cretica* seeds of all other species either disappear completely from the soil or lose their viability. Size of the seed bank for both *A. stocksii* and *C. cretica* was also substantially reduced either through predation or the loss of viability.

Most of the initial work on the seed bank dynamics was conducted on the cool, moist salt marshes. Halophytes native to those areas were reported to be highly tolerant to salinity and temperature stress while in the soil [1, 8, 35, 40], and when salinity and temperature stress were removed they germinated readily. They also maintained a larger seed bank, which plays a significant role in both evolutionary and ecological success of the marsh ecosystem. Seed bank studies reported for the arid sub-tropical communities dominated by perennial shrubs and grasses showed an entirely different pattern. Most of the shrubs are reported to produce millions of seeds; however, after dispersal seeds disappear quickly. Seeds of only some species remain in the soil including *A. stocksii*, *C. cretica*, *A. macrostachyum* and *S. fruticosa* [3, 22, 23, 25, 26]. However, except for *A. stocksii* and *S. fruticosa*, the recruitment of the species through seeds is rare. The present study and other related unpublished work being carried out indicate that seeds of those species that are represented by either large or small numbers loose their viability and therefore are not part of the seed bank. This explains the lack of recruitment through seedlings. Our data also indicate that a number of previous studies that lack information on the viability of seeds would lead to erroneous interpretations. The results presented in this study validate the hypothesis that the seed bank has a role in contributing to the fitness of some perennial shrub populations in arid habitats; however, their main role appears to be to maximize fitness by occasionally contributing fresh genotypes to the gene pool of the population, which primarily rely on duplicating the successful genotypes. Further studies on the role of seed banks under saline desert conditions are in progress.

**Table 4.** Number of seeds ($m^{-2}$) of various species in the seed bank of *Cressa cretica* community.

| Year | Species | Nov | Dec | Jan | Feb | Mar | Apr | May | Jun | Jul | Aug | Sep | Oct |
|---|---|---|---|---|---|---|---|---|---|---|---|---|---|
| **1999–2000** | *Atriplex stocksii* | 0 ± 0 | 0 ± 0 | 0 ± 0 | 884.6 ± 748 | 884.6 ± 88 | 4423.2 ± 237 | 3538.5 ± 235 | 2653.9 ± 188 | 0 ± 0 | 1769.2 ± 117 | 3538.5 ± 235 | 0 ± 0 |
| | *Cressa cretica* | 2653 ± 178 | 884 ± 88 | 0 ± 0 | 0 ± 0 | 885 ± 88 | 3539 ± 265 | 1769.2 ± 117 | 884 ± 84 | 0 ± 0 | 884 ± 83 | 3538 ± 235 | 2653 ± 188 |
| | *Cyperus arenarius* | 26539 ± 189 | 884 ± 88 | 0 ± 0 | 884 ± 88 | 0 ± 0 | 884 ± 88 | 2654 ± 182 | 7077 ± 368 | 6192 ± 188 | 16808 ± 245 | 22 116 ± 482 | 38039 ± 1408 |
| | *Salsola imbricata* | 0 ± 0 | 0 ± 0 | 0 ± 0 | 1769.2 ± 117 | 0 ± 0 | 2653.9 ± 188 | 0 ± 0 | 884.6 ± 885 | 0 ± 0 | 1769.2 ± 117 | 0 ± 0 | 0 ± 0 |
| **2001–2002** | *Atriplex stocksii* | 0 ± 0 | 885 ± 88 | 0 ± 0 | 0 ± 0 | 0 ± 0 | 3539 ± 270 | 884.6 ± 885 | 0 ± 0 | 0 ± 0 | 0 ± 0 | 0 ± 0 | 0 ± 0 |
| | *Cressa cretica* | 9731 ± 596 | 3538 ± 144 | 2654 ± 188 | 884 ± 84 | 0 ± 0 | 0 ± 0 | 0 ± 0 | 0 ± 0 | 1769.2 ± 119 | 0 ± 0 | 2653 ± 188 | 1888 ± 188 |
| | *Cyperus arenarius* | 1180 ± 44 | 4423 ± 237 | 8846 ± 527 | 2654 ± 188 | 2654 ± 188 | 3539 ± 270 | 7962 ± 339 | 1769 ± 117 | 7077 ± 368 | 1608 ± 442 | 0 ± 0 | 6194 ± 293 |
| | *Salsola imbricata* | 0 ± 0 | 0 ± 0 | 0 ± 0 | 0 ± 0 | 0 ± 0 | 0 ± 0 | 0 ± 0 | 0 ± 0 | 2654 ± 188 | 0 ± 0 | 0 ± 0 | 0 ± 0 |

**Table 5.** Number ($m^{-2}$) of viable seeds in the seed bank of *Cressa cretica* community.

| Month | *Atriplex stocksii* | | *Cressa cretica* | | *Cyperus arenarius* | | *Salsola imbricata* | |
|---|---|---|---|---|---|---|---|---|
| | 1999–2000 | 2001–2002 | 1999–2000 | 2001–2002 | 1999–2000 | 2001–2002 | 1999–2000 | 2001–2002 |
| Nov | 0 ± 0 | 0 ± 0 | 0 ± 0 | 4423 ± 197 | 1769 ± 133 | 0 ± 0 | 0 ± 0 | 0 ± 0 |
| Dec | 0 ± 0 | 885 ± 89 | 0 ± 0 | 0 ± 0 | 885 ± 89 | 0 ± 0 | 0 ± 0 | 0 ± 0 |
| Jan | 0 ± 0 | 0 ± 0 | 0 ± 0 | 2653 ± 1876 | 0 ± 0 | 2654 ± 187 | 0 ± 0 | 0 ± 0 |
| Feb | 0 ± 0 | 0 ± 0 | 0 ± 0 | 0 ± 0 | 0 ± 0 | 885 ± 89 | 0 ± 0 | 0 ± 0 |
| Mar | 0 ± 0 | 0 ± 0 | 885 ± 89 | 885 ± 89 | 0 ± 0 | 0 ± 0 | 0 ± 0 | 0 ± 0 |
| Apr | 0 ± 0 | 885 ± 89 | 1769 ± 133 | 1769 ± 133 | 0 ± 0 | 885 ± 89 | 0 ± 0 | 0 ± 0 |
| May | 885 ± 89 | 885 ± 89 | 0 ± 0 | 0 ± 0 | 0 ± 0 | 885 ± 89 | 0 ± 0 | 0 ± 0 |
| Jun | 0 ± 0 | 0 ± 0 | 0 ± 0 | 0 ± 0 | 0 ± 0 | 0 ± 0 | 0 ± 0 | 0 ± 0 |
| Jul | 885 ± 89 | 0 ± 0 | 885 ± 89 | 1769 ± 133 | 885 ±89 | 0 ± 0 | 0 ± 0 | 0 ± 0 |
| Aug | 885 ± 89 | 0 ± 0 | 0 ± 0 | 0 ± 0 | 0 ± 0 | 885 ± 89 | 885 ± 89 | 0 ± 0 |
| Sep | 885 ± 89 | 0 ± 0 | 0 ± 0 | 0 ± 0 | 0 ± 0 | 0 ± 0 | 0 ± 0 | 0 ± 0 |
| Oct | 0 ± 0 | 0 ± 0 | 0 ± 0 | 5308 ±196 | 0 ± 0 | 0 ± 0 | 0 ± 0 | 0 ± 0 |

**Table 6.** Number of seeds ($m^{-2}$) in the seed bank of *Atriplex stocksii* community.

| Year | Species | Nov | Dec | Jan | Feb | Mar | Apr | May | Jun | Jul | Aug | Sep | Oct |
|---|---|---|---|---|---|---|---|---|---|---|---|---|---|
| 1999–2000 | *Atriplex stocksii* | 3539 ± 90 | 5308 ± 176 | 10 616 ± 228 | 4423 ± 589 | 6192 ± 147 | 3539 ± 294 | 885 ± 0 | 7077 ± 361 | 2654 ± 442 | 4423 ± 38 | 5307 ± 265 | 7077 ± 361 |
| | *Cressa cretica* | 885 ± 76 | 0 ± 0 | 0 ± 0 | 0 ± 0 | 885 ± 89 | 885 ± 68 | 0 ± 0 | 885 ± 78 | 885 ± 64 | 0 ± 0 | 0 ± 0 | 0 ± 0 |
| | *Cyperus arenarius* | 0 ± 0 | 1769 ± 340 | 0 ± 0 | 0 ± 0 | 1769 ± 364 | 6192 ± 254 | 885 ± 0 | 6192 ± 216 | 1769 ± 362 | 4423 ± 241 | 15 039 ± 532 | 8160 ± 179 |
| | *Cyperus bulbosus* | 0 ± 0 | 0 ± 0 | 0 ± 0 | 0 ± 0 | 0 ± 0 | 0 ± 0 | 0 ± 0 | 0 ± 0 | 17 693 ± 228 | 0 ± 0 | 0 ± 0 | 0 ± 0 |
| | *Salsola imbricata* | 0 ± 0 | 0 ± 0 | 885 ± 0 | 0 ± 0 | 0 ± 0 | 885 ± 34 | 0 ± 0 | 0 ± 0 | 885 ± 54 | 885 ± 89 | 0 ± 0 | 0 ± 0 |
| 2001–2002 | *Atriplex stocksii* | 4423 ± 295 | 2654 ± 0442 | 3539 ± 23 | 3539 ± 45 | 15 924 ± 731 | 2654 ±442 | 53 078 ± 255 | 4423 ± 221 | 0 ± 0 | 10 616 ± 456 | 885 ± 0 | 7962 ± 423 |
| | *Cressa cretica* | 0 ± 0 | 0 ± 0 | 0 ± 0 | 0 ± 0 | 0 ± 0 | 0 ± 0 | 0 ± 0 | 0 ± 0 | 0 ± 0 | 0 ± 0 | 0 ± 0 | 0 ± 0 |
| | *Cyperus arenarius* | 2654 ± 442 | 4423 ± 442 | 4423 ± 589 | 0 ± 0 | 0 ± 0 | 0 ± 0 | 0 ± 0 | 8845 ± 654 | 1769 ± 145 | 6192 ± 147 | 2654 ± 442 | 7077 ± 525 |
| | *Cyperus bulbosus* | 0 ± 0 | 0 ± 0 | 0 ± 0 | 0 ± 0 | 0 ± 0 | 0 ± 0 | 0 ± 0 | 0 ± 0 | 0 ± 0 | 0 ± 0 | 0 ± 0 | 0 ± 0 |
| | *Salsola imbricata* | 0 ± 0 | 0 ± 0 | 0 ± 0 | 0 ± 0 | 0 ± 0 | 0 ± 0 | 0 ± 0 | 0 ± 0 | 0 ± 0 | 0 ± 0 | 0 ± 0 | 0 ± 0 |

**Table 7.** Number ($m^{-2}$) of viable seeds in the seed bank of *Atriplex stocksii* community.

| Month | *Atriplex stocksii* | | *Cressa cretica* | | *Cyperus arenarius* | | *Cyperus bulbosus* | |
|---|---|---|---|---|---|---|---|---|
| | **1999–2000** | **2001–2002** | **1999–2000** | **2001–2002** | **1999–2000** | **2001–2002** | **1999–2000** | **2001–2002** |
| Nov | 885 ± 87 | 4423 ±197 | 885 ± 87 | 0 ± 0 | 0 ± 0 | 0 ± 0 | 0 ± 0 | 0 ± 0 |
| Dec | 1769 ±134 | 2654 ±188 | 1769 ±134 | 0 ± 0 | 0 ± 0 | 0 ± 0 | 0 ± 0 | 0 ± 0 |
| Jan | 4423 ±197 | 2654 ±188 | 0 ± 0 | 0 ± 0 | 0 ± 0 | 885 ± 87 | 0 ± 0 | 0 ± 0 |
| Feb | 1770 ±1333 | 2654 ±188 | 0 ± 0 | 0 ± 0 | 0 ± 0 | 0 ± 0 | 0 ± 0 | 0 ± 0 |
| Mar | 3539 ±144 | 9732 ±206 | 885 ± 87 | 0 ± 0 | 0 ± 0 | 0 ± 0 | 0 ± 0 | 0 ± 0 |
| Apr | 2653 ±1888 | 885 ± 87 | 885 ± 87 | 0 ± 0 | 0 ± 0 | 0 ± 0 | 0 ± 0 | 0 ± 0 |
| May | 885 ± 87 | 2654 ±189 | 0 ± 0 | 0 ± 0 | 0 ± 0 | 0 ± 0 | 0 ± 0 | 0 ± 0 |
| Jun | 1770 ±133 | 3539 ±144 | 0 ± 0 | 0 ± 0 | 0 ± 0 | 0 ± 0 | 0 ± 0 | 0 ± 0 |
| Jul | 885 ± 87 | 0 ± 0 | 0 ± 0 | 0 ± 0 | 885 ± 87 | 885 ± 87 | 0 ± 0 | 0 ± 0 |
| Aug | 885 ± 87 | 2654 ±188 | 0 ± 0 | 0 ± 0 | 0 ± 0 | 0 ± 0 | 1769 ±133 | 0 ± 0 |
| Sep | 1770 ±133 | 0 ± 0 | 0 ± 0 | 0 ± 0 | 0 ± 0 | 0 ± 0 | 0 ± 0 | 0 ± 0 |
| Oct | 4423 ±197 | 4423 ±198 | 0 ± 0 | 0 ± 0 | 0 ± 0 | 0 ± 0 | 0 ± 0 | 0 ± 0 |

## References


1 Ungar IA (2001) Seed bank and seed population dynamics of halophytes. *Wetland Ecol Manag* 9: 499–510
2 Bertness MD, Gough L, Shumway SW (1992) Salt tolerance and the distribution of fugitive salt marsh plants. *Ecology* 73: 1842–1851
3 Gulzar S, Khan MA (1994) Seed banks of coastal shrub communities. *Ecoprint* 1: 1–6
4 Aziz S, Khan MA (1995) Role of disturbance of the seed bank and demography of *Leucus urticifolia* (Labiatae) population in a maritime subtropical desert of Pakistan. *Int J Plant Sci* 156: 834–840
5 Cantero JJ, Leon R, Cisneros JM, Cantero A (1998) Habitat structure and vegetation relationships in central Argentina salt marsh landscapes. *Plant Ecol* 137: 79–100
6 Shiferaw H, Teketay D, Nemomissa S, Assefa F (2005) Some biological characteristics that foster the invasion of *Prosopis juliflora* (SW.) DC at middle east Rift Valley Area, north eastern Ethiopia. *J Arid Environ* 58: 135–154
7 Wang SM, Zhang X, Li Y, Zhang L, Xiong YC, Wang G (2005) Spatial distribution patterns of the soil seed bank of *Stipagrostis pinnata* (Trin.) de Winter in the Gurbantonggut Desert of north-west China. *J Arid Environ* 63: 203–222
8 Ungar IA (1995) Seed bank ecology of halophytes. In: MA Khan, IA Ungar (eds): *Biology of Salt Tolerant Plants.* University of Karachi, Pakistan, 65–79
9 Jutila HME (1998) Effect of different treatments on the seed bank of grazed and ungrazed Baltic seashore meadows. *Can J Bot* 76: 1188–1197
10 Jutila HME (1998) Seed banks of grazed and ungrazed Baltic seashore meadows. *J Veg Sci* 9: 395–408
11 Jerling L (1984) Composition and viability of the seed bank along a successional gradient on a Baltic Sea shore meadow. *Holarctic Ecol* 6: 150–156
12 Jefferies RL, Davy AJ, Rudmick J (1981) Population biology of the salt marsh annual *Salicornia europaea* agg. *J Ecol* 69: 17–31
13 Hartman JM (1988) Recolonization of small disturbance patches in a New England salt marsh. *Am J Bot* 75: 1625–1631
14 Hopkins DR, Parker VT (1984) A study of the seed bank of a salt marsh in northern San Francisco Bay. *Am J Bot* 71: 348–355
15 Valiela I, Rietsma CS (1995) Disturbance of salt marsh vegetation by wrack mats in Great Sippewissett Marsh. *Oecologia* 102: 106–112
16 Badger KS, Ungar IA (1990) Seedling competition and the distribution of *Hordeum jubatum* L. along a soil salinity gradient. *Funct Ecol* 4: 639–644
17 Badger KS, Ungar IA (1994) Seed bank dynamics in an inland salt marsh with special emphasis on the halophyte *Hordeum jubatum* L. *Int J Plant Sci* 155: 66–72
18 Egan TP, Ungar IA (2000) Similarity between seed banks and aboveground vegetation along a salinity gradient. *J Veg Sci* 11: 189–194
19 Hutchings MJ, Russell PJ (1989) The seed regeneration dynamics of an emergent salt marsh. *J Ecol* 77: 615–637
20 Ungar IA, Woodell SRJ (1993) The relationship between the seed bank and species composition of plant communities in two British salt marshes. *J Veg Sci* 4: 531–536
21 Ungar IA, Woodell SRJ (1996) Similarity of seed bank to the aboveground vegetation in grazed and ungrazed salt marsh communities on the Gower Peninsula, South Wales. *Int J Plant Sci* 157: 746–749
22 Zaman AU, Khan MA (1992) The role of buried viable seeds in a saline desert plant community. *Bangladesh J Bot* 21: 1–10

23 Khan MA (1993) Relationship of seed bank to plant distribution in saline arid communities. *Pak J Bot* 25: 73–82
24 Khan MA (1990) The relationship of seed bank to vegetation in a saline desert community. In: DN Sen, S Mohammad (eds): *Proceedings International Seed Symposium*, Jodhpur, India, 87–92
25 Aziz S, Khan MA (1996) Seed bank dynamics of a semiarid coastal shrub community in Pakistan. *J Arid Environ* 34: 81–87
26 Gul B, Khan MA (2001) Seasonal seed bank patterns of an *Arthrocnemum macrostachyum* (Chenopodiaceae) community along a coastal marsh inundation gradient on the Arabian Sea near Karachi, Pakistan. *Pak J Bot* 33: 305–314
27 Cottam G, Curtis JT (1956) The use of distance measurements in phytosociology sampling. *Ecology* 37: 451–460
28 Khan MA, Gul B (1999) Seed bank strategies of coastal populations at Arabian Sea coast. In: ED McArthur, WK Ostler, CL Wambolt (eds): *Proceedings of the symposium on Shrubland Ecotones. Proceedings RMRS-P-11*, Ogden, UT: US. Department of Agriculture, Forest Service, Rocky Mountain Research Station USA, 227–230
29 Gul B, Weber DJ (2001) Seed bank dynamics in a Great Basin salt playa. *J Arid Environ* 49: 785–794
30 Adam MA, Marsh DM, Knox JS (2005) Importance of seed bank for population viability and population monitoring in a threatened wetland herb. *Biol Conserv* 124: 425–436
31 Jutila HME (2002) Seed banks of river delta meadows on the west coast of Finland. *Ann Bot* Fennici 39: 49–61
32 Ungar IA (1978) Halophyte seed germination. *Bot Rev* 44: 2333–2264
33 Woodell SRJ (1985) Salinity and seed germination in coastal plants. *Vegetatio* 61:223–230
34 Keiffer CH, Ungar IA (1997) The effect of extended exposure to hypersaline conditions on the germination of five inland halophyte species. *Am J Bot* 84: 104–111
35 Khan MA, Ungar IA (1997) Effects of thermoperiod on recovery of seed germination of halophytes from saline conditions. *Am J Bot* 84: 279–283
36 Dodd GL, Donovan LA (1999) Water potential and ionic effects on germination and seedling growth of two cold desert shrubs. *Am J Bot* 86: 1146–1153
37 Ungar IA (1987) Population ecology of halophyte seeds. *Bot Rev* 53: 301–334
38 Pugnaire FI, Lazaro R (2000) Seed bank and understory species composition in a semi-arid environment: The effect of shrub age and rainfall. *Ann Bot* 86: 807–813
39 Boudell JA, Link SO, Johansen JR (2002) Effect of soil microtopography on seed bank distribution in the shrub-steppe. *West N Am Natur* 62:14– 24
40 Khan MA, Ungar IA (1986) Life history and population dynamics of *Atriplex triangularis*. *Vegetatio* 66: 17–25

Biosaline Agriculture and High Salinity Tolerance
Edited by Chedly Abdelly, Münir Öztürk, Muhammad Ashraf and Claude Grignon


# Study of kochia (*Kochia scoparia*) as a forage crop

Mohammad Kafi[1] and Majid Jami-al-Ahmadi[2]

[1] *Center of Excellence for Special Crops, Ferdowsi University of Mashhad, Iran*
[2] *Faculty of Agriculture, Birjand University, Iran*

## Abstract

The production of halophytes using saline waters and soils, and feeding livestock with them, is one of the most sustainable ways of conserving desert ecosystems and food production for people living in these areas. Therefore, to study the possibility of growing kochia (*Kochia scoparia*) as a forage crop in desert environments with saline underground water, a research project was performed in Birjand, in the center of South Khorasan province, Iran. The rate and percentage of germination, radiation use efficiency (RUE), growth and ion accumulation in kochia were studied at three levels of saline irrigation water (1.5, 8.6 and 28.2 dS/m), three irrigation intervals and two plant densities (10 and 20 plants/$m^2$). The results showed that salinity negatively influenced the majority of plant's morphological and physiological indices, yet the dry matter accumulation in the highest salinity level reached 60% of plants in lower saline levels, and even moderate salinity caused a small stimulus in plant growth and yield performance. However, mostly no difference was observed with the lowest salinity level. The radiation absorption of kochia rose as leaf area index (LAI) increased, and 95% of radiation was absorbed with an LAI equal to 4.5–5. In general, increase in salinity caused a delay in early season development, and accelerated plant maturity at late season. Kochia's adjustability for vegetative growth and forage yield show no response to plant density, but the seed yield increased on increasing the plant density. In conclusion, the kochia's high production capacity, desirable digestibility and crude protein content in the presence of salinity and other desert stresses, such as high temperature and drought, make this plant suitable as a forage crop in harsh environmental conditions.

## Introduction

The phenomenon of soil salinization is a multidimensional problem that is caused by water, soil and biological factors, and resolving this problem will not be simple. Traditional approaches of agronomic management, such as designing drainage systems and land rehabilitation, using conventional cultivars and/or soil and water manipulation to improve plant products under adverse environmental condition, although they may be useful, in many cases are met with nonchalance and disinclination by farmers, due to high costs and emerging secondary problems [1, 2]. Sustainable utilization of saline water and soil resources needs reappraisal of natural resources and obsolete and/or underexploited plants [3, 4].

Forage and food production from halophytes is one of the sustainable utilization methods of soil and water resources in arid and semi-arid agro-ecosystems and nourish settled populations in these regions. Domestication of salt-tolerant plants (halophytes), which grow naturally in saline and arid vegetation sites, could be introduced to provide new crops for cultivation under environmental stresses induced by salinity and aridity. These plants could be a sensible alternative to conventional crops in many developing countries. Thus, biosaline agriculture is an inevitable necessity, particularly in arid and semi-arid regions, and has recently received more attention. In fact, the main constraint that previously prevented a better utilization of these plants was a lack of proper insight about the possibilities of using these plants and availability of fresh water [5, 6].

The *Kochia* species has recently attracted the attention of researchers worldwide. These plants are drought and salt tolerant and are grazed in their natural state by livestock [7], and may provide a good source of forage in dry regions that use saline water for irrigation [8, 9]. Farmers already cultivate *K. scoparia* as a drought-tolerant forage crop in lands where planting other crops is difficult; thus, it has been called as "poor man's alfalfa" [10]. Its rapid growth during a relative shorter period and drought tolerance indicate that *K. scoparia* has the capacity to turn into important forage especially in semi-arid regions [11]. Characteristics like relatively inexpensive production, adaptation to arid, saline and low-fertile lands, low water requirement, and resistance to insects and diseases, has caught the attention of researchers during recent years to use this plant as a forage crop [10].

The economical cultivation expansion of halophytes like kochia as a cash crop involves an understanding of biology, ecology, and superior management pattern of these plants to obtain optimal and sustainable production. The first prerequisite is understanding of the plant's capabilities. The main objectives in studying *K. scoparia* were to benefit from its halophytic nature and investigation of its cultivation possibility in regions that have a low productivity potential due to salinity or other specific environmental constraints. The possibility of domestication and sustainable utilization of kochia as a feed source for ruminants needs a multilateral study of its different characteristics under different climates and conditions. To achieve this we performed a comprehensive study of *K. scoparia*, and its responses to changes in growth medium, in the field as well as under a controlled environment. Part of our results are presented in this chapter.

## Explanation of experiments

The field studies consisted of two experiments at two different locations. In the first, the response of plants to different salinity levels in the irrigation water (1.5, 8.6 and 28.2 dS $m^{-1}$) was investigated at Birjand, a dry area with limited water resources for irrigation, particularly of fresh water. In the second, we cultivated *Kochia* at different combinations of plant densities (5,

**Table 1.** Chemical analysis of soil before and after of experiment in Mashhad and Birjand.

| | | EC | Absorbable elements (ppm) | | | | Soluble ions (meq $l^{-1}$) | | | | | | |
|---|---|---|---|---|---|---|---|---|---|---|---|---|---|
| **Location** | | **dS $m^{-1}$** | **pH** | **N** | **P** | **K** | $HCO_3^-$ | $Cl^-$ | $SO_4^=$ | $Na^+$ | $K^+$ | $Mg^{2+}$ | $Ca^{2+}$ |
| Before experiments | | | | | | | | | | | | | |
| Birjand | | 10.33 | 7.36 | 9.33 | 1.25 | 150 | 3.6 | 90 | 20.8 | 43.48 | 1.03 | 37.2 | 22.2 |
| Mashhad | | 12.4 | 7.66 | 24.5 | 28 | 420 | 4.5 | 75 | 44.5 | 67 | 1.3 | 28 | 29 |
| After experiment | | | | | | | | | | | | | |
| Birjand | S1 | 1.65 | 8.3 | 30.33 | 0.25 | 140 | 3.6 | 8.8 | 3 | 8.7 | 0.18 | 2.6 | 1.4 |
| | S2 | 13.24 | 7.73 | 11.66 | 0.25 | 145 | 1.6 | 94 | 46 | 76.1 | 0.9 | 48.2 | 27.6 |
| | S3 | 22.7 | 7.93 | 21 | 0.25 | 160 | 2.2 | 18.4 | 64 | 186.96 | 1.28 | 54.4 | 27 |

S1, S2, and S3 refer to salinity of irrigation water equal to 1.5, 8.6 and 28.2 dS $m^{-1}$, respectively. Soil texture in Birjand and Mashhad were sandy loam and silty clay loam, respectively.

**Table 2.** Chemical analysis of waters used during different experiments at Birjand and Mashhad.

| | | EC | | Soluble ions (meq $l^{-1}$) | | | | | | |
|---|---|---|---|---|---|---|---|---|---|---|
| **Location** | | **dS $m^{-1}$** | **pH** | $HCO_3^-$ | $Cl^-$ | $SO_4^=$ | $Na^+$ | $K^+$ | $Mg^{2+}$ | $Ca^{2+}$ |
| Birjand | | 5.6 | 7.57 | 3.7 | 32.5 | 19.8 | 38 | 0.18 | 11.6 | 7.0 |
| Mashhad | | 5.5 | 6.81 | – | 39.2 | – | 24.7 | 0.19 | 15.4 | 5.4 |
| Salinity | S1 | 1.5 | 6.92 | – | 8 | – | 10.8 | 0.15 | 5.8 | 1.4 |
| experiment | S2 | 8.6 | 6.3 | – | 74.4 | – | 48.7 | 0.40 | 26.6 | 12.4 |
| | S3 | 28.2 | 6.68 | – | 204 | – | 250 | 1.4 | – | 9.2 |

S1, S2, and S3 refer to salinity of irrigation water equal to 1.5, 8.6 and 28.2 dS $m^{-1}$ respectively.

10, 15, and 20 plants $m^{-2}$) and irrigation intervals (7 and 14 days) at Birjand (13°59' N, 53°32' E) and Mashhad (36°20' N, 59°35' E). Based on meteorological information, the average annual precipitation in Birjand and Mashhad is 176 and 260 mm, respectively. The chemical characteristics of applied waters and soil, before and after experiments, are represented in Tables 1 and 2.

## *Germination at different temperatures*

Kochia often grows in very dry environments; therefore, for rapid germination the limited moisture in spring is of importance [12, 13]. *K. scoparia* has a good germination in a wide thermal range between 4° and 41°C [12, 13]. Young et al. [14] studied germination of different accessions of *K. prostrata* under different temperatures and found the maximum germination of all accessions occurred in 20°C. Khan et al. [15] observed that the germination rate (GR) was the highest at higher temperatures. However, there is disagreement about the optimum temperature for kochia germination. Optimal germination has been reported at constant temperatures of 16°C [16], 17°C [17], 10° and 15°C [17], inconstant 15/30° or 20/30°C [12], and inconstant night/day temperatures of 25–30°C.

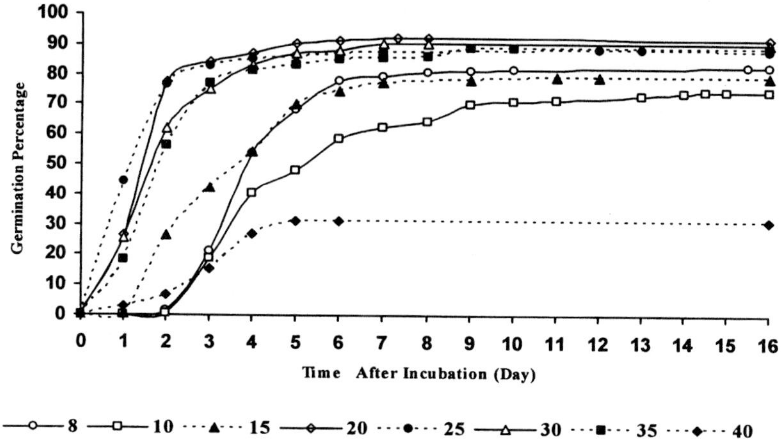


**Figure 1.** Cumulative germination percentage of Kochia seed at 24-h intervals in different temperatures. Each data point represents the average of four replications (together 160 seeds).

Considering the importance of these temperatures in research and agronomic planning, we studied *K. scoparia* germination response to different temperatures, to determine its cardinal germination temperatures. We evaluated the *Kochia* seed response to temperature at a thermal range of 10–40°C, with 5°C intervals, with an additional temperature of 8°C in growth chambers at Institute of sustainable agriculture (Instituto De Agricultura Sostenible, CSIC, Cordoba, Spain). Germination was recorded when tip of the radicle with a 5-mm length was uncoiled [14]. Germination counting was done at 12-h intervals, continuing until cumulative germinated seeds became stable in two consecutive records or when 100% germination was done.

*Germination percentage*

During first 96 h of *Kochia* germination, seeds at 20° and 25°C germinated rapidly and nearly reached their peak of germination (Fig. 1). At the lower temperatures (8, 10, and 15°C) fewer seeds germinated during the same period and final germination percentages (GP) at these temperatures were also lower, when compared with 25–35°C.

On the average, 85% of seeds germinated at temperatures of 8–35°C; at 40°C, the GP showed a significant reduction ($p<0.01$) and dropped to 31.5%. Our results are also in agreement with those of Everitt et al. [18]; however, they report that the significant reduction in germination of *K. scoparia* began from 30°C, and there were only 11% germination at 40°C, that is lower than our findings; however, our findings support their results and those of Romo and Haferkamp [19] that cold to moderate temperatures

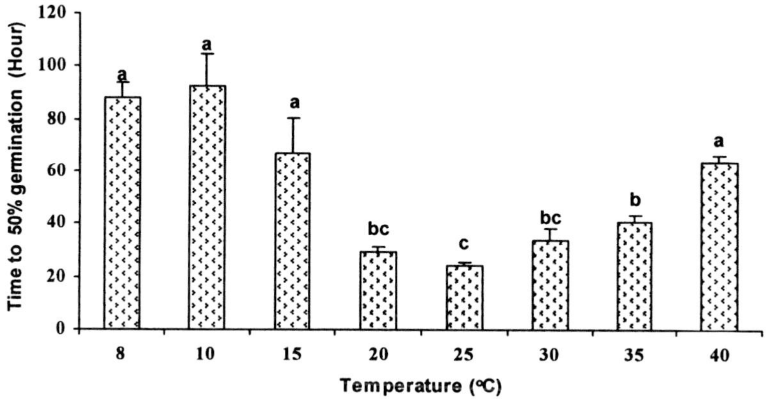


**Figure 2.** Temperature response of time taken to 50% germination. Error bars represent standard error of means for each temperature. Non-similar letters above the bars indicate significant different between temperatures at $p<0.01$ level. The coefficient of variance (CV) is 4.6%.

improve *Kochia*'s germination, while constant high temperatures constraints germination severely.

### *Germination rate*

As the daily germination trend indicates, the highest GP of 86% at 25°C occurred during the first 48 h after the start of the test, and about half of seeds finished their germination within the first 24 h of test at this temperature (Fig. 1). The inverse of time span until 50% germination (T50) was determined as an index of germination rate (GR) [5, 20, 21].

Analysis of variance showed a significant difference ($p<0.01$) in required time for 50% germination ($T_{50}$; Fig. 2). The minimum time span for 50% germination was 24 h at 25°C. Flores and Briones [8], in a study of six desert species at three temperatures (12, 20 and 26°C), also found that, as temperature increased, the beginning of germination and time span up to 50% germination decreased. Temperatures below 15°C caused a delay in germination of *Salicornia rubra* [22]. In winterfat (*Kraschenninkovia lanata*), a drought-tolerant shrub, when temperature increased, germination rate was raised and time to peak of germination decreased.

### *Cardinal temperatures*

The calculation method for cardinal temperatures is based on dependence of developmental rate to temperature. To describe cardinal temperatures we employed a frequently used intersected lines model, in which GR index and temperatures were dependent and independent variables, respectively:

$$Y = a1 + b1x \quad \text{if } T_b < T \leq T_o,\ b > 0$$
$$Y = a2 + b2x \quad \text{if } T_o < T \leq T_m,\ b < 0$$

where Y is the GR and T, $T_b$, $T_o$, and $T_m$ are average, base, optimal and maximum temperatures, respectively. In this model $T_b$ and $T_m$ are derived from the interception of each regression line with the abscissa and $T_o$ is the intersection point of two linear regression lines that describe response of GR to temperature at suboptimal and superoptimal temperatures.

Based on model, estimated $T_b$, $T_o$, and $T_m$ of kochia are 3.4°, 24°, and 49.7°C, respectively (Fig. 3). This appears to be the first report about cardinal temperatures of kochia in this field. It is evident that more work is required for these results to be generalized. In interpretation of cardinal temperatures, it should be considered that from these relations one can obtain $T_b$ and $T_o$; however, extrapolated values of $T_m$ are not constant for all GP and should be accepted carefully.

The estimation of $T_m$ from the model reflects the high tolerance of kochia to high temperatures, especially in the germination stage. It should be noted that in nature, seeds are subject to thermal fluctuation when germinating [20], thus these results under constant temperatures may differ slightly from germination in the field. The wide thermal range of kochia germination from 3.5°C ($T_b$) to 50°C, and its rapid germination ability indicate the high potential of kochia to establish in most regions of the world.

Since kochia can germinate at low soil temperatures and is highly cold tolerant in early growth stages, it can be sown at late winter; thus kochia germinates with the first moisture in spring. With this fast start to the growth season, competition with other annual plants would not be complicated. The ability of kochia to germinate at low temperatures and its fast germination in high temperatures allows farmers to use kochia as a second crop or to cultivate it twice a year (two plantings), thus allowing maximum exploitation of resources, especially in arid and semi-arid regions.

### *Germination at different levels of salinity*

The initial establishment of a plant species in saline habitats depends on their seed germination response to thermal and saline regimes, and it is the level of this response that usually determines whether a population can persists until reproductive maturity or not [5, 23]. Indeed, the seed ability to germinate under high salinity levels is an important factor for successful halophytes establishment through direct sowing [9, 24].

*K. scoparia* is a halophyte that germinates well under moderate and even high salinity level [17, 25–27]. Everitt et al. [18] reported that there was not

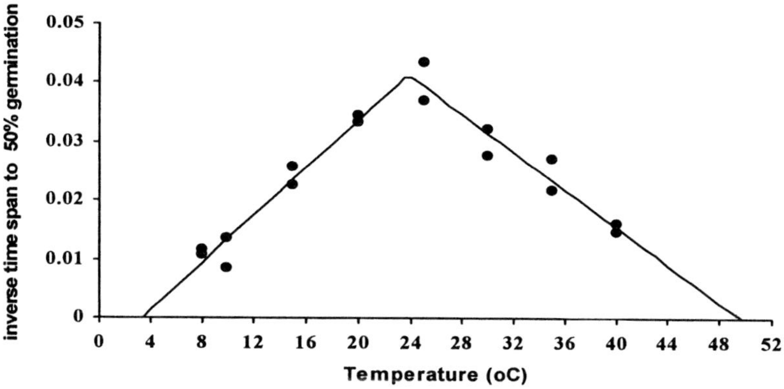


**Figure 3.** Kochia germination rate response to temperature based on inverse time taken to 50% germination index. Each data point is mean of two replications of each series of experiment. Fitted line between points has been obtained using non-linear triangular regression, and is the basis of determination of cardinal temperatures.

any significant effect of salinity induced by six different salts on *K. scoparia* germination up to 20 dS $m^{-1}$, although germination decreased significantly as salinity increased; however, even at 40 dS $m^{-1}$ there was 40% and 45% germination with sodium chloride and calcium chloride, respectively. Steppuhn and wall [27] used equal NaCl and $CaCl_2$ volumes to generate salinity concentrations from zero to 30 dS $m^{-1}$ and reviewed the relation between maximum cumulative *Kochia* germination with salinity of substrate. In their study, germination reductions due to salinity begun at concentrations lower than 20 dS $m^{-1}$ and reduction rate between 12 and 30 dS $m^{-1}$ was in average of –3.3% per dS $m^{-1}$ increase in salinity. According to Khan et al. [22], the highest germination (95%) occurred in distilled water, which decreased to 14% in 30 dS $m^{-1}$.

We studied seed germination characteristics of *Kochia scoparia* under unequal combination of NaCl and $CaCl_2$. For this, the saline solutions were prepared by adding appropriate amounts of NaCl and $CaCl_2$ in a 2:1 molar ratio to distilled water, to obtain an electrical conductivity (EC) range of 0–20 dS $m^{-1}$, with 5-dS $m^{-1}$ intervals. After adding saline solutions, petri dishes were put in a second plastic container and, to avoid evaporation, this second container was encased in a plastic film. As the lowest number of seeds germinated was 8, in one of 20 dS $m^{-1}$ replications, the time taken for 8 seeds to germinate was recorded as an arbitrary index of GR.

## *Germination percentage*

There was a reverse relation between the peak of cumulative germination and salinity of the medium (Fig. 4). In addition to the reduction of the total num-

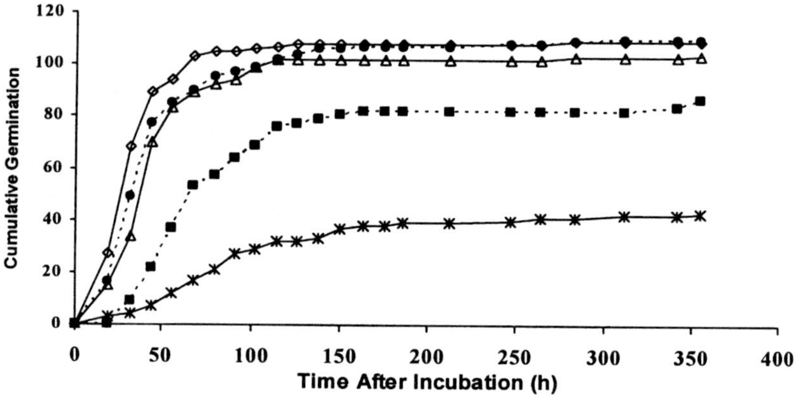


**Figure 4.** The trend of cumulative Kochia seed germination at 0(◇), 5(•), 10(△), 15(▪), and 20(✲) dS m$^{-1}$ salinity levels.

**Table 3.** Effects of salinity on germination properties of *Kochia scoparia.*

| Time to 50% germination (h) | Time to germination of 8 seeds (h) | Germination percentage | Average germination seeds/petri dish | Salinity dS/m |
|---|---|---|---|---|
| 27.8 b** | 20 b* | 90.84 a** | 27.25 a** | 0 |
| 34.25 b | 25.3 b | 91.67 a | 27.5 a | 5 |
| 37.8 b | 30.25 b | 86.67 a | 26 ab | 10 |
| 60.30 a | 54.5 b | 72.5 ab | 21.75 b | 15 |
| 77.8 a | 143.3 a | 35.84 b | 10.75 c | 20 |

In each column, non-similar letters indicate significant difference at 0.05 (*), or 0.01 (**) probability level.

ber of germinated seeds, enhancement of salinity caused an increase in the time needed for germination. The peak of germination occurred within the first 100 h of experiment, but a few seeds still germinated even at 14 days after the start of tests (Fig. 4). This indicates a high genetic diversity of kochia seed lots, which maybe a reason of its success in growing in harsh environments and also becoming a successful weed in agroecosystems.

Salinity up to 10 dS m$^{-1}$ did not impose any considerable effect on seed germination, but on passing this threshold, GP and GR decreased significantly ($p<0.01$, Fig. 4, Tab. 3). An extreme reduction occurred at 20 dS m$^{-1}$, but still more than 35% of seeds germinated at this level of salinity. In the study of Khan et al. [22], 14% germination was seen for Kochia at 30 dS m$^{-1}$. Regardless of inequality of units used by researchers to explain salinity concentration, these reports clearly show that each species has very particular germination requirements and response to salinity, and that each species also

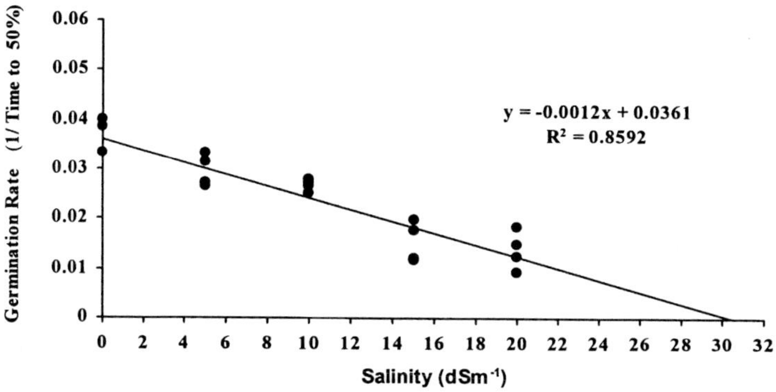


**Figure 5.** Germination rate (inverse of time taken to 50% germination) of *K. scoparia* under different salinity levels. Each series of point belong to four replications of each treatment.

displays a distinct germination, based on the location and dominating climate conditions during growth and seed production.

### *Germination rate*

Increases in salinity caused a reduction of seed GR and seeds required a longer period to complete their germination at high salinity levels (Tab. 3). Figure 5 shows that as salinity increased, seed GR (the inverse of time taken to 50% germination) decreased linearly; the time required to this stage varied from 28 h in distilled water to 78 h at 20 dS $m^{-1}$. On the other hand, whereas in controls about half of seeds completed their germination within the first day of the test, it took up to 3 days in a 20 dS $m^{-1}$, clearly showing a delay in germination due to salinity enhancement. However, there was no significant difference between salinity levels for GR, up to 10 dS $m^{-1}$ salinity. In the study of Khan et al. [22], as salinity increased, the average required time to reach germination stability (the end of germination) also increased, from 4 days in distilled water reached to 9 days in 24 and 30 dS $m^{-1}$.

Fitting a line between data points of GR and extrapolating it until it crosses the horizontal axis, the threshold of germination tolerance appears to be slightly more than 30 dS $m^{-1}$. It seems that above this salinity level GR approaches to zero; however, according to Steppuhn and wall [27], there was still 14% germination at 30 dS $m^{-1}$. Of course, they used equal volume of NaCl and $CaCl_2$ in preparing saline solution and this may be the reason for difference between the results. It should be noted that, in present study, the effects of concentrations higher than 20 dS $m^{-1}$ were not evaluated. In addition, the temperature used here (20°C) differed from temperature used in their study [27] (15°C) and in a study concerning interaction between salinity and temperature [15]; this temperature difference may contribute in variation in results.

Considering the lack of any significant reduction in GR and GP up to 10 dS $m^{-1}$, it may be possible to cultivate *K. scoparia* at farms using saline water with the same salt concentration, provided that the soil is wetted until establishment of seedlings so that soil water potential remains at a level that does not interfere with seeds.

Therefore, for planting in the early spring, two issues should be considered: if air temperature (prior, soil temperature) is high, it is evident that GP is more important than GR. In this situation, one can sow more seeds to insure a good crop stand. If temperature is low or water is salty, GR would be reduced and other factors such as microorganisms may affect the seed; thus in this situation irrigation should be done, which should not be saline enough to reduces GR severely. Taking these points into account, a good plant establishment could be expected using saline water or soil.

## Growth and development

### *Vegetative growth*

There were no significant differences in respect to leaf area index (LAI) between different salinity levels, particularly during later growth stages (Fig. 6a), and the maximum difference was observed at the maximum LAI, which was reached about 100 days after plantation (DAP). From Figure 6a, it can be concluded that salinity caused the trend of increasing plant leaf area to be slowed; however, it seems that Kochia can tolerate salinity, approaching a maximum LAI in the presence of salinity.

Considering the high correlation between produced dry matter and leaf area [28], which also was observed here ($r^2=0.88$, $p>0.01$), the variation in total dry matter appears to largely follow the changing trend of LAI (Fig. 6a, b). Moderate salinity increased dry matter accumulation (Fig. 6b). Stimulation of growth with moderate salinities has also been reported in *K. indica* [7]. According to these results, it appears that *K. scoparia* needs salts for its optimum growth and that low to moderate salinity levels stimulate its growth. Thus, contrary to a pervious assumption that Kochia is a voluntary halophyte [25], it seems rather to be an obligated halophyte species.

Increase of salinity to 28.2 dS $m^{-1}$ caused a noticeable alteration in partitioning coefficients, in favor to continuity of photosynthetate allocation to leaf production. Other reports (e.g., [29, 30]) also emphasized increase of leaf to stem ratio in response to increase of salinity, especially in halophytes. These and others [6, 31], all agree on this point that salinity exerts more restriction for stem growth than leaves.

Because of the halophytic nature of Kochia, it may be more realistic to consider the increase of leaf thickness as a factor contributing to tolerance

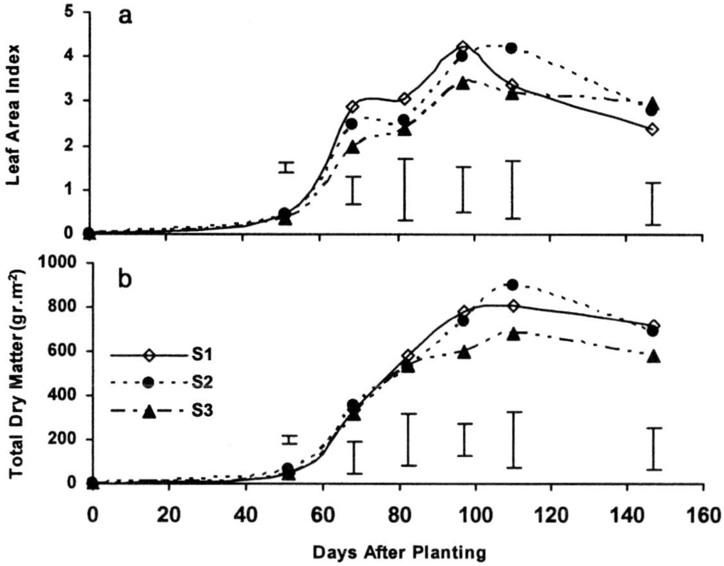


**Figure 6.** Changing trend of leaf area index (LAI) (a) and total dry matter (b) in response to different salinities of irrigation water. S1, S2, and S3 represent a salinity of 1.5, 8.6, and 28.2 dS m$^{-1}$, respectively. The vertical bars indicate the least significant difference (LSD) at 0.05 probability.

mechanism of the plant to higher salinities. As maximum leaf to stem weight ratio showed a direct correlation with electrolytes leakage from leaves ($r^2=0.48$, $p<0.01$), the continuity of leaf production at the highest salinity levels (Fig. 6a) is a response to more salt accumulation in leaves, which can damage plant cells, leading to defoliation of aged leaves. In this situation, plant continues to produce new leaves to maintain a high photosynthesizing area and control intensive salt accumulation at growth points [21, 26]. This will cause reduction of salt concentration inside other tissues and evasion of salinity damage, allowing photosynthesis to continue at the cost of some reduction in dry matter accumulation.

### *Reproductive growth*

Kochia is a short-day plant, which produces flowers and seeds in summer [12, 32]. Reproduction of Kochia species takes place with both self- or out-crossing [13]. The length of daylight that causes flowering induction varies from 13 to 15 h between different Kochia populations [12]. In our trials, plants in each location flowered and matured rather simultaneously between different treatments (±7 days). In Birjand, flowering started only after a lapse of 68 DAP [1083 thermal unit (TU) accumulation]; however, peak of pollination occurred ~14 days later when 1348 TU had accumulated. Flowering was terminated at 110 DAP (1824 TU), thus the duration of flowering was ~42 days in Birjand. In Mashhad, flowering started at 75 DAP, with only 737.4 TU accumulation, which is much lower than in Birjand. The extended flowering period of Kochia is probably mostly due to the high diversity between individuals, which is affected by environmental condition *per se*. This is confirmed by the high variation seen between individuals in growth index measurements.

The final harvest for seed (ripening) was done at 174 DAP (2986 TU) and 168 DAP (2404 TU) in Birjand and Mashhad, respectively. It has been stated that any plant needs a constant number of TU to reach a given developmental stage [16], but growth period differences between Birjand and Mashhad were notable. A precise classification of Kochia growth stages based on plant phonology could help to describe Kochia responses to environment conditions.

*Salinity and light utilization characteristics*

Kochia benefits from the $C_4$ photosynthesis pathway [18] and, therefore, is not a shade-tolerant plant. The only available information about its light characteristics were presented by Fischer et al. [20]. They reported K values of 0.65, 0.36, and 0.33 for Kochia, wheat, and barley, respectively. In their study radiation absorption by the canopy of Kochia rose linearly as LAI and plant height increased. Kochia maintained its high rate of photosynthesis at PAR values above 400 mM $m^{-2}$ $s^{-1}$, but at PAR values below this threshold, its $CO_2$ loss due to dark respiration was greater and its net assimilation rate was lower than that of cereals plants. This result shows clearly that wheat and barley tolerate shade better than Kochia.

To calculate the canopy light extinction coefficient, we measured incident radiation using a linear solarimeter (Sunscan, ΔT) between the hours of 11.00 to 14.00 over the solar noon. Since the diffuse component of solar radiation has a better uniform distribution than the direct one, all measurements were done with artificial shading. Charles-Edwards and Lawn [25] stated that if the measured radiation at solar noon is used to determine radiation interception, the ratio of absorbed radiation would be underestimated up to 10%. Thus, $F_{abs}$ was corrected for the whole day using equation presented by Charles-Edwards et al. ([7], Eq. 1):

$$Q_{day} = 2 \times Q_{noon} / [1 + Q_{noon}] \quad [1]$$

The amounts of radiation absorbed by canopy were calculated using a constant K value between salinity treatments, assuming 45% of intercepted radiation is photosynthetically active.

Since all radiation measurements were done at solar noon, its value indicates the relation between natural logarithm (Ln) of the canopy $F_{abs}$ at solar noon with LAI. The slope of this relation is the light extinction coefficient (K) value of Kochia at solar noon, and is 0.59. This value is lower than that reported K of 0.65 by Fischer et al. [20] for Kochia. However, in their experiment K was calculated under competitive condition with wheat and barley. It should be noted that measurements of radiation above and bottom of canopy in our experiment were done with elimination of direct radiation by artificial shading because, according to Sinclair and Horie [33], ignoring this component may generate a considerable error in calculations .

The amount of radiation intercepted by canopy and its transmitted fraction change with time and both reach to peak around solar noon [18]. Thus, the daily variation in flux intensity of incident radiation above the canopy ($I_0$) and the fraction transmitted through the canopy ($I/I_0$) should be considered in calculating of K. After correcting the amount of radiation absorption by the canopy using Eq. (1), the K of Kochia for the whole day increased to 0.75, which was higher than the value reported by Fischer et al. [20]. This value was used in further calculations and was assumed constant in all treatments. Begue et al. [28] also stated that $F_{abs}$ at noon is lower than its value at morning or afternoon and, to obtain real K values, the daily variation of K with change in solar angle should be considered in calculations.

The fraction of radiation intercepted by canopy increased with LAI and 95% of incident radiation was absorbed at a LAI of 4.5–5. Figure 6a indicates that plants reached to this LAI in about 100 DAP, i.e., in mid to late flowering period. This time coincides almost with early summer and after that the total daily radiation that reaches the plant canopy starts to decrease. The lack of synchrony between maximum incoming seasonal radiation and maximum LAI will lead to an inefficient absorption and utilization of light by the canopy.

### *Radiation absorption*

Figure 7 indicates the estimated absorbed radiation during the growth season at different salinities. It seems likely that plants at lower salinities closed their canopies earlier and that the extension rate of leaf area at early season was reduced as salinity increased. At high salinity level (28.2 dS $m^{-1}$) plants approached their maximum $F_{abs}$ later in the growth season. The important point is that the peak of $F_{abs}$ at this salinity level coincided with the time at which the incident radiation to the canopy started to decrease (Fig. 7). Therefore, a lower amount of absorbed radiation would be expected in this treatment. The higher $F_{abs}$ in higher salinity in the late season were due to its higher LAI in this time (Fig. 6a). Reduction of $F_{abs}$ during the late season is a consequence of normal leaf desiccation.

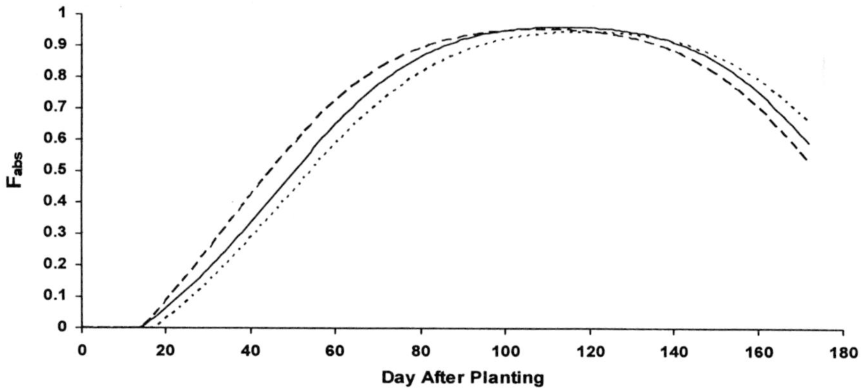


**Figure 7.** The predicted effects of salinity on the fraction of absorbed radiation of Kochia during growth season. Salinity levels are 1.5 (patchy line), 8.6 (solid line), and 28.2 (dotted line) dS $m^{-1}$.

**Table 4.** Radiation use efficiencies of kochia in different salinity levels.

| **EC (dS.m$^{-1}$)** | **RUE (gr.MJ$^{-1}$ PAR)** | $R^2$ |
|---|---|---|
| 1.5 | 2.41 | 0.86 |
| 8.6 | 2.48 | 0.87 |
| 28.2 | 2.20 | 0.85 |

## *Radiation use efficiency*

Fitting a linear regression between data point for each salinity level showed that the increase in salinity level from 1.5 to 8.6 dS $m^{-1}$ caused a slight increase in RUE that decreased again with higher salinity; however, there was not much difference in RUE between different salinity levels (Tab. 4). The moderate salinity level, which had the highest RUE, is the treatment that showed the highest above-ground biomass accumulation, which is a consequence of the stimulating effect of moderate salinity on growth of halophytes such as Kochia [7, 30, 31].

It had been observed that plants exposed to salinity assimilate carbon with a higher efficiency as a result of carbon hydrate effects on osmotic adjustment, so that the increase in efficiency of carbon fixation and assimilation leads to an improved osmotic adjustment process [34]. However, it is evident that this is only true if the salinity level does not exceed the plant tolerance threshold and its capacity for osmotic adjustment.

Our results indicated that Kochia has a reliable tolerance to salinity, so that an increase in salinity in the irrigation water from 1.5 to 28.2 dS $m^{-1}$, which can cause plant death in many species, leads to only a negligible decrease in

**Table 5.** Effects of different levels of salinity of irrigation water on yield and yield attributes of *K. scoparia*.

| Salinity (dS $m^{-1}$) | Plant dry weight (g) | Forage dry matter (%) | Forage dry yield (kg $ha^{-1}$) |
|---|---|---|---|
| 1.5 | 37.12 (± 4.02)* | 34.20 (± 1.87) | 9670.6 (± 564.8) |
| 8.6 | 47.64 (± 5.75) | 33.58 (± 2.68) | 9907.0 (± 1221.8) |
| 28.2 | 26.7 (± 3.00) | 26.11 (± 1.00) | 6182.1 (± 687.5) |

*Numbers inside parenthesis indicate Standard Error of Means (SEM).

RUE. Moreover, the moderate salinity level of 8.6 dS $m^{-1}$ induced stimulation in plant growth and improvement of RUE. Therefore, *K. scoparia* is not only a saline-tolerant species, it also shows a good response to moderate levels of accumulated salts in soils, and could be used successfully in saline and desert environments to produce forage for domestic animals in these harsh areas, which have already been faced with forage shortage.

### *Salinity and dry matter production*

In our study moderate salinity (8.6 dS $m^{-1}$) caused a slight stimulation of growth and an increase in plant dry weight. At 28.2 dS $m^{-1}$ plant dry weight decreased significantly (Tab. 5). These findings are in agreement with those of Ashour et al. [35], who underlined the stimulating effect of moderate salinity on growth and yield of halophytic grasses. The larger plant dry weight reduction at higher salinities can be attributed to several factors, such as leaf area loss, increased respiration, turgor reduction, rising osmotic adjustment costs and salt uptake and excretion [26].

Average forage dry matter percentage at first cutting (26.7%) was different from second cutting (35.9%) and the effects of salinity were obvious only at the second harvest (Fig. 8), i.e., dry matter content increased with the progress of the growth season. Plants in the highest salinity conserved more water content to withstand the negative effects of salinity. Thus, it can be concluded that control of plant water is part of the adaptive process in *K. scoparia* for osmotic adjustment, especially as plant ion concentration and water content together determine total osmotic pressure, which have a physical relation with salinity of external medium of roots [16].

Increase in salinity from 1.5 to 8.6 dS $m^{-1}$ increased the dry forage yield slightly, but with a further salinity increase to 28.2 $dsm^{-1}$ the yield reduced significantly (Tab. 5); however, plants under this high salinity level produced a forage yield equal to 64% of plants at the lowest salinity level (a 36% reduction). These results are in agreement with the findings of Noaman and El-Haddad [30] with six halophyte species and Steppuhn et al. [27] with four halophyte species. The more negative effects of salinity on stem growth and increase of leaf to stem ratio in response to salinity enhancement have also

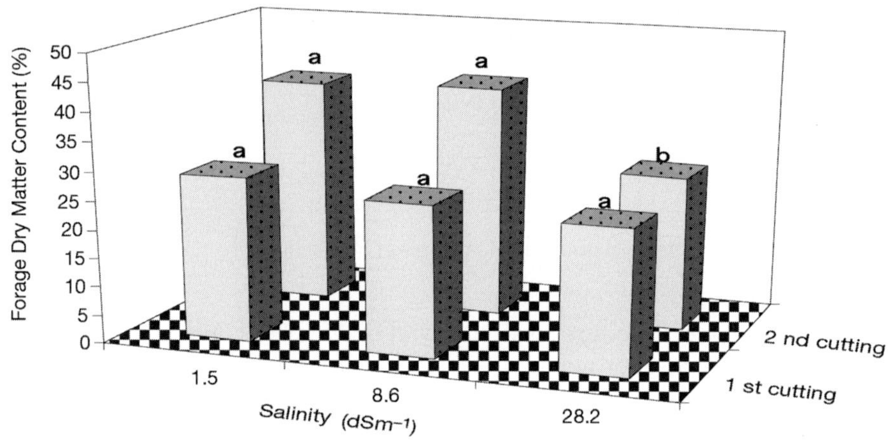


**Figure 8.** Salinity effects on dry matter content (%) of produced Kochia forage, harvested at two successive cutting at 82 and 145 days after plantation (DAP). Non-similar letters above bars indicate significant difference at 0.01 probability level.

been emphasized for *Atriplex griffithi* var. *stocksii* and *Suaeda fruticosa* [15, 31].

The total forage yield (sum of two cuttings) in Mashhad and Birjand were 19 697.3 and 10 524.2 kg ha$^{-1}$, respectively. Undersander et al. [10] reported a forage yield between 2 and 8 ton ha$^{-1}$ dry matter [10], which are lower than the obtained yield in our studies. Our harvested yields are similar to those of Sherrod [11] who harvested Kochia from a wild green stand at three different times (pre-, mid-, and post-flowering) and observed a linear dry matter yield increment with progress in maturity, and at final harvest 11 326.7 kg ha$^{-1}$ dry matter was obtained, whereas Foster (1980, cited in [17]) in New Mexico harvest 26 ton ha$^{-1}$ on a dry weight basis from a well-irrigated and fertilized stand of *K. scoparia* with four harvests during the growth season; however, this report seems superlative. These reports indicate a great variation between obtained yields of *K. scoparia*, which is a result of its high genetic variation [13, 36, 37] and differences between landraces or wild collections. Another effective factor on these yield fluctuations is the acclimation of Kochia to different environmental conditions, which was also observed in our study and caused different growth behavior and yield of Kochia in two different locations (Mashhad and Birjand).

It seems that *K. scoparia* has a very high yield potential, and in this respect is competitive with most of conventional forage crops. However, it should be kept in mind that these yields in our study have been produced by plants that resulted from seeds that were not subjected to any breeding program. Despite this, the yield in Mashhad was comparable with an alfalfa yield, and yields obtained in Birjand were not lower than those for most forage crops under the same conditions in saline areas. We believe that under these conditions, particularly after breeding and domestication, *K. scoparia* could compete with any conventional forage crop.

### *Seed yield*

Seed yield is a complex trait, which is a function of distinct yield components that are influenced greatly by weather conditions. Kochia plants show a dramatic reproductive flexibility that depending upon the environment and competition conditions, as seed production may vary greatly between single plants from only 5 seeds per plant under stressful conditions to more than 50000–100000 seeds per plant under favorable conditions. Nevertheless, a single Kochia plant usually produces about 14600 seeds per year [12, 13]. Kochia seed yield has been reported to be ~1500–2000 kg ha$^{-1}$ [10]. Cox et al. (1969, cited in [7]) estimated that average seed yield of *K. scoparia* is equal to 2200 kg ha$^{-1}$.

In our study, increase in water salinity from 1.5 to 28.2 dS m$^{-1}$ caused a seed yield reduction, from 2701 to 2198 kg ha$^{-1}$. Nevertheless, this non-significance effect of salinity on seed yield in our study contradicts the results of Noaman and El-Haddad [30], who reported that at high salinity levels seed yields of six halophytic grasses generally decreased more than vegetative growth. Thus, under saline stress, Kochia plants, although reducing a portion of vegetative growth, appear to always allocate a constant photosynthetate volume to seed production and this trait is not influenced by external factors.

## Conclusions

The extended temperature range for Kochia germination, from a base temperature of 3.5°C to maximum of 50°C, and its ability to germinate at salinities of up to 30 dS m$^{-1}$ are indications of its high establishment potential in most of dry and saline areas. Kochia's capability for quick germination, emergence and growth will favor its establishment inside saline seedbeds. Field trials indicated that Kochia displays a good tolerance to salinity after establishment, so that an increase in salinity of irrigation water from 1.5 to 28.2 dS m$^{-1}$ is accompanied by only a 36% yield reduction. Indeed, moderate salinity level of 8.6 dS m$^{-1}$ stimulated plant growth and improved RUE and quality properties of produced forage.

In general, quick germination and growth of Kochia and its desirable drought and salinity tolerance indicate that it can be regarded as a valuable forage plant when there is a shortage of conventional forage plants, particularly in arid and semi-arid regions. Kochia can also provide a good source of forage in areas with a deficit in irrigation or using saline waters, and, due to its high nutritional value, can constitute an important component of ruminant diets in these areas. Cultivation of Kochia using saline waters for rehabilitation of saline areas that have been left barren can be regarded as an approach to sustainable and low-input agriculture, which is needed to enhance genetic diversity of agricultural products, and can help in food production for people settled in these regions.

## References

1 Pessarakli M (1994) *Handbook of plant and crop stress*. Marcel Dekker, New York

2 Siadat H, Bybordi M, Malakouti MJ (1997) Salt-affected soils of Iran: A country report. Proceeding of International symposium on "*Sustainable Management of Salt Affected Soils in the Arid Ecosystems*". Cairo, Egypt

3 Glenn EP, Swingle RS, Riley JJ, Mota CU, Watson MC, Squires VR (1992) North American halophytes: Potential use in animal husbandry. In: VR Squires, AT Ayoub (eds): *Halophytes as a resource for livestock and for rehabilitation of degraded lands*. Kluwer Academic Publishers, Dordrecht

4 Lieth H, Lohmann M (2000) *Cash crop halophytes for future halophyte growers*. Institute of Environmental Systems Research, University of Osnabrück, Osnabrück

5 Aflakpui GKS, Gregory PJ, Froud-Williams RJ (1998) Effects of temperature on seed germination rate of *Striga hermonthica* (Del.) Benth. *Crop Protect* 17: 129–133

6 Fuehring HD, Finkner RE, Oty CW (1985) Yield and composition of kochia forage as affected by salinity of water and percent leaching. Available at: http://wrri.nmsu.edu/publish/techrpt/abstracts/abs199.html (accessed 4 Jun 2005)

7 Zahran MA (1993) Juncus and kochia: Fiber- and fodder-producing halophytes under salinity and aridity stress. In: M Pessarakli (ed): *Handbook of plant and crop stress*. Marcel Dekker, New York

8 Florcs J, Briones O (2001) Plant life-form and germination in a Mexican inter-tropical desert: Effects of soil water potential and temperature. *J Arid Environ* 47: 485–497

9 Miyamoto S, Glenn EP, Singh NT (1992) Utilization of halophytic plants for fodder production with brackish water in subtropic deserts. In: VR Squires, AT Ayoub (eds): *Halophytes as a resource for livestock and for rehabilitation of degraded lands*. Kluwer Academic Publishers, Dordrecht

10 Undersander DJ, Durgan BR, Kaminski AR, Doll JD, Worf GL, Schulte EE (1990) Alternative Field Crops Manual: Kochia. Available at: http://www.hort.purdue.edu/newcrop/afcm/kochia.html (accessed 3 Dec 2000)

11 Sherrod LB (1971) Nutritive value of *Kochia scoparia*. I. Yield and chemical composition at three stages of maturity. *Agron J* 63: 343–344

12 Clarke LD, West NE (1969) Germination of *Kochia americana* in relation to salinity. *J Range Manag* 22: 286–287

13 Mullenix W (1998) Kochia (Kochia spp.) biology outline and bibliography. Available at: http://www.agron.iastate.edu/~weeds/WeedBiolLibrary/kochiabiblio.html (accessed 3 Dec 2000)

14 Young JA, Evans RA, Stevens R, Everett RL (1981) Germination of *Kochia prostrata* seed. *Agron J* 73: 957–961

15 Khan MA, Gul B, Weber DJ (2000) Germination response of *Salicornia rubra* to temperature and salinity. *J Arid Environ* 45: 207–214

16 Kader MA, Jutzi SC (2004) Effects of thermal and salt treatments during imbibition on germination and seedling growth of sorghum at 42/19°C. *J Agron Crop Sci* 190: 35–38
17 Eberlin CV, Fore ZQ (1996) Kochia biology. Available at: http://mandakzerotill.org/book11.kochia.html (accessed 12 Dec 2002)
18 Everitt JH, Alaniz MA, Lee JB (1983) Seed germination characteristics of *Kochia scoparia*. *J Range Manag* 36: 646–648
19 Romo JT, Haferkamp MR (1987) Forage kochia germination response to temperature, water stress, and specific ions. *Agron J* 79: 27–30
20 Fischer AJ, Messersmith CG, Nalewaja JD, Duysen ME (2000) Interference between spring cereals and *Kochia scoparia* related to environment and photosynthetic pathways. *Agron J* 92: 173–181
21 Smart AJ, Schacht WH, Moser LE (2001) Predicting leaf/stem ratio and nutritive value in grazed and nongrazed big bluestem. *Agron J* 93: 1243–1249
22 Khan MA, Ungar IA (2001) Seed germination of *Triglochin maritime* as influenced by salinity and dormancy relieving compounds. *Biologia Plantarum* 44: 301–303
23 Khan MA, Gul B, Weber DJ (2001) Influence of salinity and temperature on germination of *Kochia scoparia*. *Wetlands Ecol Manag* 9: 483–489
24 Tobe KX, Li, Omasa K (2000) Seed germination and radicle growth of a halophyte, *Kalidium capsicum* (Chenopodiaceae). *Ann Bot* 85: 391–396
25 Charles-Edwards DA, Doley D, Rimmington GM (1986) *Modelling plant growth and development.* Academic Press, Australia
26 Kafi M, Mahdavi Damghani M (2000) *Mechanisms of environmental stress resistance in plants* (in Persian). Ferdowsi University Press, Mashhad
27 Steppuhn H, Volkmar KM, Miller PR (2001) Comparing canola, field pea, dry bean, and durum wheat crops grown in saline media. *Crop Sci* 41: 1827–1833
28 Begue A, Desprat JF, Imbernon J, Baret F (1991) Radiation use efficiency of pearl millet in the Sahelian zone. *Agric Forest Meteo* 56: 93–110
29 Waller SS, Britton CM, Schmidt DK, Stubbendieck J, Sneva FA (1983) Germination characteristics of two varieties of *Kochia prosterata* (L.) Schrad. *J Range Manag* 36: 242–245
30 Noaman MN, El-Haddad E (2000) Effects of irrigation water salinity and leaching fraction on the growth of six halophyte species. *J Agric Sci (Cambridge)* 135: 279–285
31 Khan MA, Ungar IA, Showalter AM (2000) Effects of salinity on growth, water relations and ion accumulation of the subtropical perennial halophyte, *Atriplex griffithi* var. *stocksii*. *Ann Bot* 85: 225–232
32 Unknown (2001) Kochia. Available at: http://www.wa.gov/agr/weedboared/weed-info/kochia.html (accessed 13 Nov. 2003)
33 Sinclair TR, Horie T (1989) Leaf nitrogen, photosynthesis, and crop radiation use efficiency: A review. *Crop Sci* 29: 90–98
34 Wyn Jones RG, Gorham J (1983) Osmoregulation. In: OL Lange, PS Nobel, CB Osmond, H Ziehler (eds): *Encyclopedia of plant physiology*. New Series, vol. 12C. Physiological Plant Ecology. Springer, Berlin, 35–38
35 Ashour, NI, Serag MS, Abd El-Haleem AK, Mekki BB (1997) Forage production from three grass species under saline irrigation in Egypt. *J Arid Environ* 37: 299–307
36 Smith DT, Wiese AF, Cooley AW (1975) Postemergence control of Kochia and Russian thistle in early spring. *Agron J* 67: 752–754
37 Bhardwaj SN, Saini Munshising AD, Singh KD (1988) Contribution of area, thickness and conductance of leaf in biomass production in upland cotton (*Gossypium hirsutum*). *Indian J Agric Sci* 58: 100–103

Biosaline Agriculture and High Salinity Tolerance
Edited by Chedly Abdelly, Münir Öztürk, Muhammad Ashraf and Claude Grignon


# Biodiversity of plant species and adaptation to drought and salt conditions. Selection of species for sustainable reforestation activity to combat desertification

Cristina M. Monteverdi[1], Marco Lauteri[2] and Riccardo Valentini[1]

[1] *Department of Forest Sciences and Resources, University of Tuscia, Via S. Camillo De Lellis s.n.c., 01100, Viterbo, Italy*
[2] *CNR, Institute of Agro-Environmental and Forest Biology, Via G. Marconi 2, 05010, Porano (TR), Italy*

## Abstract

Soil salinity is a threat for many agriculture and forest communities. Particularly ecosystems surrounding desert and coastal areas need to be preserved and restored for their intrinsic ecological value and vulnerability. Knowledge on physiology and ecology of plant tolerance, resistance and adaptation to salt exposure is fundamental to face land degradation and related desertification processes. Studies of physiological adaptive traits are important to select suitable species for a sustainable management of rural and forest environments. Salinity affects productivity of plant species. Moreover the capacity to tolerate salinity is widely variable at intra- and inter-specific level. It is known that variations of photosynthetic performances and related parameters (e.g., leaf area) are mainly responsible of growth reduction under salt stress. Generally salinity limits $CO_2$ availability at the carboxylation sites. A reduced assimilation rate is often accompanied by a more than proportional decrease in transpiration rate. Thus, salt-stress conditions usually determine an increase of instantaneous water-use efficiency (WUE). The investigation of salinity tolerance in natural environments is not easy because of climatic factors and field heterogeneity of salty soil. Moreover, because of the multiple and complex physiological responses to salinity, it is very important to find a suitable index to properly integrate the different physiological processes involved. Carbon isotope discrimination (Δ) in plant material has been demonstrated by several studies to be a promising approach that integrates physiological processes on different time scales under both drought and salt conditions. Δ is directly related to the ratio of intercellular to atmospheric partial pressure of $CO_2$ ($p_i/p_a$) and negatively related with WUE. Moreover, it is a relatively low-cost, easy and non-invasive technique to investigate ecophysiological traits of plant species in controlled and field conditions. A brief overview and perspectives in applying Δ in ecophysiological studies, related to salinity and drought tolerance, are shown.

## Introduction

Ecosystem vulnerability and degradation due to unsustainable land-use patterns represent major warnings of this century. Fundamental ecosystem functions, such as biodiversity and primary productivity, are severely endangered.

Land degradation processes and desertification are quickly developing in many regions. As a consequence, in the near future many countries will not be able to manage agricultural and forestry resources in a sustainable way.

Soil salinization is one of the most important processes in land degradation and represents a major environmental hazard. About 41% of the world lands is considered subjected to desertification and land salinization because of excessive exploitation of the land. This percentage is increasing.

Data from the FAO/UNESCO indicate that about 397 million ha of the world lands are affected by soil salinization and that sodic soils cover 434 million ha. These areas are not necessarily arable lands, but include all salt-affected areas of the world (FAO 2000, www.fao.org). Concerning arable lands, it has been estimated that 3 ha/min of surface are lost in the world because of increase in soil salinity (FAO 2000, www.fao.org).

Salt accumulates in coastal soils, in places where the water table is shallow and the evapotranspirative demand is high, in depressed or in intensively irrigated areas, managed with traditional and not sustainable techniques where drainage is insufficient. Soil salinity is a threat for many agriculture and forest communities. Particularly ecosystems surrounding desert and coastal areas need to be preserved and restored for their intrinsic ecological value and vulnerability.

Knowledge on the physiology and ecology of plant tolerance, resistance and adaptation to salt exposure is fundamental to face land degradation and related desertification processes. In saline environments plant productivity is limited due to various adverse effects of salt on plant growth [1, 2]. Salt stress may determine different negative effects on plants, including water potential reduction, disturbances in ion homeostasis and toxicity. Generally, salinity largely affects higher plants, both halophytes and non-halophytes. Moreover, the capacity to tolerate salinity is widely variable among and within species [3, 4]. Often non-halophytes show dramatic growth reduction, although exposed at only low salt concentrations [5].

The basis of the decline in plant growth under saline conditions is not completely understood. There are different theories concerning salinity effects on plant growth. According to Gale [6] and Schwarz and Gale [7], the growth reduction under saline conditions could be related to increased respiration rates resulting from higher energy requirements. Other authors argue that the reduction in productivity is related to combined effects of salinity on photosynthesis and carbon allocation [5, 8, 9]. Another cause is addressed to the reduction of leaf area involved in $CO_2$ assimilation. It has been shown in many works that salinity affects carbon uptake *via* effects on leaf expansion [10–12].

According to Flowers et al. [13], Greenway and Munns [14], Munns and Termaat [15], Brugnoli and Lauteri [5], modifications of ion balance, water status, mineral nutrition, stomatal conductance, photosynthetic $CO_2$ assimilation, carbon allocation and utilization are the main physiological causes determining a lower growth rate. Within this pool of causes, variations of many photosynthetic performances and related parameters (e.g., leaf area) are mainly responsible of the growth reduction under salt stress [14].

Increasing soil salinity is known to negatively affect photosynthesis through stomatal and non-stomatal effects. Both of these may be involved in the reduction of photosynthetic $CO_2$ assimilation rate on a leaf area basis [5, 16–18].

## Stomatal effects

Salinity limits $CO_2$ availability at the carboxylation sites. Salt-stress conditions induce stomatal closure so that photosynthetic rate is reduced because of substrate limitation [5, 19–25]. A reduced assimilation rate is often accompanied by a more than proportional decrease in transpiration rate. Thus, salt stress usually determines an increase of instantaneous water-use efficiency (WUE). The capacity of WUE regulation is addressed by several authors as a relevant mechanism of salt tolerance in plants. Decreased water consumption per unit of biomass accumulated implies a reduced amount of salt absorbed by plants and a reduced energy cost for compartmentation or extrusion of salt itself. Salt stress affects plant water relations by reducing water potentials. Decreased osmotic potentials and increased turgor pressures are common responses in plants exposed to salt [26–28]. A high WUE in salt-dryland plant species has important physiological and ecological implications.

An ecological role of WUE for adaptation to saline conditions has been established by recent studies (Brugnoli and co-workers, unpublished), which indicate how plant communities typical of saline ecosystems generally show higher WUE than communities colonizing less saline-exposed habitats.

## Non-stomatal effects

Apart from the reduction in photosynthetic rate due to the reduction of stomatal conductance, many reports highlight salt-stress effects in terms of non-stomatal limitations of photosynthesis. Common effects concern the reduction in activity of ribulose-1,5-bisphosphate carboxylase (Rubisco) or limitations of the capacity for RuBP regeneration due to reduced electron transport [19, 23, 29–31]. Under chronic salt-stress conditions, non-stomatal limitation of photosynthesis can be simply related to a decreased allocation of resources to enzymes of carbon fixation despite effective damage to the photosynthetic apparatus [32].

During gas exchange measurements sub-stomatal $CO_2$ partial pressure ($p_i$) could be overestimated because of patchy photosynthesis due to stomatal heterogeneity. Terashima et al. [33] and Brugnoli and Lauteri [5] showed heterogeneous stomatal apertures under various stress conditions including salinity in leaves characterized by mesophyll compartmentation. The overestimation of $p_i$ might lead to an incorrect evaluation of stomatal and non-stomatal effects of salinity on photosynthesis.

Modern techniques based on stable carbon isotope analysis (for reviews see [34–36]) allow the separation of stomatal from false non-stomatal responses. Carbon isotope discrimination (Δ), a complex physiological parameter representing the depletion of the heavy stable isotope $^{13}C$ in plant tissues with respect to its abundance in the atmospheric $CO_2$, is directly related to the ratio of intercellular to atmospheric partial pressure of $CO_2$; this ratio ($p_i/p_a$) represents the equilibrium between the availability and the requirement of $CO_2$ at the leaf level, which is the set point for gas exchange activity [37]. In this way, the carbon stable isotope technique represents a reliable tool to estimate the intercellular $CO_2$ partial pressure. It is interesting to note that the $p_i/p_a$ parameter is also negatively related with WUE, so that a mechanistic negative relationship between Δ and WUE follows. According to this theory, Δ analysis allows an assimilation-weighted estimation of both $p_i/p_a$ and intrinsic WUE integrated over different time scales, depending on the tissues or metabolites being analyzed ([38–41] and Monteverdi and Lauteri, unpublished).

In conclusion, several experiments have shown that salinity largely affects photosynthesis because of stomatal effects. Nevertheless, non-stomatal effects, like reduced photosynthetic capacity and reduced assimilation rate and linear electron flow at $p_i$ representative of operational conditions, could sometime have an effect. These effects were always associated with reduction of the enzymes of the photosynthetic carbon reduction-oxygenation cycle. As a consequence of both limitations, photosynthetic carbon gain is reduced to an extent depending on the levels of salt stress and plant tolerance.

## Physiological indicators of salinity tolerance

In the last few decades many studies have attempted to highlight the complex physiological mechanisms involved in plant tolerance to salinity. The huge amount of findings has led to an understanding of several physiological responses to salinity that define qualitative and quantitative differences between salt-sensitive and salt-tolerant species. Despite this, the availability of reliable physiological indicators is still an unreached target for breeders. Tolerance to salinity, in fact, involves multiple biochemical pathways (e.g., the pathways determining the plant capacities of water retention and acquisition, of chloroplast protection and of ion balance). This physiological complexity matches the huge variability of plant responses to salinity at both the inter- and intra-specific level. Tolerance to salinity must be considered as a complex phenotypic performance, determined by several quantitative traits whose expression is controlled by the genotype–environment interaction [3, 42]. For these reasons, the analysis of salinity tolerance in natural demes, populations and biocenoses is not an easy matter. Climatic unpredictability and field heterogeneity often exacerbate the constraints in performing ecophysiological studies on plant adaptation to salinity. However, Δ analysis in plant material has been proposed in several studies as a reliable physiological approach both under controlled and field conditions. In fact Δ represents a complex physiological parameter integrating complex plant performances over different time scales, like those expressed in the presence of abiotic stresses (e.g., salinity and drought).

Furthermore, Δ has been proposed in several works as a candidate indicator for the screening of salt-tolerant species and genotypes. Generally, an increase of salinity is correlated to a decrease in Δ of plant carbon pools. Such a decrease in Δ in response to increase in salt concentration has been shown in many species: cotton [5], common bean [5], wheat [43], sugarcane [44], barley [24, 25], sunflower (Brugnoli and Lauteri, unpublished), and poplar (Monteverdi and Lauteri, unpublished). The enrichment of $^{13}C$ in plant dry matter (i.e., the decrease of Δ) under salinity conditions indicates a long-term decrease of the photosynthetic operational set point, $p_i$. In this case, a decreased $CO_2$ partial pressure at the carboxylation sites reveals leading stomatal limitations of photosynthesis rather than non-stomatal ones. A reduction of photosynthetic capacity ascribed to detrimental effects of salt would lead to higher operational set points, with a reverse effect on Δ expression (increase in Δ). Thus, Δ integrates the effects of both stomatal and non-stomatal limitations, and the suitability of Δ as a physiological indicator of salt tolerance depends on the kind of prevailing effect. For example, Jiang et al. [25] compared the tolerance of 14 barley isogenic lines. They observed a significant variation in Δ on leaves of the different barley genotypes together with a general decrease of Δ values with increasing salt concentration. However, Δ was not the most suitable parameter for selecting the most tolerant barley lines, likely because of the simultaneous and confounding occurrence of both stomatal and non-stomatal effects. On the contrary, the most tolerant isogenic lines were highlighted by the highest values of stomatal conductance and photosystem II efficiency.

A recent study on rice [4] has also shown a high variability of Δ among different isogenic lines exposed to salt-stress conditions. Contrary to the barley case study, tolerant isogenic lines of rice showed higher Δ than susceptible lines. This indicates that, in rice, the variability in tolerance was positively linked with variability in stomatal conductance, and that possible non-stomatal limitations affected the different genotypes uniformly. As a consequence a negative relationship between Δ and salt symptoms was verified in rice. Furthermore, the increase in WUE (as indicated by a decrease in Δ [45]) showed by the susceptible rice genotypes in the same study suggests higher levels of stress in these isogenic lines. This elevated sensitivity to salt could result from an excessive storage of salts in the cell vacuole and chloroplast cytoplasm, with detrimental effects on photosynthesis. A higher Δ in the tolerant lines could be also explained by a decreased content of non-structural carbohydrates, with a less negative $\delta^{13}C$ than the remaining biomass [38] or by an increasing content of lipids with more negative values of $\delta^{13}C$ [46].

Frequently, saline lands are also affected by drought so that plant growth in such environments is restricted not only by salt content but also by water limitations. Consequently, selective pressures could have determined, in plants, linked adaptive strategies to cope with both drought and salinity. Shaheen and Hood-Nowotny [47] determined the combined effect of drought and salinity on Δ in wheat cultivars. They found that Δ varied significantly among the different cultivars depending on both salinity and watering treatments. Salinity induced a linear and significant decrease in Δ of the tested cultivars. Under water-deficit conditions, salt-tolerant cultivars showed positive correlation between harvest index and Δ. The authors stated that Δ is a useful trait in screening wheat cultivars with improved harvest index in saline areas.

The present review points out that high water-use efficiency is not an *a priori* indicator of adaptive capacity to drought and salinity conditions. Different studies indicated that low WUE is often associated with high Δ, high $CO_2$ assimilation and high productivity. Thus, looking for high Δ values can provide adequate results in selection of genotypes adapted to both dry and saline environments. For example, Δ has been reported as a powerful indicator of drought adaptivness in comparative studies on forest tree populations from contrasting habitats: *Pinus pinaster* [48], *Castanea sativa* [41], and *Nothofagus* spp. [49]. It should be noted, however, that Δ integrates complex physiological performances and that a more detailed knowledge on the physiological causes underlying Δ expression should be taken into account for each case study.

Recent research on halophytes (*Tamarix jordanis*, *Tamarix aphylla*) and moderately tolerant glycophytes (*Populus alba*), subjected to different salinity levels, reported that relatively higher biomass production (the main proof of tolerance) was generally related with relatively higher values of Δ accompanied by relatively higher photosynthetic rates. On the contrary, the susceptible species showed lower biomass production correlated with higher reduction of both stomatal conductance and $CO_2$ assimilation (Monteverdi and Lauteri, unpublished data). These results suggest better adaptive strategies to salinity for the more productive species: a higher capacity for osmoregulation related to a high quantity of organic osmolytes, salt ion compartmentation into safe cellular spaces (vacuoles) and a more efficient salt extrusion at the leaf level. In particular, *T. jordanis* showed higher adaptedness to chronic salt-stress conditions than *T. aphylla*, maintaining quite constant and high values of Δ and of photosynthetic parameters for all salt treatments in a long-term experiment. The higher adaptedness of *T. jordanis* was also confirmed by the analysis of total osmolytes at leaf level. This evidence suggests a higher capacity for osmoregulation and of salt ion compartmentation into vacuoles in *T. jordanis* rather than in *T. aphylla*.

## Conclusions

In conclusion, salinity is one of the most serious threats to both agricultural and natural landscapes in many part of the world. Salinity problems are aggravated in arid and semi-arid regions. To face the daunting rate of land degradation and over-exploitation of resources, planning of reforestation measures, based on ecological and physiological knowledge of salinity-related problems, is strongly needed. In particular, selection of species equipped with adequate adaptedness to the environmental characteristics of the different bioregions requires the knowledge of adaptive mechanisms and the availability of physiological indicators that can be used in comparative and ecological surveys. Due to the complexity of genetic traits underlying the adaptive mechanisms in tolerant species, it is particularly important to point out physiological characters with the intrinsic capacity to integrate the multiple physiological responses under stress conditions. Reaching such a target could strongly aid the identification and characterization of suitable plant species in relation to their adaptedness, productivity and capacity to restore salt and drought affected or degraded lands. Even though much work is still needed, a comprehensive physiological parameter, Δ, operated during the photosynthetic process, seems a promising proxy of plant adaptedness to complex abiotic stresses like drought and salinity. However, it should be noted that using Δ to acquire fundamental information on tolerance mechanisms requires a combination with detailed physiological studies to separate and understand the factors that control the expression of Δ itself. Finally, any plan of selection and introduction of salt- and drought-resistant species should be combined with economic and social programs to accomplish an ecological and sustainable management of salt-affected lands.

### *Acknowledgements*

The authors are grateful to Italian Ministry of Environment, Land and Sea (MATTM) and to Euro-Mediterranean Centre for Climate Change (CMCC) for funding and supporting the high quality researches on plant adaptation to saline environments.

## References

1 Allakhvererdies SI, Sakamoto A, Nishiyama Y, Inaba M, Murata N, (2000) Ionic and osmotic effects on NaCl-induced inactivation of photosystem I and II in *Synechoccus* sp. *Plant Physiol* 123: 1047–1056

2 Borsani O, Valpuesta V, Botella MA (2003) Developing salt tolerant plants in a new century: A molecular biology approach. *Plant Cell Tissue Organ Cult* 73: 101–115

3 Ashraf M, Harris PJC (2004) Potential biochemical indicators of salinity tolerance in plants. *Plant Sci* 166: 3–16

4 Shaheen R, Hood-Nowotny RC (2005) Carbon isotope discrimination: Potential for screening salinity tolerance in rice at the seedling stage using hydroponics. *Plant Breeding* 124: 220–224

5 Brugnoli E, Lauteri M (1991) Effects of salinity on stomatal conductance, photosynthetic capacity, and carbon isotope discrimination of salt tolerant (*Gossipium hirsutum* L.) and salt-sensitive (*Phaseolus vulgaris* L.) C3 non-halophytes. *Plant Physiol* 95: 628–635
6 Gale J (1975) The combined effect of environmental factors and salinity on plant growth. In: A Poljakoff-Mayber, J Gale (eds): *Plants and Saline Environments*. Springer, Berlin, 186–192
7 Schwarz M, Gale J (1981) Maintenance respiration and carbon balance of plants at low levels of sodium chloride salinity. *J Exp Bot* 32: 933–941
8 Ball MC (1981) *Physiology of photosynthesis in two mangrove species: Responses to salinity and other environmental factors*. PhD thesis, Australian National University, Canberra
9 De Jong TM (1978) Comparative gas exchange of four California beach taxa. *Oecologia* 51: 343–351
10 Papp JC, Ball MC, Terry N (1983) A comparative study of the effects of NaCl salinity on respiration, photosynthesis, and leaf extension growth in *Beta vulgaris* L. (sugar beet). *Plant Cell Environ* 6: 675–671
11 Wang Y, Nil N (2000) Changes in chlorophyll, ribulose biphosphate carboxylase-oxygenase, glycine betaine content, photosynthesis and transpiration in *Amaranthus tricolor* leaves during salt stress. *J Hort Sci Biotechnol* 75: 623–627
12 Parida AK, Das AB (2005) Salt tolerance and salinità effects on plants: A review. *Ecotoxicol Environ Safety* 60: 324–349
13 Flowers TJ, Troke PF, Yeo AR (1977) The mechanisms of salt tolerance in halophytes. *Annu Rev Plant Physiol* 28: 89–121
14 Greenway H, Munns R (1980) Mechanisms of salt tolerance in non-halophytes. *Annu Rev Plant Physiol* 31: 149–190
15 Munns R, Termaat A (1986) Whole plant responses to salinity. *Aust J Plant Physiol* 13: 143–160
16 Levitt J (1980) Responses of plants to environmental stress. In: *Water, Radiation, Salt and Other Stresses*, vol. II. Academic Press, New York
17 Bethke PC, Drew MC (1992) Stomatal and non stomatal components to inhibition of photosynthesis in leaves of *Capsicum annum* during progressive exposure to NaCl salinity. *Plant Physiol* 99: 219–226
18 Kao WY, Tsai TT, Shih CN (2003) Photosynthetic gas exchange and chlorophyll a fluorescence of three wild soybean species in response to NaCl treatments. *Photosynthetica* 41: 415–419
19 Downton WJS, Grant WJR, Robinson SP (1985) Photosynthetic and stomatal responses of spinach leaves to salt stress. *Plant Physiol* 77: 85–88
20 Ball MC, Farquhar GD (1984) Photosynthetic and stomatal response of two mangrove species, *Aegiceras corniculatum* and *Avicennia marina* to long term salinity and humidity conditions. *Plant Physiol* 74: 1–6
21 Ball MC, Farquhar GD (1984) Photosynthetic and stomatal response of the grey mangrove *Avicennia marina* to transient salinity conditions. *Plant Physiol* 74: 7–11
22 Seemann JR, Critchley C (1985) Effects of salt stress on the growth, ion content, stomatal behaviour and photosynthetic capacity of salt sensitive species, *Phaseolus vulgaris* L. *Planta* 164: 151–162
23 Reddy PS, Ramanjulu S, Sudhakar C, Veeranjaneyulu K (1998) Differential sensitivity of stomatal and non-stomatal components to NaCl or $Na_2SO_4$ salinity in horsegram, *Macrotyloma uniflorum* (Lam.). *Photosynthetica* 35: 99–105
24 Isla R, Aragues R, Royo A (1998) Validity of various physiological traits as screening criteria for salt tolerance in barley. *Field Crop Res* 58: 97–107

25 Jiang Q, Roche D, Monaco TA, Durham S (2006) Gas exchange, chlorophyll fluorescence parameters and carbon isotope discrimination of 14 barley genetic lines in response to salinity. *Field Crops Res* 96: 269–278
26 Meloni DA, Oliva MA, Ruiz HA, Martinez CA (2001) Contribution of proline and inorganic solutes to osmotic adjustment in cotton under stress. *J Plant Nutr* 24: 599–612
27 Khan MA (2001) Experimental assessment of salinity tolerance of *Ceriops tagal* seedlings and saplings from the Indus delta. *Pak Aquat Bot* 70: 259–268
28 Romeroranda R, Soria T, Cuartero J, (2001) Tomato plant-water uptake and plant-water relationships under saline growth conditions. *Plant Sci* 160: 265–272
29 Seemann JR, Sharkey TD (1986) Salinity and nitrogen effects on photosynthesis, ribulose-1,5-bisphosphate carboxylase and metabolite pool size in *Phaseolus vulgaris* L. *Plant Physiol* 82: 555–560
30 Murata N, Mohanty PS, Hayashi H, Papageorgiou G (1992) Glycinebetaine stabilizes the association of extrinsic proteins with the photosynthetic oxygen-evolving complex. *FEBS Lett* 296: 187–189
31 Murota K, Ohshita Y, Watanabe A, Aso S, Sato F, Yamada Y (1994) Changes related to salt tolerance in thylakoid membranes of photoautotropically cultured green tobacco cells. *Plant Cell Physiol* 35: 107–113
32 Brugnoli E, Björkman O (1992) Growth of cotton under continuous salinity stress: Influence on allocation pattern, stomatal and non-stomatal components of photosynthesis and dissipation of excess light energy. *Planta* 187: 335–347
33 Terashima I, Wong SC, Osmond CB, Farquhar GD (1988) Characterisation of non-uniform photosynthesis induced by abscisic acid in leaves having different mesophyll anatomies. *Plant Cell Physiol* 29: 385–395
34 Farquhar GD, Hubich KT, Condon AG, Richards RA (1988) Carbon isotope fractionation and plant water-use efficiency. In: PW Rundel, JR Ehleringer, K Nagy (eds): *Applications of stable isotope ratio to ecological research*. Springer, New York, 21–40
35 Brugnoli E, Scartazza A, Lauteri M, Monteverdi MC, Maguas C (1998) Carbon isotope discrimination in structural and non-structural carbohydrates in relation to productivity and adaptation to unfavourable conditions. In: H Griffitths (ed): *Stable isotopes: Integration of biological, ecological and geochemical processes*. Bios Scientific Publishers, Oxford, 133–146
36 Brugnoli E, Farquhar GD (2000) Photosynthetic fractionation of carbon isotopes. In: RC Leegood, TD Sharkey, S von Caemmerer (eds): *Photosynthesis: Physiology and Metabolism – Advances in Photosynthesis*, vol. 9. Kluwer Academic Publisher, Dordrecht, 399–434
37 Ehleringer JR (1993) Carbon and water relations in desert plants: An isotopic perspective. In: JR Ehleringer, AE Hall, GD Farquhar (eds): *Stable isotope and plant carbon-water relations*. Academic Press, San Diego, 155–172
38 Brugnoli E, Hubick KT, Von Caemmerer S, Wong SC, Farquhar GD (1988) Correlation between the carbon isotope discrimination in leaf starch and sugars of C3 plants and the ratio of intercellular and atmospheric partial pressure of carbon dioxide. *Plant Physiol* 88: 1418–1424
39 Lauteri M, Brugnoli E, Spaccino L (1993) Carbon isotope discrimination in leaf soluble sugars and in whole-plant dry matter in *Helianthus annuus* L. grown under different water conditions. In: Ehleringer JR, Hall AE, Farquhar GD (eds): *Stable isotope and plant carbon-water relations.* Academic Press, San Diego, 93–108
40 Scartazza A, Lauteri M, Guido MC, Brugnoli E (1998) Carbon isotope discrimination in leaf and stem sugars, water-use efficiency and mesophyll conductance during different developmental stages in rice subjected to drought. Aust *J Plant Physiol* 25: 489–498

41 Lauteri M, Pliura A, Monteverdi MC, Brugnoli E, Villani F, Eriksson G (2004) Genetic variation in carbon isotope discrimination in six European populations of *Castanea sativa* mill. Originating from contrasting localities. *J Evol Biol* 17: 1286–1296
42 Pareek A, Singla S L, Grower A (1997) Salt responsive protein/genes in crop plants. In: PK Jaiwal, RP Singh, A Gulati (eds): *Strategies for improving salt tolerance in higher plants*. Oxford and IBH Publication, New Delhi, 365–391
43 Ansari R, Naqvi MS, Khanzada NA, Hubick KT (1988) Carbon isotopes discrimination in wheat cultivars under saline conditions. *Pak J Bot* 30: 87–93
44 Frederick CM, Plaut Z, Saliendra ZN (1994) Carbon isotope discrimination, gas exchange and growth of sugarcane cultivars under salinity. *Plant Physiol* 104: 521–526
45 Farquhar GD, Ehleringer JR, Hubik KT (1989) Carbon isotope discrimination and photosynthesis. *Ann Rev Plant Physiol* 11: 503–537
46 Robinson SP, Downton WJS, Millhouse JA (1983) Photosynthesis and ion content of leaves and isolated chloroplasts of salt-stressed spinach. *Plant Physiol* 73: 238–242
47 Shaheen R, Hood-Nowotny RC (2005) Effect of drought and salinity on carbon isotope discrimination in wheat cultivars. *Plant Sci* 168: 901–909
48 Guehl, JM, Nguyen-Queyerens A, Loustau D, Ferhi A (1995) Genetic and environmental determinants of water-use efficiency and carbon isotope discrimination in forest trees. In: M Bonnet-Masimbert, H Sandermanns (eds): *The EUROSILVA contribution to forest tree physiology*. Editions Colloques de l'INRA, Paris, 297–321
49 Read JJ, Farquhar GD (1991) Comparative studies in *Nothofagus* (Fagaceae) leaf carbon isotope discrimination. *Funct Ecol* 5: 684–695

Biosaline Agriculture and High Salinity Tolerance
Edited by Chedly Abdelly, Münir Öztürk, Muhammad Ashraf and Claude Grignon


# Potentiality of salt marshes in Mediterranean coastal zone of Egypt

Hassan M. El-Shaer and Mohamed H. El-Morsy

*Desert Research Center, 1 Mathaf El Mataria St., P. O. Box 11753, Mataria, Cairo, Egypt*

## Abstract

The Northern Coastal region of Egypt extends around 1000 km long and 30 km inland. Its major characteristics as an arid Mediterranean climate are limited rainfall that varies between 80 and 250 mm/year where a relative insignificant rain-fed agriculture is practiced. Rangelands and livestock production are the main common natural elements in the region. Socio-economic characteristics are based on tribal systems of nomads and agro-pastoralists. The local economy depends on livestock sales and trade of agricultural products. Salt marshes plant species are dominant and grow naturally and intensively in the saline environments. Such plant species could have great potentialities for many human and animal usages. The production and nutritive values of these plants are affected by different human and environmental factors that cause severe deterioration. Many salt marsh plants communities are of great interest to grazers in the Mediterranean coastal zones of Egypt because of their prolonged production period, as they constitute good grazing in the dry season in summer and autumn, particularly for sheep, goats, camels and wildlife. However, the real potentialities of such salt march plants have not been fully exploited and evaluated for different purposes. The aim of this study is to survey, briefly, the salt marshes species, its habitats, potentialities and the ecological factors affecting their distribution in different areas of the Mediterranean coastal zones of Egypt.

## Introduction

The Mediterranean coastal zone (MCZ) of Egypt extends from Sallum to Rafah for about 970 km and is geographically divided into three sections: Western, Middle and Eastern. The native range vegetation cover consists of sparse stands of shrubs, sub-shrubs and herbaceous perennials, with a winter-spring cover of ephemerals (short-cycle annuals) of varying density. Natural saline areas of different kinds cover considerable areas in the region with altitudes ranging from –300 m below sea level to almost 200 m above sea level. These saline areas are inhabited by halophytes, some of which dominate the well-recognized widespread plant communities. About 50% of the total indigenous flora of Egypt (800 species) are represented in this region [1], but floristic diversity has been considerably reduced in recent times due to overuse of the vegetation. Indigenous plant communities are distributed among different types of habitats each of which is identified by a collection of species of similar ecological requirements that forms an indicator group.

Salt marshes are organized into zones parallel to the shore line, and the number of zones in any one locality is dependent on the topographic features of the littoral belt. The plant communities may be classified according to the growth form of the dominant species into five groups: succulent half-shrub types, non-succulent half-shrub types, grassland types, succulent shrub types, and non-succulent shrub types [1, 2].

Indigenous salt marshes are generally considered as being among the productive ecosystems in Egypt. Their high primary production has been, from the beginning, presumed to be due to adjacent coastal waters. Recently there has been a sharp increase in losses of arable lands due to salinization. This is especially hard for the farmers in the Mediterranean regions where irrigation farming is very common and many fields have reached a level of soil salinity that prevents farmers from raising common crops. This is a worldwide problem in dry and semi-dry regions like Egypt. The loss in soil fertility in certain regions of Egypt are so severe that rural populations are starving, leaving their villages or requesting permanent living support through outside aid. In recent years several studies have been performed into the development of sustainable utilization of halophytes and other salt marshes [3–5]. The objective of this article is to demonstrate, in brief, the current situation, constraints and potentiality of salt marshes in the MCZ of Egypt for further regional development.

## Physical environment

The MCZ of Egypt occupies a portion of the semi-arid belt south of the Mediterranean Sea. The northwestern coast zone of Egypt (Western section) extends from Alexandria to the Libyan boarder; it is approximately 480 km long and 25–50 km wide. It is a region of warm coastal deserts, with the warmest summer month having a mean temperature less than 30°C, and the coldest winter month having a mean temperature above 10°C; although occasional short rainstorms occur in winter, most days are sunny and mild.

The orientation of the coast, the proximity of the Mediterranean Sea, and the orography are the main factors to contribute to the climate of this region. The proximity of the sea has a direct effect on air temperature and humidity, and consequently on evaporation and condensation, but does not increase the amount of rainfall. The MCZ of Egypt receives noticeable amounts of rainfall, especially in winter. The rainy months are October to February. In summer, no or little rain is recorded, while in autumn, occasional heavy rain may occur. The climate at the Mediterranean coast of Egypt meteorological station indicates that dry periods extend from February to November, and it is obvious from the diagram that the precipitation curves underlie the mean temperature curves throughout the year except for a short time in winter from December to February.

Solar radiation is very strong from March to end of October, whereas relative cloudiness prevails from November to January. Air relative humidity is generally high in summer (>70%) and winter (60–70%). Diurnal variations in temperature and relative humidity may be favorable in summer for the condensation of water vapor as dew.

The coastal salines are prominent and form an important ecosystem in the region. Extensive Sabkhas are present at levels slightly above or below sea level. The coastal salt marshes comprise different series: They may be formed by sand bars, salty flat areas, coastal swamps subject to flooding and spray by seawater, or down streams wadis into which rain water has poured.

## North coastal areas of Egypt

The North Coastal areas of Egypt are composed of two major sub-zones: the Northwest Coast (NWC) and the North Coastal Areas of Sinai.

### *Northwest Coast*

The NWC of Egypt forms a belt about 20 km deep, which extends for about 500 km between Amria (20 km west of Alexandria) and El Salloum near the borders with Libya.

#### Main climatic features

The NWC is characterized by a dry Mediterranean climate, with average high and low temperatures of 18.1°C and 8.1°C in the winter and 29.2°C and 20°C in the summer, respectively.

Rainfall in the NWC ranges between 105.0 mm/year at El Salloum and 199.6 mm/year at Alexandria. Data from eight stations situated near the coastline show that most of the rainfall (70% or more) occurs within the winter months (November to February), mostly during December and January.

There is significant variation in rainfall from one location to another, which is attributed mainly to the orientation of the coast at these locations. The prevailing rainfall gradient from the north shows that the average mean decreases sharply from 150 mm near the coast to 50 mm at 20–70 km inland. The NWC area has the highest average wind speed in Egypt in the winter where wind speed could reach 18.5 km/h. Wind speed drops gradually inland.

### Soil and water resources

Soil types and properties are highly influenced by geomorphic factors. The main soil units could be summarized as follows: (1) Coastal sand dunes, (2) soils of the lagoonal depressions, (3) consolidated dunes, (4) deep sand and clay loam soils, (5) moderate to limited depths of sandy to clay loam, (6) wind blown formations, and (7) soils of the alluvial fans and outwash plains over the plateau.

Water resources are mainly that of rainfall, groundwater resources are limited and usually of low quality especially with respect to varied salinity content.

### Human resources

Human resources are diversified in distribution, background, level of education and occupations. Agro-pastoralists with tribal traditions are the main dwellers of these rangelands. Lately wide-spread construction of tourist villages and summer resorts have spread all along the Mediterranean Sea Coast, bringing human resources affiliated to these villages and summer resorts. Urban dwellers occupy several cities and small towns along the main International Road to the Libyan borders.

## *North East Coast of Sinai*

### Climatic features

North Sinai areas are characterized by the dry Mediterranean climate type with relatively rainy, cool winters and dry hot summers. Air temperatures are similar to those of the NWC, but with large variations diurnally, seasonally and geographically. The annual wind speed is around 14.0 km/h and the prevailing wind direction is northwest and north. The amount of rainfall in Sinai decreases from the northeast towards the southwest. The greatest rainfall is found in Rafah (304 mm/year) in the northeast. The annual average along the Mediterranean coast amounts to 120 mm/year. Rainfall decreases in the uplands to the south to about 32 mm/year.

Rainfall occurs in Sinai mainly during the winter season (November to March) and during spring or fall. Rainfall is practically absent from May to October. Along the Mediterranean Coast, 60% of the rain occurs in the winter, while 40% falls during the transitional seasons. Due to differences in water availability, growing seasons differ in the different parts of the Governorate of North Sinai.

## Soil and water resources

The desert soils of Northern Sinai are of three different origins: aeolian, alluvial and soil formed *in situ*. The latter is related to land form and is found in the plateau region of Wadi Al Arish on either calcareous or volcanic parent material. The majority of alluvial soils were formed under recent climatic conditions. They constitute the present wadi beds and they are characterized by a granulometric differentiation according to flood intensities and sedimentation times. As a consequence, soil upstream of the wadis is coarser in texture than the soils further downstream. In the dune area, the soils generally differ from those in the gravel plain. The dune area is dominated by soils with almost no signs of soil forming processes. Saline soils are found exclusively in the coastal zone.

The Tina plain in the west was formed by alluvial Nile deposits as a natural extension of the old Nile valley. The soils are heavy textured with high salinity contents due to the water-logging condition attributed to the near-sea and low-lying location. Water resources are varied, i.e., rainfall water with possible runoffs if the rainfall exceeds 10 mm per rain storm. When runoff occurs, wadi beds begin to carry water depending on the amount and duration of rainfall. It is estimated that 60% of the mean rainfall in Sinai is lost to evapotranspiration.

Groundwater in Sinai may be classified into two types. Shallow groundwater occurs mainly within the weathered layer of igneous and metamorphic rocks, quaternary rock, recent deposits such as wadi fill or sediments and sand dunes. Deep groundwater mainly occurs as semi-confined aquifers of per-Quaternary formation. Groundwater resources in the North Coastal area are limited in nature and in general of low quality.

A third water resource is being introduced to the area, namely the "Al Salam Canal", which will convey mixed Nile and agricultural drainage water across the Suez Canal to reclaim 400 000 feddans in Northern Sinai.

## Human resources

Populations are mainly of tribal. Socio-economic characteristics are based on tribal systems of nomads and depend on rain-fed agricultural products, livestock and local crafts.

## **Cover vegetation in the MCZ of Egypt**

The North Mediterranean Coastal Belt is the richest part of Egypt in flowering plants; most of these species are annual weeds that flourish during the rainy season, giving the area a temporary gray grassland aspect. During the longer dry periods, the characteristic woody shrubs and perennial herbs constitute the scrub vegetation of the area, scattered sparsely in parts and grouped in denser distinct patches in favored habitats. An account on the plant communities in the different landforms is given below.

*Coastal sand dunes*

The most common plant communities inhabiting sand dunes in the north coast of Egypt are: *Ammophila arenaria*, *Euphorbia paralias*, *Ononis vaginalis*, *Crucianella maritime*, *Stipagrostis scoparia*, *Convolvulus lanatus*, *Artemisia monosperma*, *Cornulaca monacantha*, *Artemisia monosperma*, *Retama reatam* and *Iris mariae*.

*Swamp and saline habitats*

The most common leading species in this habitat are: *Phragmites australis*, *Typha* sp., *Scirpus littoralis*, *Salicornia herbacea*, *Juncus maritimus*, *Halocnemum strobilaceum*, *Limoniastrum monopetalum*, *Salsola tetrandra*, *Suaeda pruinosa*, *Suaeda fruticosa*, *Frankenia laevis*, *Limonium pruinosum*, *Haplophyllum tuberculatum*, *Sporobolus pungens*, *Traganum nudatum*, *Reaumuria hirtilla*, *Orlaya maritime*, *Atriplex halimus*, *Arthrocnemum macrostachyum*, *Suaeda aegyptiaca*, *Juncus arabicus*, and *Anabasis setifera*.

*Uncultivated desert areas*

There are several communities that dominate the region; the most important and common ones are: *Thymelaea hirsuta - Gymnocarpos decanderum*, *Plantago albicans - Asphodelus microcarpus* and *Anabasis articulata - Haloxylon salicornicum*.

**Cover vegetation of salt marsh and its potentialities**

The salt marsh vegetation in the MCZ of Egypt has common features: simplicity of structure, limited number of species and notable differences in cover due to minor differences in ground level. Many halophytes are able to form phytogenic mounds, which vary in size according to the plant species and the prevailing environmental conditions.

There are many salt marshes forming the lakes of the Mediterranean Sea costal belt of Egypt [6] such as: Bardawil Lake, Manzalah lake, Mariout lake, and Buroulos lake.

Soil salinity, moisture content, calcium carbonates and the human disturbance may be considered the most important environmental factors that affect the distribution and the abundance of halophytic vegetation and other habitats of the salt marshes. There is a significant variation in the distribution of the halophytic species with elevation. The halophytic plant communities surrounding the salt marshes along the Mediterranean coast of Egypt often exhibit large zonal pattern of species composition [1]. The plant species found in the salt marshes of Bardawil Lake are similar but not identical to those in the delta lakes and the western Mediterranean salt marsh habitat of Egypt [7, 8]. Five groups of halophytic plants have been identified [9] in Bardawil Lake salt marshes, which are dominated by: *Salsola tetrandra*, *Nitraria retusa*, *Zygophyllum album*, *Halocnemum strobilaceum*, *Arthrocnemum macrostachym*, *Suaeda aegyptiaca*, and *Sarcocornia fruiticosa*.

In salt marshes habitat of the Mediterranean coast of Egypt comprises five communities dominated by *Zygophyllum aegyptium*, *Inula crithmoides*, *Arthrocnemum macrostachyum*, *Juncus acutus* and *Halocnemum strobilaceum* [2].

The Coastal salt marshes have unique vegetation tolerant of the multiple stresses and frequent disturbances related to tidal inundation [6, 10]. The family *Chenopodiaceae* showed a clear relationship between topological index and elevation. Ramadan [11] studied the species distributions at habitats with different salinity levels, i.e., fresh water, transitional water quality, low-saline soil and high-saline soil and found the following habitats:

- mainly fresh slightly brackish waters: the species of *Echinocloa stagnina*, *Panicum repens*, and *Phragmites australis*.
- saline grounds: *Arthrocnemum macrostachyum*, *Atriplex portulacoides*, *Juncus acutus*, *Halocnemum strobilaceum*, *Sarcocornia fruticosa*, and *Phragmites australis*.
- fresh brackish water or wetlands with shallow surface water of moderate to fairly high salinity: *Phragmites australis*-*Typha domingensis*, *Ceratophyllum demersum*, *Echinocloa stagnina*, *Eichhornia crassipes* and *Panicum repens*.
- fresh brackish-fairly saline water: *Inula crithmoidies*, *Azolla filiculoides*, *Lemna gibba*, *Phragmites australis*, *Cyperus laevigatus* and *Typha domingensis*.
- highly saline grounds: *Halocnemum strobilaceum*, *Salsola longifolia*, *Sarcocornia fruticosa* and *Phragmites australis*.
- highly saline grounds also: *Sarcocornia fruticosa*, *Tamarix nilotica*, *Juncus acutus*, *Salsola longifolia*, *Atriplex portulacoides*, *Halocnemum strobilaceum*, and *Phragmites australis*.

Rushes and reeds constitute an important component of the wetlands in Egypt and coastal and inland reed swamps. Rushes are represented by two *Juncus* species: *J. acutus* and *J. rigidus*. The reeds are represented by *Phragmites* and *Typha*. The *Phragmites* tolerates the highest soil salinity followed by *J. rigidus*, *J. acutus* and *Typha domingensis*. *J. rigidus* tolerates soil salinity better than *J. acutus* [6, 12]. Reeds and rushes provide a habitat (e.g., nesting sites, substratum, feeding materials) for a wide range of organisms, from invertebrates to birds. They are very important as producers in nutrient-rich freshwater habitats such as the River Nile, Manzala, and Burulius Lakes and the irrigation and drainage canals. Moderate growths of macrophytes can help oxygenate the water, assisting the survival of fish and invertebrates. They are also important to processes such as nutrient cycling within freshwater ecosystems [1].

The common reed *Phragmites* is widespread in Egypt, and is an economically valuable resource for making fishing rods, weaver's spools, musical instrument mouth pieces, roofing, fencing, and basket weaving, as well as for animal fodder and fuel.

There are two *Typha* species in Egypt: *Typha domingensis* and *Typha elephantina*. The first *Typha* is recorded wherever marshy conditions prevail [13], especially along the margins of drains, the main Nile branches and the northern lakes of Egypt [14]. These plants can be environmentally and economically important. They can provide a highly desirable habitat type for fish. These plants can serve as nutrient 'scrubbers' in polluted aquatic systems, thereby playing a key role in nutrient cycling.

The rush *J. acutus* show a wide range of distribution in the marsh and saline habitats compared with *J. rigidus*, which is mainly dominant in saline habitats. The most common associates with both species are obligate halophytes, e.g., *Halocnemum strobilaceum*, *Inula crithmoides*, *Arthrocemum macrostachyum*, *Suaeda vera* and *Atriplex portulacoides*. *J. rigidus* is more tolerant to aridity and soil salinity [15].

*Juncus* species produces fiber and if successfully managed can provide good quality paper. They could represent a non-conventional crop to be cultivated on salt-affected soils of Egypt [15]. They are used for making mats, baskets and sandals [16]; mats are used for manufacturing cheese (El Gcben el Areesh) and also as prayer mats. *Juncus* species could be used as a raw material in various chemical industries, e.g., drug, oils, etc. However, *J. rigidus* is superior to *J. acutus* for the following reasons [12]: (1) The salt tolerance and desalination effects of *J. rigidus* are higher than those of *J. acutus*; (2) *J. rigidus* are resistant against fungal infection, an advantage that increases its vegetative yield; and (3) the seeds of *J. rigidus* are, relatively, richer in their chemical constituents than those of *J. acutus*.

The feasibility of growing halophytes on salt marshes can be maximized with plant species that in addition to its primary product can also provide 'indirect' and economical benefits. Factors affecting the input costs, such as establishment costs of the plants and the cost of water, and the value of product (feed quality in case of forage species), in addition to indirect benefits could certainly increase the value of halophytes to be used commercially.

**Economic benefits and uses of halophytes**

There are many usages of halophytes in Egypt [4], e.g.:

- Halophytes as fuel wood and coal: *Tamarix* spp. and *Acacia* spp. trees are the most important ligneous plants as a good source for firewood and coal.
- As timber: *Tamarix* spp., *Halexylon* spp. and *Prosopis* spp. are considered the main sources of timber in Egypt for construction of roofs and some furniture.
- As fodders: Several halophytic range plants represent the main source of animal feeds in Sinai and the NWC of Egypt [3].
- As medicine: The ash of some halophytes is sometimes used in the folk medicine, especially for wounds, scabies and burns, e.g., *Zygophyllum* spp. [17].
- As windbreaks: *Tamarix aphylla* and *Prosopis* spp. have been used for a long time as windbreaks in the desert areas.

*Traditional uses of salt marshes in MCZ of Egypt*

In view of the thorough knowledge of the characteristics of the halophytes and their distribution, the different traditional uses of the halophytes in the region are given below in brief:

- Use of range plants as animal fodders: The nomadic pastoralism in the region long ago made the Bedouins aware of the characteristics of halophytes. Halophyte usage as animal fodders can provide a good source of animal feeds under arid and semi-arid areas [3, 5, 18], particularly the palatable and nutritious plant species grown naturally in this region (Tab. 1) [5]. Based on their nutritive values, many of these halophytes such as *Nitraria retusa* and *Suaeda fruticosa*, in particular during good rainy seasons, are considered as good quality fodders since they contain sufficient nutrients for animals not only for maintenance but also for productive purposes [3]. The production and nutritive values of these plants are affected by different human and environmental factors that can cause severe deterioration. However, several halophytic plants used as animal fodders are limited by high salt content, e.g., *Atriplex* spp., *Haloxylom salicornicum*, *Halocnemum strabilaceum*, and *Salsola tetranda*, which represents some constraints to their use as illustrated in Table 2. Although most of these plant species contain high concentrations of different minerals, these concentrations are still in safe ranges and do not cause severe harm to animals. Sufficient knowledge is now available to consider implementation of saline land

**Table 1.** Chemical composition of common salt marsh plant species in Mediterranean coastal zone of Egypt during spring 2006 (% on dry matter basis)

| Plant species | DM | Ash | CP | EE | CF | NFE |
|---|---|---|---|---|---|---|
| *Aeluropus lagopoides* | 38.06 | 12.84 | 12.20 | 7.54 | 32.15 | 35.27 |
| *Anabasis articulate* | 16.16 | 38.63 | 11.32 | 4.82 | 19.00 | 26.23 |
| *Arthrocnemum glaucum* | 24.60 | 26.59 | 15.45 | 8.15 | 10.17 | 39.64 |
| *Atriplex halimus* | 16.39 | 27.51 | 15.04 | 6.79 | 19.00 | 31.66 |
| *Chrysanthemum coronarium* | 38.65 | 20.65 | 11.57 | 13.33 | 25.80 | 28.64 |
| *Halocnemum strobilaceum* | 42.01 | 32.55 | 8.45 | 6.86 | 11.60 | 40.54 |
| *Juncus rigidus* | 32.11 | 15.21 | 16.89 | 8.43 | 31.05 | 28.41 |
| *Limoniastrum monopetalum* | 15.84 | 22.60 | 20.33 | 8.35 | 17.00 | 31.73 |
| *Lygeum spartum* | 33.06 | 7.33 | 17.83 | 9.73 | 13.65 | 51.46 |
| *Mesembryanthemum nodiflorum* | 13.22 | 46.44 | 11.82 | 8.41 | 4.36 | 28.97 |
| *Phragmites australis* | 54.30 | 13.85 | 13.75 | 9.07 | 26.75 | 36.57 |
| *Salicornia fruticosa* | 22.76 | 41.62 | 13.14 | 6.83 | 14.75 | 23.66 |
| *Suaeda vermiculata* | 21.60 | 30.63 | 11.26 | 7.20 | 23.55 | 27.36 |
| *Tamarix nilotica* | 25.98 | 20.22 | 18.89 | 8.45 | 14.71 | 37.74 |
| *Thymelaea hirsute* | 53.64 | 8.43 | 13.76 | 9.80 | 26.75 | 41.25 |
| *Zygophyllum album* | 11.4 | 28.80 | 7.51 | 9.53 | 11.25 | 42.91 |

From [5]
DM, dry matter; CP, crude protein; EE, ether extract; CF, crude fiber; NFE, nitrogen free extract

management schemes incorporating trees and shrubs in dry land and irrigated areas [3].

- Use as timber: Among the halophytes used by the Arabs as timber are the different *Tamarix* spp. that have been used for construction of roofs and for manufacturing furniture.
- Use for fuel wood and coal: *Tamarix* trees and shrubs are the most important ligneous plants of the halophytes, which are used as a good source for firewood and coal.
- Use as windbreaks and amenity plantations: *Tamarix aphylla* has been used for a long time as a windbreak in the desert areas where there is plantation. It is one of the most successful plants used as a windbreak in the region where soil conditions are not favorable for planting non-saline-tolerant plants; *Tamarix* is still the plant of choice.
- Use for tannin: *Rhizophora mucronata* (Oundo) is used for tanning and dying clothes.
- Use for cleaning clothes and utensils: The dried and powdered leaves of *Seidlitzia rosmarinus* is a famous plant used for this purpose. Known as Shnan in Arabic, Bedouins make use of this for washing clothes, hair, utensils, etc. and to make a lather-like soap. Among the other plants used for this purpose is *Hammada elegans*.

**Table 2.** Concentrations of some minerals of common halophytes in Egypt

| **Plant species** | **Ca** | **P** | **Na** | **K** | **Mg** | **S** | **Zn** | **Cu** | **Fe** | **Mn** |
|---|---|---|---|---|---|---|---|---|---|---|
| | % | % | % | % | % | % | **ppm** | **ppm** | **ppm** | **Ppm** |
| *Atriplex halimus* | 1.69 | 0.32 | 3.91 | 0.57 | 0.32 | 0.17 | 64 | 10 | 503 | 51 |
| *Nitraria retusa* | 1.96 | 0.22 | 5.35 | 0.66 | 0.36 | 0.14 | 32 | 11 | 567 | 62 |
| *Haloxylom salicornicum* | 4.00 | 0.15 | 5.01 | 1.74 | 0.33 | 0.07 | 73 | 18 | 603 | 80 |
| *Haloc. strabilaceum* | 2.45 | 0.14 | 7.00 | 1.65 | 0.39 | 0.10 | 99.7 | 17 | 621 | 87 |
| *Salsola tetranda* | 3.98 | 0.16 | 5.65 | 1.45 | 0.59 | 0.12 | 44 | 8.88 | 664 | 79 |
| *Suaeda fruticosa* | 2.11 | 0.41 | 4.06 | 1.29 | 0.30 | 0.20 | 55 | 13 | 674 | 88 |
| *Tamarix aphylla* | 3.73 | 0.16 | 2.75 | 0.78 | 0.43 | 0.12 | 38 | 13 | 274 | 60 |
| *Tamarix mannifera* | 1.44 | 0.01 | 2.60 | 0.8 | 0.46 | 0.09 | 38.5 | 16 | 291 | 52 |
| *Zygophyllum album* | 3.47 | 0.14 | 2.84 | 0.9 | 0.64 | 0.09 | 43.3 | 7.78 | 393 | 52 |

From [19, 20]

- Use as a medicine: The ash of some halophytes is usually used in the folk medicine in the region, especially for wounds, burns and scabies. The ash of *Typha* spp. *domingensis* was traditionally used for wounds. The ash of Harm (*Zygophyllum*) is used for treating scabies in camels. The effect of the ash in these cases may be due to its salt content as well as being aseptic after burning the plant material.

## Conclusion

Indigenous salt marshes with their naturally grown and intensively farmed flora are considered as being among the productive ecosystems in Egypt. Halophytic rangelands and livestock production are the main common natural elements in the MCZ of Egypt. The local economy of the region depends on livestock sales and trade of some agricultural products. Some of the plant species could have great potential for many human and animal usages. The utilization of halophytes for rehabilitation and reclamation of salt-affected lands has proven to be feasible if certain precautions are considered.

There is a need for development of proper agromangement and conditions to maximize the productivity of these known economical halophytic species. Also, considerable efforts are underway toward genetic improvement of the crop characteristics, just as there is with any crop. Finding ways to make marginal land and saline wasteland productive will provide land of a better quality to be farmed in more rational ways to lessen the risk of degradation.

## References


1 Zahran MA, Willis AJ (1992) *The vegetation of Egypt*. Chapman and Hall, London

2 Kassas M, Zahran MA (1967) On the ecology of the Red Sea littoral salt marsh. *Egypt Ecol Monogr* 37: 792–316

3 El Shaer HM (1999) Conservation and improvement of natural rangelands as animal feed resources in the north western coast of Egypt. *The third Conference on Desertification and Environmental Studies Beyond the Year 2000, 30 Nov–4 Dec 1999*. Center for Desert Studies, King Saud University, Kingdom of Saudi Arabia

4 El Shaer HM (2006) Halophytes as cash crops for animal feeds in arid and semi-arid regions. In: M Öztürk, Y Waisel, MA Khan, G Görk (eds): *Biosaline agriculture and salinity tolerance in plants*. Birkhäuser, Basel, 117–128

5 Gab Allah MSA ( 2006) *Ecological and chemical studies on some common Rangeland Plants in Egypt and Libya*. MSc. Thesis. Institute of African Research and Studies, Cairo university, Egypt

6 Shaheen SE (1998) *Geoenvironmental studies on Bardawil lagoon and its surroundings, North Sinai, Egypt*. Ph.D Thesis, Faculty of Science, Mansoura University, Mansonra, Egypt

7 Montasir AH (1937) Ecology of Lake Manzala. *Egyptian Univ Bull Fac Sci* 2: 1–50

8 Khedr AA (1999) Floristic composition and phytogeography in a Mediterranean deltaic lake (Lake Burollos). *Egypt Ecologia Mediterranea* 25: 1–11

9 Khedr AA, Zahran MA (2002) The salt marsh visitation of lake Bardawil, North Sinai, an overview. *Proceedings of the International Symposium on "The Optimum Resources Utilization in Salt-Affected Ecosystems in Arid and Semi-Arid Regions"*, 8–10 April 2002, Cairo, Egypt, 339–345

10 Bouma TJ, Nielsen KL, Vanhal J, Koutstaal B (2001) Root system topology and diameter distribution of species from habitats differing in inundation frequency. *Funct Ecol* 15: 360–369

11 Ramadan AA (2002) Population dynamics and multivariate analysis of vegetation in Lake Manzala. *Egypt Pak J Biol Sci* 5: 842–852

12 Serag MS, Zahran MA, Khedr AA (2002) Ecology and economic potentialities of the dominant salt-tolerant reeds and rushes in the Nile delta. *Proceedings of the International Symposium on "The Optimum Resources Utilization in Salt-Affected Ecosystems in Arid and Semi-Arid Regions"* , 8–10 April 2002, Cairo, Egypt, 245–256

13 Tackholm V (1974) *Students' Flora of Egypt*. Cairo University Press, Cairo, Egypt

14 Serag MS, Khedr AA, Zahran MA, Willis AJ (1999). Ecology of some aquatic plants in polluted water courses, Nile delta, Egypt. *J Union Arab Biol* 9 (B): 85–97

15 Zahran MA (1993) Juncus and Kochi fiber-and fodder-producing halophytes under salinity and aridity stress. In: M Pessarakli (ed): *Handbook of plant and crop stress.* Marcel Dekker, New York, 505–528

16 Tackholm V, Drar H (1954) *Flora of Egypt.* Bulletin Fac. Sci. III. University of Cairo, Cairo

17 Batanouny KH (1994) Halophytes and halophytic plant communities in the Arab region: Their potential as a rangeland resource. In: VR Squires, AT Ayoub (eds): *Halophytes as a resource for livestock and for rehabilitation of degraded lands*. Kluwer Academic Publishers, Dordrecht, 139

18 Squires VR, Ayoub AT (1994) Halophytes as resource for livestock and for rehabilitation of degraded lands. *Proceeding of the International Workshop on "Halophytes for reclamation of saline wasteland and as a resource for livestock: Problems and prospects"*, Kluwer Academic Publishers, London

19 El Shaer HM (1986) Seasonal variation of the mineral composition of the natural pasture of Southern Sinai grazed by sheep. *Proceedings of the 2nd Egyptian–British Conference on Animal and Poultry Production*, 26–28 August 1986, Bangor, Gwynedd, UK
20 Attia-Ismail SA, Fayed AM, Fahmy AA (2003) Some mineral, and nitrogen utilization of sheep fed salt plant and monensin. *Egy J Nutr Feeds* 6: 151–161

Biosaline Agriculture and High Salinity Tolerance
Edited by Chedly Abdelly, Münir Öztürk, Muhammad Ashraf and Claude Grignon


# Studies on the halophyte desert vegetation in the Northern Caspian Region (Caspian Lowland and Mangyshlak)

Irina N. Safronova

*Komarov Botanical Institute RAS, St.-Petersburg, 197376, Russia*

## Abstract

Halophyte vegetation is a characteristic feature of the temperate desert of Turan, in the Northern Caspian Region – the westernmost part of Turan. It includes the Caspian Lowland and the Mangyshlak. The Caspian Lowland belongs to the steppe and desert regions. The boundary between these Regions runs along the Kuma River on the south-west of the Caspian Lowland (45° N) and further – along the Kuma-Manych Depression and the eastern slope of the Ergeni Height (45° E), reaches 47° N and than turns to the north-east, crosses the Volga river at 48°N and runs to the east along this latitude. The Mangyshlak lies completely within the desert region. It is situated to the south of the Caspian Lowland at the eastern shore of the Caspian Sea between 45° and 42° N. Halophytes play a great role in desert vegetation of the Northern Caspian Region. They form communities that are confined to salted soils and to salted variants of zonal soils. The halophytes comprise a significant number of species, including the families *Chenopodiaceae*, *Poaceae* and *Asteraceae*, together with *Tamaricaceae* and *Limoniaceae*. The most numerous is the group of dwarf semishrub halophytes. Some shrub species and two semishrubs are common. Some species of halophilous perennial grasses and one annual grass are characteristic of the desert vegetation. Many species of the annual saltworts also occur. The halophytes can be divided into three groups: (1) those restricted to the solonchak (salt marshes); (2) those restricted to solonetz and takyr; and (3) those with their distribution connected with the salted variants of the zonal soil types. The latitudinal (zonal) and longitudinal (regional) regularities are revealed in distribution of halophytic desert vegetation. Halophyte communities in the desert zone are of a great importance as pastures. There are well adapted to environmental conditions and suitable for phytomelioration of pastures with low productivity.

## Introduction

The halophytic vegetation is characteristic feature of the temperate desert of Turan. Many scientists have studied its ecology, distribution, composition, and the salinity tolerance of its species ([1–18], and others). Here only the halophyte vegetation of the Northern Caspian Region, the Westernmost part of Turan, is presented.

The Northern Caspian Region includes the Caspian Lowland and the Mangyshlak. The Caspian Lowland borders the Caspian Sea on the west from

44°40' N on the west shore to 45° 20' N on the east shore. The northern limits extend from 50°20' N (around Volga River – north of the town of Kamyshin) to 51°20' N (town of Uralsk). The western limit passes at 45° E along the Ergeni Height protruding to the west up to 43° E in the Kuma-Manych Depression. Mangyshlak is situated to the south of this lowland at the eastern shore of the Caspian Sea between 45° and 42° N. It is washed by the Caspian Sea on the north and west and bordered by the Usturt Plateau on the east and by Bay Kara-Bogaz-Gol on the south. The Caspian Lowland belongs to the steppe and desert regions, whereas the Mangyshlak lies completely within desert. The boundary between steppe and desert regions goes along the Kuma river on the south-west of the Caspian Lowland (45° N) and further-along the Kuma-Manych Depression and the eastern slope of the Ergeni Height, reaching 47° N and then turns to the north-east, crossing the Volga river at 48° N and goes to the east along this latitude [19].

The climate of the area studied is extremely continental with dry hot summers and cold winters, and winds are rather constant. The seasonal and daily temperatures change significantly. Mean temperature of July is 24° to 26°C and the mean temperature of January is –5° to –9.5°C. Annual precipitation is 150–200 mm, which is distributed unevenly. Evaporation is very high 800–900 mm. The snow cover is approximately 10 cm deep. Dust storms are recorded five to six times in a month, at a wind speed of 10–12 m/s [20–23].

The relief is rather flat as a whole. The most part of it lies below the world sea level. The absolute height goes down gradually from 15–20 m above sea level on the north to –27 m below sea level at the Caspian Sea shore [24]. The relief of the Mangyshlak is much more complicated. Its northernmost part – the Buzachi peninsula, North Mangyshlak Depression, the Tyub-Caragan peninsula – represents a plain, central part being strongly dissected with mountain ranges like Karataushik, Northern and Southern Aktau, Western and Eastern Karatau. The vast undulated Mangyshlak Plateau spreads to the south of the Mountain Mangyshlak. On the south it is separated by the large depressions of Kaundy and Basgurly from the same plain Kenderli-Kayasan Plateau. The Eastern Mangyshlak is characterized by the complicated inversion relief. On the southeast, it joins the large depression of Karin-Zharyk, which comes out to the bay of Kara-Bogas-Gol.

The dry climate determines the development of the peculiar soil types and salinity is a characteristic feature of desert soils because the salts are not washed out of soils. In the Turan Desert two types of soils are distinguished: brown and gray-brown ones. They are represented by the different variants (loam, sandy-loam, sand, etc) and by the different genera (eroded, primitive, solonetz-solonchak, etc.). Solonetz, solonchak, takyrs, takyromorphic desert soils occupy vast areas. Solonetz are the salted soils with compact surface and a tight clay horizon of column or prismatic structure in the lower part of soil profile. Solonchak is salted soil with a high content of soluble salts in the surface layer. Takyr is soil having rather compact surface, which cracks into polygons of regular form of 8–10 cm in diameter with a crust of 2–5 cm in depth; the whole profile is characterized by the high content of carbonates, the salts are represented by chlorides and sulfates in different proportion. The crust does not usually contain salts. Takyrs develop on alluvial and proluvial plains due to periodical flooding. Takyromorphic desert soils are middle–heavy loam; the whole profile contains carbonates; soluble salts (mainly chlorides and sulfates) are present almost permanently but may be washed out up to some depth. These soils develop on alluvial, lake and proluvial deposits [25].

## Vegetation

The Turan Desert is subdivided into three longitudinal subzones [18, 26]. The northern subzone is characterized by the domination of communities of dwarf semishrub wormwoods on the vegetation cover of a placor environment (*Artemisia lerchiana*, *A. semiarida*, *A. sublessingiana*); in the middle subzone coenoses of *Artemisia terrae-albae* and perennial saltwort (*Anabasis salsa*, *Salsola arbusculiformis*) prevail; the southern subzone is dominated by communities of *Artemisia kemrudica* and *Salsola gemmascens*. The middle subzone is borders on the southern one to the south of 43° N in the Mangyshlak [17, 27]. Halophytes play a great role in desert vegetation of the Northern Caspian Region. They form communities confined to saline soils and to salted variants of zonal soils. A significant number of halophytic species belonging to the families *Chenopodiaceae*, *Poaceae* and *Asteraceae*, together with *Tamaricaceae* and *Limoniaceae* occur. The highest number belongs to the group of dwarf semishrub halophytes: *Anabasis salsa*, *A. aphylla*, *A. brachiata*, *Arthrophytum lehmannianum*, *Atriplex cana*, *A. verrucifera*, *Camphorosma monspeliaca*, *Halimocnemis karelinii*, *H. villosa*, *Halocnemum strobilaceum*, *Halogeton arachnoides*, *H. glomeratus*, *Halopeplis pygmaea*, *Halostachys belangeriana*, *H. caspica*, *Halothamnus glauca*, *H. subaphylla*, *Iljinia regelii*, *Kalidium caspicum*, *K. foliatum*, *Kochia prostrata*, *Limonium suffruticosum*, *Nanophyton erinaceum*, *Reaumurea fruticosa*, *Salsola dendroides*, *S. gemmascens*, *S. laricina*, and *S. orientalis*. Many species of the annual saltworts also occur: *Atriplex tatarica*, *Bassia hyssopifolia*, *Ceratocarpus arenarius*, *C. utriculosus*, *Climacoptera affinis*, *C. aralensis*, *C. brachiata*, *C. crassa*, *Corispermum aralo-caspicum*, *C. hyssopifolium*, *Echinopsilon sedoides*, *Girgensohnia oppositiflora*, *Kirilowia eriantha*, *Kochia iranica*, *Petrosimonia brachiata*, *P. brachyphylla*, *P. glaucescens*, *P. hirsutissima*, *P. litwinowi*, *P. oppositifolia*, *P. triandra*, *Salicornia europaea*, *S. perennans*, *Salsola foliosa*, *Suaeda acuminata*, *S. altissima*, and *S. confusa*. The halophilous dwarf semishrub wormwoods are represented by four species: *Artemisia halopohila*, *A. gurganica*, *A. pauciflora*, and *A. santonica*. Shrub species of genus *Tamarix* and *Haloxylon aphyllum* are also common. There are only two semishrub species: *Krascheninnikovia ceratoides* and *Salsola arbusculiformis*. Some species of halophilous perennial grasses (*Achnatherum splendens*, *Aeluropus pungens*, *Leymus ramosus*,

*Puccinellia distans*, and *P. sclerodos*) and one annual grass (*Eremopyrum triticeum*) are also characteristic of this desert vegetation.

The latitudinal (zonal) and longitudinal (regional) features are clearly observed in the distribution of halophytic desert vegetation. Some of these formations play essential role in the physiognomy of vegetation of the region. The halophytes may be divided into three groups: (1) those restricted to solonchak; (2) those restricted to solonetz and takyr; and (3) those with distribution concentrated in the salted variants of the zonal soil types.

## Hyperhalophytic deserts on solonchak

Phytocoenoses of dwarf semishrub saltworts (Mediterranean *Halocnemum strobilaceum*, Irano-Turanian *Kalidium capsicum* and *Salsola dendroides*, Kazakhstanian-Irano-Turanian *Suaeda physophora*) are most common and spread everywhere, in all three latitudinal subzones [15, 28]. In the northern subzone, the communities of such dwarf semishrub saltwort as Kazakhstanian-Irano-Northturanian *Halimione verrucifera*, Kazakhstanian-Northturanian *Atriplex cana*, Pontic-Kazakhstanian-Irano-Northturanian *Limonium suffruticosum* and dwarf semishrub wormwood *Artemisia santonica* are characteristic. There are several noticeable halophilous grasses (*Aeluropus pungens*, *Puccinellia distans*) and shrubs (Mediterranean *Tamarix ramosissima* and *T. laxa*). Today the huge areas on the West of the Caspian Lowland are occupied by the coenoses of annual saltworts (*Climacoptera crassa*, *C. brachiata*, *Suaeda acuminata*, *S. altissima*, and *Salicornia perennans*). They are connected not only with solonchak themselves but also with salinification of brown zonal soils owing to the strong anthropogenic pressure (overgrazing, ploughing, melioration). Along the eastern shore of the Caspian Sea from north to south floristic composition changes gradually: in the middle subzone *Artemisia santonica* is replaced by *Artemisia halophila*, *Halimione verrucifera* disappears, and the role of annual saltworts and grasses decreases.

In the Mangyshlak, numerous deep depressions have naked sors (solonchak with crust salty stratum on the surface) surrounded by chubby solonchak which are occupied by coenoses of *Halocnemum strobilaceum*, *Kalidium caspicum*, *Artemisia halophila*, and *Limonium suffruticosum*, with an admixture of *Atriplex cana*, *Climacoptera brachiata*, *Halimocnemis karelinii*, and *Frankenia hirsuta*. A similar vegetation spreads out on salted plains along the seashore and along the solonchak streams in mountain massifs, but single bushes of *Tamarix elongata* and *T. gracilis* also occur here together with East Mediterranean *Achnatherum splendens*. On the sand shell rocks of the salty littoral plains, coenoses of *Aeluropus littoralis*, *Puccinellia distans*, *Limonium suffruticosum*, *Artemisia halophila*, *Suaeda dendroides*, *Anabasis salsa* and *Salsola dendroides* are distributed together with *Camphorosma lessingii*. The dry streams are occupied by coenoses of *Salsola dendroides* and *Anabasis aphylla*. In the southern subzone, the vegetation of solonchak is similar to a large extent with that of middle deserts. The difference is in the presence of Irano-Turanian-Central Asian *Halostachys caspica* and Irano-Southturanian-Central Asian *Salsola gemmascens* communities, which are absent in the middle desert. At the same time, the *Atriplex cana* and *Aeluropus littoralis* coenoses disappear and Irano-Turanian *Reaumurea fruticosa* becomes more dominant. *Halocnemum strobilaceum* is more widely distributed in both habitats, the littoral plains and sor depressions, and along the streams *Tamarix elongata* is well developed. *Salsola gemmascens* participates in diverse communities with *Halocnemum strobilaceum*, *Reaumurea fruticosa*, and *Kalidium caspicum*.

### Halophytic deserts on solonetz and takyrs

Solonetz are characteristic of the northern and middle subzones. Takyrs appear in the middle strip but they are spread more widely southward where they reach a great size.

*Artemisia pauciflora* Web. is halophilous dwarf semishrub of the desert steppe, and is a Norththuranian-Kazakhstanian element. Its range occupies a large area from the Don river on the west (42° E) to Lake Zaisan on the east (85° E); its northern boundary goes along 52–53° N, and the southern one along 47–48° N [29]. It usually occurs in the vegetation of the steppe region, and is found only in the northern subzone of the desert region. *Artemisia pauciflora* is restricted to solonetz and outcrops of Tertial salted clays throughout almost the whole range [17]. The size of communities varies from 1 $m^2$ up to several kilometers. In the desert region in communities of *Artemisia pauciflora*, the halophilous dwarf semishrubs (*Kochia prostrata*, *Atriplex cana*, *Anabasis salsa*) represent common co-dominants; in spring there is a well-developed synusia of ephemeroides (*Poa bulbosa*, *Tulipa biflora*, *T. schrenkiana*, *Ferula caspica*, *F. nuda*, and *Ornithogalum fisheri*) and ephemers (*Ceratocephalus orthoceras*, *Alyssum desertorum*, *Eremopyrum triticeum*, and *E. orientale*). Diverse communities of perennial saltwarts with sparse cover are also found on the solonetz and takyrs. They are composed of the same species that are spread on the plains with salted zonal soils (*Anabasis salsa*, *A. brachiata*, and *Nanophyton erinaceum*).

### Halophytic deserts on the salted zonal soils (brown and gray-brown)

*Anabasis salsa* (C. A. Mey.) Benth. ex Volkens is a dwarf semishrub of the steppe-desert; it is a halophyte, and represents a Kazakhstanian-Northturanian element. It occupies vast area from 44° E on the west (from Ergeni Height) to 92° E on the east (in Dzungaria). The northern limit of distribution is 54° N, and the southern 40° N [30]. In the Eurasian Steppe Region *Anabasis salsa* is confined mainly to solonetz, solonchak and salt rock. In the Turan Desert, it is one of the dominant ("landscape") plants. *Anabasis salsa* grows on plains and hills on soils of different mechanical texture (clays, loams, sandy loams) and of different salt content (strongly salted soils, solonetz, solonchak), and on outcrops of salted clays. This species is uncommon in the northern subzone in the western part of the Caspian Lowland. To the east from the Ural river *Anabasis salsa* is a permanent component of complexes on plains with strongly salted soils, giving way only to wormwood communities of *Artemisia lerchiana*, *A. semiarida*, and *A. sublessingiana*. It is the most significant plant in vegetation within the middle desert subzone, especially between the Caspian and Aral seas (in the Mangyshlak and the Usturt Plateau) where its communities occupy a huge area. In the southern subzone the role of *Anabasis salsa* is different from that in the middle one. It occurs only in the northern part of subzone as a dominant or a co-dominant (usually with *Salsola gemmascens*) species on the eroded and shallow scree soils at the slopes of rolling hills, on plains with salted loamy scree soils, on takyromorphic soils and on takyrs. On eroded gypsum soils its communities include several dwarf semishrubs with about the same abundance – *Salsola gemmascens*, *S. orientalis*, *Artemisia kemrudica*, and *Nanophyton erinaceum* [17]. The *Anabasis salsa* formation in the Mangyshlak includes 96 species: 90 of them in the middle subzone, 50 in the southern one. The separate coenoses include 2–29 species. General cover percentage in plant communities is 20–30%. Plant communities are poor in species and have few perennials: *Anabasis aphylla*, *Salsola orientalis*, and *Rheum tataricum*. Synusia of annuals is noticeable with a mean cover of 5–10% or 30–50% in separate years. Usually 2–4 annual species predominate. On strongly salted loamy soils, solonetz and takyrs communities of *Anabasis salsa* are very poor. In the middle subzone, in habitats with soils of lighter mechanical texture (small content of screen, outcrops of stones), *Artemisia terrae-albae* and *Salsola orientalis* become the co-dominants of *Anabasis salsa*, on takyrs – *Artemisia gurganica* (only in the Western Mangyshlak) or *Anabasis aphylla*. On strongly eroded soils, *Nanophyton erinaceum* or *Anabasis brachiata* are abundant in communities of *Anabasis salsa*, and on salted sandy loams – *Arthrophytum lehmannianum*. The most constant annual species are *Eremopyron orientale*, *Ceratocephala testiculata* (in middle and southern subzones), *Lappula spinocarpos*, *Leptaleum filifolium*, *Veronica amoena* (more abundant in the middle subzone), and *Diptychocarpus strictus* (more often and abundant in the southern). So, there is certain the North-South trend in abundance of *Anabasis salsa* in plant communities.

*Anabasis brachiata* Fisch. et C. A. Mey. ex Kar. et Kir. is a desert halopetrophilous perennial, and a Turanian endemic. In the western geographical range this species penetrates to the Apsheron peninsula and reaches the Kura River, covering the whole territory between the Caspian Sea and the Aral Sea. In the north it reaches the Lake Inder (the left bank of the Ural river), in the south it is bordered by the western spur of the Kopet-Dag Mountains [31]. It is widespread in both the middle and southern subzones [17]. The species is restricted to outcrops of the mix-colored salted clays and plains with the salted scree gray-brown soils. The general cover in its communities is 5–30%, and species diversity is fairly low, with mono-dominant coenoses not being rare. The complexes of perennial saltworts (*Anabasis salsa*, *Nanophyton erinaceum*, *Salsola gemmascens*, and *S. orientalis*) are characteristic of the southern subzone with the rolling-hill relief, and shallow gray-brown soils contain much gypsum. Among them a noticeable role is played by *Anabasis brachiata*. Takyrs of various size are common elements of landscape, and sparse grouping of *Anabasis brachiata* and *A. eriopoda* (along with those of *Anabasis salsa* and *Nanophyton erinaceum*) are characteristic of small takyrs. Such coenoses also occupy huge areas of bozyngens.

*Nanophyton erinaceum* (Pall.) Bunge is a steppe-desert halopetrophilous dwarf semishrub, and a Turano-Gzungarian element [32], with a geographical range extending from 51° E to 94° E and from 53° N to 40° N, including the steppe and desert regions [29, 30]. Its phytocoenotic optimum lies in the Turan Desert, where this species is confined to the outcrops of various salted rocks, plains and slopes of hills with salted loam scree soils, the scree-loam solonetz and takyrs. *Nanophyton erinaceum* does not occur in the northern subzone in the Caspian Lowland because of the absence of suitable habitats. It is widespread in the Mangyshlak in both the middle and the southern subzones [17]. In the middle subzone, *Nanophyton erinaceum* is especially characteristic in the Mountain Mangyshlak, where it is restricted to gentle slopes of hills, salted rocks, and salted plains between hills. Communities of *Nanophyton erinaceum* prevail on shale, clays, sandstones, and others. Their vegetation is sparse with a cover of 5–10%. They are poor in species. Sometimes there are no other perennial species, with only ephemers like *Alyssum turkestanicum*, *Eremopyrum orientale*, and *Trigonella arcuata* being present with a noticeable abundance. In the ravines, the coenoses of *Nanophyton erinaceum* are more closed with abundant *Poa bulbosa* and *Catabrosella humilis* or *Artemisia terrae-albae* and *A. gurganica*. In the hilly areas (melkosopochnik) with salty soils, a difference is observed between northern and southern slopes: ephemeroids *Catabrosella humilis* and *Poa bulbosa* as well as annuals *Veronica amoena* and *Trigonella arcuata* are abundant only on the northern slopes; *Eremopyrum orientale* and *Anisantha tectorum* are distributed on the southern slopes; *Eremopyrum orientale* and *Roemeria hybrida* are spread on the tops. On the solonchak chalk plains (between hills) the communities of *Nanophyton erinaceum* participate in halophytic complexes together with the communities of *Artemisia halophila*, *Aeluropus littoralis*, and *Atriplex cana*. Coenoses of *Nanophyton erinaceum* occur often in the southern subzone. Here they are components of numerous complexes. On the plains, bozyngens (convex or flat form of land surface with the soils enriched by gypsum) occur constantly with groupings, in which the lichen cover of *Xanthaparmelia camtschadalis* is well developed. These groupings are formed usually by *Anabasis brachiata*, but sometimes by *Nanophyton erinaceum*, as well as some other species (*S. gemmascens*, *S. orientalis*, *Anabasis salsa*, and *Nanophyton erinaceum*). Takyrs, as bozyngens, are permanent elements of the landscape on the south of Mangyshlak. They are of various length (from 30–50 m up to some kilometers) and depth. The small takyrs are occupied by various sparse groupings, notably *Nanophyton erinaceum*. In the *Nanophyton erinaceum* formations 39 species have been recorded, with a participation of 8–26 species in the community structure. General cover is 5–35%.

*Arthrophytum lehmannianum* Bunge is a desert halopetrophytic dwarf semishrub, a Turanian element. Its western distribution limit passes through the Mangyshlak [29]. The areas of its coenoses are not particularly large, but show fairly wide distribution in the Mangyshlak in the middle subzone [17], being abundant in the north-east part of the Karagie Depression, on the northern side of Mangyshlak Plateau, in the Eastern Mahgyshlak, in the north part of the Karin-Zharyk Depression; and on the northern slopes of the Southern Aktau Range. Its communities are restricted to flat rolling hills and to small stony outcrops of salted sandstone and chalk. In the Mangyshlak there are 73 species in the *Arthrophytum lehmannianum* formation, but this varies between 11 and 26 species in different communities. General cover is 25–50%. The dwarf semishrub wormwood *Artemisia terrae-albae* participates in all coenoses. *Salsola orientalis* and *Anabasis salsa* are also characteristic in its communities. The co-dominant species may be *Nanophyton erinaceum*, but this is not so common. Occasionally *Atraphaxis replicata* shrub and *Convolvulus fruticosus* and *Krascheninnikovia ceratoides* semishrubs are found, and the tufted grasses *Stipa caspia* and *Catabrosella humilis* occur most often. Psammophilous sedge *Carex physodes* are also frequent. Perennial herbs are represented by only 5 species and their role is inconspicuous; in addition, 20 annual species occur, but only 3–4 of them are abundant. *Eremopyrum orientale*, *Meniocus linifolius*, *Ceratocephala testiculata*, *Ceratocarpus utriculosus*, and *Leptaleum filifolium* are species that are also constantly found.

*Salsola arbusculiformis* Drob. is a steppe-desert halopetrophilous dwarf semishrub, representing a Turanian element. Its reaches the western limit of its geographical range in the Mangyshlak, and is widely distributed in the north Turanian deserts eastward of the Aral sea and in the northern part of the Usturt Plateau, with an isolated locality in the central part of the Usturt Plateau and two isolated localities in the Mangyshlak (central part of the Northern Aktau Range and the Eastern Karatau Range). Phytocoenotic optimum of this species is in the middle subzone [17, 33]. In the Eastern Karatau its communities are confined to salty violet sandstones of the Permian and Trias age. The cover is 70%. In these coenoses the ephemeroid *Poa bulbosa* and two wormwood species (*Artemisia gurganica* and *A. terrae-albae*) always participate. The shrubs *Atraphaxis replicata* and *Caragana grandiflora* and semishrub *Convolvulus fruticosus* appear on stony areas of slopes near outcrops of rocks. Grasses such as *Agropyron fragile* and *Stipa caspia* also occur in such habitats. Sometimes a small difference in community structure is observed between slopes exposed to the north and the south: *Poa bulbosa* and *Artemisia terrae-albae* are more abundant on the slopes with northern exposure, whereas annuals (mainly *Anisantha tectorum*) and *Artemisia gurganica* are more frequent on the south-exposed slopes. In the Northern Aktau Range *Salsola arbusculiformis* is restricted to carbonate limestone. On the gentle slopes with northern exposure, it forms two types of plant communities that are poor in species: with co-domination of or without wormwood with abundant ephemers (*Eremopyrum orientale*, *Meniocus linifolius*, and *Lepta-leum filifolium*) or ephemeroides (*Poa bulbosa* and *Catabrosella humilis*). On eroded areas of slopes *Salsola arbusculiformis* forms communities with abundant perennial saltworts (*Nanophyton erinaceum*, *Salsola orientalis*, and *Anabasis brachiata*). In the Mangyshlak, the formation of this species includes 75 species, and the number of species in communities varies between 11 to 33 species. General cover is 60–80%, of which 55–60% is given by *Salsola arbusculiformis*.

*Artemisia gurganica* (Krasch.) Filat. is a desert halophilous dwarf semishrub, and is an endemic species of the Mangyshlak and the Usturt Plateau [34]. Its phytocoenotic optimum lies in the middle subzone in the Mangyshlak, where communities of *A. gurganica* spread often on salted soils of different mechanical texture and on solonetz. In the strip of southern deserts they are confined to takyrs [17]. General cover is 30–40%, but may range from 50% to 70% in the years favorable for annuals. In the formation of *A. gurganica* 119 species have been recorded, the separate communities number 5–35 species. In pelitophytic deserts, this wormwood always differs in the abundance of the halophytic dwarf semishrubs *Anabasis aphylla* and *A. salsa.* In the hemipetrophytic and petrophytic variants on salty sandstone and limestone, grasses participate such as the long-vegetating ones (*Achnatherum caragana*, *Stipa caspia*, *S. caucasica*, *S. lessingiana*, some places *S. sareptana*) and the short-vegetating ones (*Catabrosella humilis*, *Poa bulbosa*) as well as shrubs (*Atraphaxis replicata*, *Caragana grandiflora*, *Rhamnus sintenisii*) and semishrubs (*Convolvulus fruticosus*, *Krascheninnikovia ceratoides*). Annuals are constant components in these coenoses, and their species diversity in these formations is very high at around 54 species. *Alyssum turkestanicum*, *Eremopyrum orientale*, *Lappula spinocarpus*, *Trigonella arcuata*, and *Veronica amoena* are always abundant. In the communities of *Artemisia gurganica*, 36 species of perennial herbs have been recorded (mostly petrophytic ones): *Astragalus ustiurtensis*, *Crambe edentula*, *Haplophyllum versicolor*, *Lagochyllus acutilobus*, *Sylene cyri*, *Tragopogon ruber*, etc. Their abundance is usually low. Communities of *Artemisia gurganica* occur only in the West Turan, especially in the middle subzone in the Mangyshlak that offers peculiar features to its vegetation.

*Salsola gemmascens* Pall. is a desert halopetrophilous dwarf semishrub, and is an Irano-Southturanian element. The western distribution limit of this species is in the Mahgyshlak (52°30' E), the eastern in the Issyk-Kul Depression (78°30' E), and the northern in the Usturt Plateau (44°40' N) [28, 29]. In the southern deserts of the Mahgyshlak *Salsola gemmascens* is one of the dominating ("landscape") plants throughout the area. It is restricted to the shallow scree salted soils with high gypsum content, takyrs, and solonchaks [17]. It frequently grows on the scree-loam soils, forming hemipetrophytic and petrophytic communities that are mono-dominant or oligo-dominant (with participation of halophytic dwarf semishrubs *Anabasis salsa*, *Salsola orientalis*, *Artemisia kemrudica*, *Nanophyton erinaceum*, *Anabasis brachiata*, and *A. eriopoda*). General cover is 30–45%, decreasing on takyrs and habitats with a high content of scree to 5–10%; in years favorable for ephemers it may achieve 70% cover. This plant formation has 56 species. The separate communities include 6–19 species, where dwarf semishrubs (*Artemisia kemrudica*, *Anabasis salsa*, *Salsola orientalis*, *Anabasis brachiata*) are abundant. The highest species diversity is observed among the annual plants (*Eremopyrum orientale*, *Lappula spinocarpos*, *Leptaleum filifolium*, *Ceratocephala testiculata*, *Strigosella scorpioides*, etc.). Perennials are unique. The communities of *Salsola gemmascens* are used as differential ones to distinguish the line between the middle and southern subzones of the desert zone since they play noticeable role in vegetation only to the south of 43°N.

## Conclusion

We can say that halophyte communities in the desert zone are of a great importance as pastures. They are well adapted to the environmental conditions and suitable for phyto-amelioration of pastures with low productivity.

### *Acknowledgements*

I am very grateful to Prof. Chedlly Abdelly for the invitation to participate in the Conference on Biosaline Agriculture and High Salinity Tolerance in Tunisia and thank the Organizing Committee as well as the Russian Foundation for fundamental investigation (RFFI, grant 04-04-49302) for their help. My special thanks go to my research associate Polozova Tamara (Dept. of Vegetation Geography and Cartography, Komarov Botanical Institute, Russian Academy of Sciences, St.-Petersburg) for translation of this manuscript into English.

## References

1 Keller BA (1940) Vegetation of saline soils of USSR. In: BK Shishkin (ed): *Vegetation of USSR*, vol. 2. Academic Press, Moscow, Leningrad, 481–521

2 Nechaeva NT (1956) Wormwood-saltwort pastures of North-West Turkmenistan. In: *Proceedings of the Institute of stock-raising of AS of Turkmenistan*, vol. 1. Academic Press, Ashgabad, 18–127

3 Shakhov AA (1956) *Salt tolerance of plants*. Academic Press, Moscow

4 Lavrenko EM (1962) *Main features of botanical geography of deserts of Eurasia and North Africa*. Academic Press, Moscow, Leningrad
5 Bilyk GI (1963) *Vegetation of saline soils of Ukraine*. Academic Press, Kiev
6 Rodin LE (1963) *Desert vegetation of the Western Turkmenia*. Academic Press, Moscow, Leningrad
7 Barsegjan AM (1966) Saltwort and solonchak vegetation of the Ararat plain. In: *Flora and vegetation of Armenia*. Academic Press, Erevan
8 Bochantsev VP (1969) Genuis *Salsola* (short history of its evolution and distribution). *Bot J* 7: 989–1001
9 Pratov U (1970) *Chenopodiaceae of the Fergana valley*. Academic Press, Tashkent
10 Bakhiev A (1979) *Vegetation indicators of saline soils and ground waters in the delta of Amu-Darya*. Academic Press, Tashkent
11 Kubanskaya ZV (1980) *Halophyte deserts of Kazakhstan*. Academic Press, Alma-Ata
12 Bykov BA (1981) Main specific features of halophyte flora and vegetation of the Middle Asia and Kazakhstan. In: *Transactions of Ac Sci Kazakh Republic, Series biological*, vol. 1. Alma-Ata, 1–9
13 Nikolskaya NI (1975) *Botanical-geographical analysis of vegetation of solonchak and solonetz in the Kasakhstan melkosopochnik as example*. Dissertation Thesis, Biological Sciences, Leningrad
14 Nikolskaya NI (1984) Vegetation of solonchak in the vegetation map of USSR, s1:4000000, for high schools. In: *Geobotanical mapping 1984*. Nauka, Leningrad, 20–25
15 Nikolskaya NI (1985) Regularities of distribution of solonchak vegetation in steppe and desert regions in USSR. *Bot J* 70, 3: 332–340
16 Akzigitova NI (1982) *Halophyte vegetation of the Middle Asia and its indicator features*. FAS Uzbekistan, Tashkent
17 Safronova IN (1996) *Deserts of Mangyshlak (outline of vegetation)*. St.-Petersburg
18 Rachkovskaya EI, Volkova EI, Khramtsov VN (eds) (2003) *Botanical Geography of Kazakhstan and Middle Asia (Desert Region)*. St-Petersburg
19 Safronova IN (2002) Phytoecological mapping of the Northern Caspian Region. In: *Geobotanical mapping 2001–2002*. St-Petersburg, 44–65
20 Trofimov IA (1995) Natural pasture of the Chernye Zemli. In: IS Zonn, VM Neronov (eds): *Biota and environment of the Kalmykia republic*. Moscow, Elista, 53–83
21 Schuchkina VP, Barmin AN, Pilipenko VN, Pirogovcki MN, Fedorovich VV, Bucharizin PI (1996) Nature of the Astrakhan Region. In: Pjatin VA (ed): *Nature and history of the Astrakhan region*. Pedogogic Institute, Astrakhan, 7–124
22 Pjatin VA (ed) (1997) *Atlas of Astrakhan region*. Federal service of geodesy and cartography of Russia, Moscow
23 Kunaev AM (ed) (1982) *Atlas of Kazakh SSR*. Moscow
24 Doskach AG (1970) *Natural division of the Caspian semidesert*. Nauka, Moscow
25 Borovskij VM (ed) (1974) *Soils of the Mangyshlak peninsula*. Nauka, Alma-Ata
26 Ladygina GM, Rachkovskaya EI, Safronova IN (eds) (1995) *Vegetation map of Kazakhstan and Middle Asia (Desert Region)*. S1:2500000. "Ekor", Moscow
27 Safronova IN (2004) On phytocoenotical mapping of the Caspian Desert Region. *Ann Bot Nuova Serie*: 83–93
28 Lavrenko EM (1965) Provincial subdivision of Centralasian and Irano-Turanian subregions of Afro-Asian desert region. *Bot J* 50, 1: 3–15
29 Musaev IF (1969). Maps of geographical ranges edificator plants of Turan. In: *Geographical ranges of USSR Flora*, vol. 2. University Press, Leningrad, 120–167

30 Lavrenko EM, Nikolskaya NI (1963) Ranges of several centralasian and northturanian species of desert plants and the problem on the botanical-geographic boundary between Middle and Central Asia. *Bot J* 48, 12: 1741–1761
31 Musaev IF (1976) Geography of species of genus *Anabasis* L. In: *Geographical ranges of USSR Flora*, vol. 3. University Press, Leningrad, 112–143
32 Rachkovskaya EI (1993) *Vegetation of gobian deserts in Mongolia*. Nauka, St-Petersburg
33 Sokolov SY, Svjazeva OA, Kubli VL (1980) *Salsola arbusculiformis* Drob. In: *Ranges of trees and shrubs of USSR*, vol. 2. Nauka, Leningrad, 18
34 Filatova NS (1984) Species of genus *Artemisia* L. (Asteraceae) subgenus Seriphidim (Bess.) Peterm. in USSR Flora. In: *Novelties of taxonomy of higher plants,* vol. 21. Nauka, Leningrad, 155–185

Biosaline Agriculture and High Salinity Tolerance
Edited by Chedly Abdelly, Münir Öztürk, Muhammad Ashraf and Claude Grignon


# Halophyte utilization for biodiversity and productivity of degraded pasture restoration in arid regions of Central Asia and Russia

Nariman Shamsutdinov[1] and Zebri Shamsutdinov[2]

[1] *All-Russian Res. Institute of Hydraulic Eng. & Land Reclamation, Moscow, Russia*
[2] *All-Russian Fodder Research Institute, Lobnya, Moscow, Russia*

## Introduction

Approximately 15% of the territories of the former USSR are arid lands that are mostly situated in the republics of Central Asia and in the south of the Russian Federation. In the agricultural lands of the Central Asian region the portion of irrigated lands is no more than 4%. The remaining 96% are non-irrigated lands of largely forage pastures (over 300 million ha) in deserts, semi-deserts and dry steppes.

Arid territories of the Russian Federation territory cover about 71.3 million ha. They extend over five natural-agricultural zones with different degrees of aridization: extra arids (deserts), 5.6 million ha; strong arids (semi-deserts), 7.5 million ha; middle arids (semi-deserts), 21.2 million ha; lax arids (southern parts of steppes), 37.0 million ha; sub-arids (North part of steppes and South Forrest steppes), 37.8 million ha; and sub-humid periodically arid zones (North part of Forrest steppe), 37.4 million ha. Together with sub-arid and sub-humid areas arid territories of Russia achieve 146.5 million ha [1].

Out of the total area of natural pastures of Russia, 73% or 50 million ha are found in arid territories of Russia. Arid lands occur on the territory of 11 areas of the Federation, i.e., the Republic of Daghestan, the Republic of Kalmykia, the Astrakhan, Volgograd, Samara, Saratov, Orenburg and Rostov Regions, the Republic of Bashkortostan, and the Krasnodar and Stavropol Territories. However, among these, the principal non-farming region of Russia with vast areas of natural pastures that support great populations of cattle is the Republic of Kalmykia.

Cattle breeding has been practiced in the northwestern part of the Circum-Caspian area, comprised mostly by the territory of Kalmykia, for five thousand years. The natural resource potential of the Republic of Kalmykia fully satisfied the requirements of cattle grazing. During the 17th to 19th centuries the pastures here supported more than 200 thousand head of large-horned cattle,

100 thousand horses, approximately 1 million fat-rumped sheep and 20 thousand camels–all Kalmykian breed. The raising of such agricultural animals of the Kalmykian breed on the basis of a pasture technology enabled maintaining sustainable condition of semi-desert and steppe ecosystems. In the 20th century radical changes in the lifestyle, with the loss of elite steppe cattle breeders as a result of wars, repressions, forced emigration, and deportation of the Kalmyk people, which lasted for 13 years, had led to a loss of the traditional and economic culture, experience of pasture cattle breeding, and disappearance of some varieties of agricultural animals of the Kalmykian breed.

The territory of Kalmykia embodies the first man-made desert of Europe, which covers approximately 1 million ha. It is facing a serious problem. The saiga population is endangered. The Kalmykian fat-rumped sheep has disappeared from the state register of domestic agricultural animals, and only 250 head of camels, 8.8 thousand horses remain in the steppe. The available large-horned cattle (147.4 thousand head) of the Kalmykian breed have partially disappeared and remaining stocks have lost their genetic features as well as quality by abandoning the traditional methods of their management. The way-out from the established environmental and economic situation is revival of historical experience of land management on the basis of pasture cattle rearing specialized on breeding of four species of agricultural animals of the Kalmykian breed. Breeding of grazing cattle based on raising big-horned beef cattle, sheep, horses and camel is the most rational way for an environmentally justified and economically beneficial development of the arid territories in the south of Russia.

At present out of 7473.1 thousand ha of land in Kalmykia, 88.8% are agricultural lands, out of which 81% covers natural pastures. More than half of natural pasturelands in the Republic are degraded and efforts are needed for restoration of their productivity and biodiversity. As a result of long-time investigations, the All-Russian Research Institute of Hydraulic Engineering and Land Reclamation, All-Russian Forage Research Institute, and the Kalmykian Research Institute of Agriculture have developed adaptive systems and methods for ecologically oriented management of pastures in arid regions in Southern Russia, including:

- Ecologically justified systems of pasture ecosystem management on the basis of pasture rotation and feedlot-based animal grazing
- Effective technologies for restoration of biodiversity and productivity of degraded pasture ecosystems based on utilization of dominant halophyte species of semishrubs and perennial grasses
- Technologies of ecological restoration.

## Genetic halophyte resources in the flora of Russia

Halophyte flora in Russia comprises more than 500 (512) species [2], which belong to 255 genera and 55 families. The share of halophyte species in the

**Table 1.** Quantitative data of leading halophyte families of Russia.

| Family | Number of | | % of the total number of halophytes | |
|---|---|---|---|---|
| | Species | Genera | Species | Genera |
| *Asteraceae* | 77 | 36 | 15.10 | 14.12 |
| *Poaceae* | 62 | 35 | 12.16 | 13.73 |
| *Chenopodiaceae* | 48 | 22 | 9.41 | 8.63 |
| *Fabaceae* | 34 | 15 | 6.67 | 5.88 |
| *Cyperaceae* | 30 | 5 | 5.88 | 1.96 |
| *Ranunculaceae* | 23 | 10 | 4.51 | 3.92 |
| *Brassicaceae* | 21 | 14 | 4.12 | 5.49 |
| *Lamiaceae* | 21 | 13 | 4.12 | 5.10 |
| *Apiaceae* | 19 | 15 | 3.73 | 5.88 |
| *Caryophyllaceae* | 18 | 11 | 3.53 | 4.31 |
| *Polygonaceae* | 15 | 5 | 2.94 | 1.96 |
| *Scophulariaceae* | 13 | 6 | 2.55 | 2.35 |
| *Liliaceae* | 11 | 6 | 2.16 | 2.35 |
| *Rosaceae* | 10 | 5 | 1.96 | 1.96 |
| *Rubiaceae* | 10 | 2 | 1.96 | 0.78 |

total number of species in the Russian flora is not large. However, their significance for selection of salt-resistant species of forage halophytes and for the development of saline lands is great. One can distinguish 15 leading families containing the greatest number of halophytes. The leading families containing the greatest number of halophytes are: *Asteraceae*, *Poaceae*, *Chenopodiaceae*, *Fabaceae*, *Cyperaceae*, *Ranunculaceae*, *Brassicaceae*, *Lamiaceae*, *Apiaceae*, *Caryophyllaceae*, *Polygonaceae*, *Scophulariaceae*, *Liliaceae*, *Rosaceae*, and *Rubiaceae*, which cover 4/5 (412) halophytes of the Russian flora. The greatest number of halophytes is found in the *Asteraceae* family (15.1%) followed by such families as *Poaceae* (62 species), *Chenopodiaceae* (48), *Fabaceae* (34), *Cyperaceae* (30), *Ranunculaceae* (23), *Brassicaceae* (21), *Lamiaceae* (21), *Apiaceae* (19), *Caryophyllaceae* (18), *Polygonaceae* (15), *Scophulariaceae* (13), *Liliaceae* (11), *Rosaceae* (10) and *Rubiaceae* (10) (Tab. 1).

Ability of halophytes to function normally and to form a high forage and medicinal mass when growing on saline soils is connected with their specific environmental, physiological and biochemical peculiarities [3–5]. All halophytes growing on saline soils need to somehow adapt to the limiting conditions of the natural environment that determine their growing and the geography of halophyte distribution over arid territories. The moisture deficit-related osmotic and toxic action of salts on plants and the physiological dryness of soils due to a high ion level in a soil solution is the first factor in this connection [3]. In the arid regions there is a deficit of moisture in saline soils accessible to organisms, which is caused by high osmotic pressure of a soil solution as well as high daytime temperatures and low air humidity. This leads to quick loss of moisture to evaporation. Halophytes need to maintain a balanced water budget: if an organism contains water in an amount sufficient for life sustenance, its absorption should be equal to its losses. Maintenance of a balance is complicated by two circumstances: chronically limited inflow of water and an aggravated permanent tendency to overuse of moisture as a result of its evaporation and other losses.

The most important anatomical-morphological and physiological-biochemical peculiarities of halophytes that ensure their normal functioning and development under saline conditions are: salt accumulation, salt release and salt resistance. Considering these features [6] halophytes are divided into three environmental-physiological groups: salt-accumulating (euhalophytes), salt-releasing (cryptohalophytes) and salt-resistant (glycohalophytes) [7]. Out of 512 halophyte species growing on the territory of Russia, the euhalophytes and mesoeuhalophytes include 99 species, cryptohalophytes (mesohalophytes) 71 species, and glycomesophytes (oligohalophytes) 340 species. Halophytes of the Russian Federation are not homogeneous by their salt resistance: 340 species are oligohalophytes, 71 mesohalophytes, 57 mesoeuhalophytes and 42 euhalophytes. Among oligohalophytes the greatest number of plants belong to the *Asteraceae* (16%) and *Poaceae* (11%) families. Among mesohalophytes the greatest number of species like oligohalophytes also belongs to the *Asteraceae* (18%) and *Poaceae* (14%). Among mesoeuhalophytes the greatest number of species belongs to the *Asteraceae* (18%) and *Chenopodiaceae* (19%) families. As for euhalophytes, here more than 47% of the species belong to the *Chenopodiaceae* family.

One of the most important features of the adaptive potential of halophytes is the life forms of the plants. Under saline conditions various life forms of halophytes may be met. For selection purposes, various life forms represent both a resource and reserve of selection taking into account soil-climatic conditions, and the degree and nature of soil salinity. Analysis of genetic resources of halophytes growing on the territory of Russia shows that they belong to different life forms. Among oligohalophytes, the greatest number of species (254) are perennial plants; shrubs comprise no more than 2–5 species. On the basis of Caspian genofund, the selection efforts are evolving with forage halophytes aimed at the development of salt-resistant, productive species for their application in environmental restoration of degraded agricultural landscapes.

## Ecologically justified systems of pasture ecosystem utilization

Natural pastures are the initial base and material basis of productivity of beef cattle, stud horse and camel breeding in arid zones of Russia. The key issue of environmentally sustainable pasture management in semi-deserts and dry steppes is the size of withdrawal and frequency of overgrazing of phytomass of pasture ecosystems [8–12]. Without causing damage to restoration processes it

**Table 2.** Scheme of double-field 10-year pasture rotation for sagebrush-grass pastures with change of spring and summer grazing in 5 years.

| Grazing cycle | First field in pasture rotation | | Second field in pasture rotation | |
|---|---|---|---|---|
| | Main season | Repeated season | Main season | Repeated season |
| First 5 years | Spring | Autumn | Summer | Winter |
| Second 5 years | Summer | Autumn | Spring | Winter |
| Next 5 years | Spring | Autumn | Summer | Winter |
| Next 5 years | Summer | Autumn | Spring | Winter |

is possible to withdraw 65–75% of the annual increment of pasture plants. Withdrawal of this amount of overground forage mass is the level that is balanced by the eating intensity of forage plants on pastures by animals. The appropriate level of eating, out of the annual increment of pasture plants by animals, forms favorable conditions for vegetative and seed restoration of plants, creates prerequisites for annual reproduction of phytomass and excludes a possibility of disturbance of ecological links in the plant community and, as a result, ensures sustainability of the whole pasture ecosystem [10, 11]. Taking into consideration the admissible norms of phytomass withdrawal and proceeding from the data of field experiments on the effect of grazing and overgrazing on the condition of forage plants and pasture productivity, different pasture rotations have been proposed. Table 2 gives a principal scheme for the rational management of a semishrub-grass type of pasture on the basis of a double-field 10-year pasture rotation with changing of spring and summer pastures each 5 years.

According to the proposed scheme of pasture rotation, in 10 years an overgrazing cycle in spring and summer is repeated. Such a scheme is common for sheep raising farms with semishrub-grass pastures. This can be detailed in each particular case with regard to natural and economic conditions in each zone. For pastures on sandy soils, a pasture rotation scheme that envisages rotation of a spring season with a winter one and summer with autumn is recommended. The overgrazing season in such schemes changes every 2 or 3 years. The principles for the most complete utilization of forage resources is based on combined grazing of different varieties of animals, in particular sheep and camels. Camels, unlike sheep, eat a wider range of species, even those plant species that are not eaten and poorly eaten by sheep. In addition, camels eat upper branches of the shrub crown that are not accessible to sheep. By eating of the upper parts of shrubs the camels do not compete here with sheep. At the same time withdrawal of the upper part of shrubs leads to their rejuvenescence, which ensures their better growth and improved productivity in the following year.

Sheep feed mostly on grass, semishrubs and also shrubs up to 1.2 m high, while camels graze on leaves and sprouts of trees, and shrubs up to 1.5–3 m high. This is the main “forage belt”, i.e., the forage ecological niche of a camel. They also eat those plant species that are not eaten or poorly eaten by big beef cattle and sheep. Thus, the feeding characteristics of sheep and camels make it possible to utilize different ecological niches without competition. This needs to be taken into consideration in development of adaptive systems for pasture cattle breeding in arid regions of the country.

Therefore, a rational system of arid pasture utilization on the basis of different schemes of pasture rotation and site-feedlot grazing of animals that creates better conditions for seed and vegetative renewal of plants, maintenance of biodiversity and preservation of their productivity on a high level is a very important principle in the system of ecologically oriented management of pastures.

## Technology of ecological restoration and productivity improvement of degraded pasture ecosystems

The adaptive technologies of ecological restoration of productivity of degraded pastures include:

- Creation of pasture ecosystems for spring-summer utilization. Such pasture ecosystems are created in regions where the natural forage lands are characterized by low productivity in the summer season. Here xerohalophyte semishrubs, such as *Kochia prostrate* (L.), *Camphorosma lessingii* Litv, *Salsola orientalis* S.G. Gmel, and *Eurotia ceratoides* Losinsk., and perennial grasses, such as *Agropyron sibiricum*, *A. desertorum*, *A. pectinatum*, and *Festuca rupicola*, are used in a ratio of 70% and 30%. An average yield of spring-summer pastures is 1.0–1.5 t/ha of dry forage mass (in unfavorable years it is not lower than 0.6–0.8 t/ha), with a yield of natural pastures (control) being 0.15–0.30 t/ha.
- Creation of perennial autumn-winter pasture ecosystems. Here forage halophyte shrubs like *Haloxylon aphyllum* (Mink.) Iljin, *Salsola paletzkiana* Litv., *Aellenia subaphylla* (C.A. Mey), semishrubs, *Kochia prostrate* (L.) Schrad, *Salsola orientalis* S.G. Gmel, *S. gemmascens* Pall, *Artemisia*, and *Ephedra strobilacea* (Bunge) and perennial grasses (*Agropyron sibiricum*, *A. desertorum*, *A. pectinatum*, and *Festuca rupicola*) in a ration of 25%, 70% and 5% are used. The autumn-winter pastures are characterized by a high sustainable productivity; in regions with an annual precipitation of 170–250 mm the yield of dry forage mass on such pastures is equal to 1.0–1.2 t/ha, while in regions with an annual precipitation of 160–250 mm it is 1.5–2.0 t/ha.
- Creation of perennial pasture ecosystems for year-round utilization is most advisable in various regions of semi-deserts and dry steppes. They are formed by using halophyte and xerophyte forage shrubs (20%), semishrubs (65%) and grasses (15%), which are consumed by sheep in different seasons. These pastures are suitable for any season of a year; their yields vary from 1.2 t/ha to 2.6 t/ha of dry forage mass.

## Technology of ecological restoration of barren sands

The processes of degradation of pasture ecosystems located on light-textured soils has led to formation of barren sands. Long-term investigations of the Kalmyk Research Institute of Agriculture, the Chernozemelsky Trial Station together with the Institute of Agricultural and Forest Reclamation (VNI-IALMI) have enabled a practical utilization of the methods for restoration of pasture ecosystems on barren sands to be developed. The technology of this method comprises several biogeocenotically justified stages:

- First stage (1st year): sowing of the main sand-fixation plant – giant ryegrass (giant sea lyme grass *Leymus racemosus* (Lam.) Tzvel. with a sowing rate 6–8 kg/ha in the period from mid-August to November
- Second stage (2nd year): planting in March-April of seedlings of *Calligonum aphyllum* (Pall.) Gurke. between rows of ryegrass to a depth of 0.6–0.7 m. In this case the spacing between rows of *Calligonum* seedlings is 1.5–2.0 m. In the same period old world winter fat is planted
- Third stage (3rd year): final fixation and formation of typical zonal sand habitat full-member pasture ecosystems by sowing of *Kochia prostrata*, mostly sandy ecotypes, white sagebrush, Siberian wheat grass and other perennial and annual grasses. The creation of pasture ecosystems on barren sands applying the described technology ensures restoration of a typical zonal pasture ecosystem with a complete grass cover and productivity of 1.5–2.6 t/ha of energy-saturated dry pasture forage [13].

## Conclusion

Pasture cattle breeding, practiced by Kalmyk tribes in the southeast of the Russkaya Plain, is a natural model of ethnos existence applied to new conditions and improved under new soil and climatic conditions. Pasture cattle breeding is a life-supporting system for the population in arid territories under the natural and socioeconomic conditions formed due to the effect of interaction of basic components of a nomadic way of life: natural ecosystem – nomadic people – aboriginal cattle. A nomadic system is based on a seasonal utilization of pasture forage and regular search for new pastures. The nomadic system is rather rational, perfect and necessary for life sustenance of communities and feeding of cattle. A comparison of nomadic systems and utilization of seasonal pastures by nomadic peoples – Mongols, Buryats, Kalmyks, Kazakhs reveal their common features. This may be explained by similarity of material culture and economic activities, a history of development of these peoples and also similarity of their habitats. At present the scientifically validated systems of ecologically oriented management of pastures are being developed in arid regions of Russia based on application of various schemes of pasture rotations and adaptive systems of phytoreclamation of degraded pasturelands.

*Acknowledgements*

We gratefully acknowledge the on-going support of the Federal Agency of Science & Innovations (projects #MD-569.2007.4, #02.512.11.2178), projects of the RFBR (Russian Foundation for Basic Research) (#07-05-97605, #07-05-13596), whose support made this study possible.

## References

1 Zvolinky VP, Zonn IS, Trofimov IA, Shamsutdinov ZSh (1998) *The lands and agro-climatic resources of Russian arid territories*. PAIM, Moscow
2 Shamsutdinov ZSh, Savchenko IV, Shamsutdinov NZ (2001) *Halophytes of Russia, their environmental assessment and utilization*. Edel-M, Moscow
3 Valter G. (1975) Vegetation on the earth. *Environmental-physiological description*, vol. 2. Mir, Moscow
4 Zalensky OV (1977) *Environmental-physiological aspects of photosynthesis studies*. Nauka, Leningrad
5 Balnokin YuV (1986) *Ion homeostasis of halotolerant algae*. Author's abstract of biology doctor dissertation. Moscow: 51
6 Richter AA (1927) Physiological foundations of resistance of southeastern plants. To the issue of salt resistance. *J Trial Agron Southeast* 3: 16
7 Genkel PM, Shakhov AA (1945) Environmental significance of a water regime of some halophytes. *Bot J* 30: 154–166
8 Larin IV, Beguchev PP, Rabotnov TA, Leont'eva IP (1975) *Meadow cultivation and pasture farming*. Kolos, Leningrad
9 Pokaninov LB (1994) *Organization of the pasture conveyor for sheep in dry steppe conditions of Kalmykia*. The abstract of the candidate thesis for biological sciences. TSChA, Moscow
10 Shamsutdinov ZS (1995) Word experience of biological melioration and prospective of their utilization in sustainable development of pasture farming in West Caspian region. In: SI Zonn (ed): *Biota and environment in Kalmykia*. Korkis, Moscow, 106–157
11 Shamsutdinov ZS (2000) Modern strategy of rational strategy of pasture farming in arid districts of Russia. In: VA Tyulydyukov (ed): *Development of scientific ideas of academician N.T. Andreev*. TSChA, Moscow, 76–95
12 Shamsutdinov ZS, Nazaryk LA, Ionis YI, Chamidov AA, Kozlov NN, Pilipko SV, Shamsutdinova EZ, Zvolinsky VP, Muchortov VI, Tsagan-Mandziev NL et al (2000) *The methodical indicating for mobilization of vegetative resources and introduction of fodder plants*. Rosselychozacademiya, Moscow
13 Lachko OA, Tsagan-Mandjiev NL (1986) *Typical technology of pasture creation and sands fixation in North-West Caspian region*. Dzangar, Elista

Biosaline Agriculture and High Salinity Tolerance
Edited by Chedly Abdelly, Münir Öztürk, Muhammad Ashraf and Claude Grignon


# Selection of a halophyte that could be used in the bioreclamation of salt-affected soils in arid and semi-arid regions

Mokded Rabhi, Ons Talbi, Abdallah Atia, Chedly Abdelly and Abderrazak Smaoui

*Laboratoire d'Adaptation de Plantes aux Stress Abiotiques, Centre de Biotechnologie de Borj Cédria, Technopole de Borj Cédria, B. P. 901, 2050 Hammam-lif, Tunisie*

## Abstract

Vegetative bioremediation or bioreclamation of salt-affected soils is an economic solution mainly for developing countries since chemical additions are becoming increasingly expensive. However, to be efficient, this approach needs sufficient irrigation. In this investigation, we evaluated the ability of some halophytes to desalinize a saline soil under non-leaching conditions with the aim of selecting appropriate species that could be used for this purpose in arid and semi-arid regions where precipitation is too low to leach salts from the rhizosphere. Three perennial species were used in this experiment: *Arthrocnemum indicum* (Willd.) Moq., *Suaeda fruticosa* Forsk., and *Sesuvium portulacastrum* L. Seedlings were grown on a saline soil under greenhouse conditions and irrigated with tap water for 170 days. Irrigations were carried out with almost no leaching. Soil salinity was significantly reduced in halophyte-grown soil as compared to the control. Plants were able to decrease the soil electrical conductivity by absorbing soluble salts, mainly sodium ions. Among the three studied species, *Sesuvium portulacastrum* L. was the most productive and was able to accumulate in shoots nearly 30% of the sodium content of each pot over the 170 days. Thus, *Sesuvium portulacastrum* L. seems to be the most promising species for saline soil desalination in arid and semi-arid regions.

## Introduction

Growth and yield of most crops cultivated on sodic and saline sodic soils are adversely affected because of poor physical properties and fertility problems in such soils [1]. Nevertheless, the vegetative bioremediation of these soils is possible without using chemical additions, which are getting increasing more expensive [2, 3]. In fact, many sodic and saline sodic soils contain at varying depths a low soluble source of $Ca^{2+}$ [3]. These ions are able to replace excess $Na^+$ from the cation exchange sites [4]. Thus, released sodium ions can be removed from the root zone through leaching irrigation [5]. Vegetative bioremediation of these calcareous sodic and saline sodic soils depends on several aspects: the $CO_2$ partial pressure within the root zone, root proton release (in $N_2$-fixing species), improvement of soil physical properties, and the $Na^+$ content of the shoot, which is removed through harvesting [6]. However, to be effective, this approach requires sufficient irrigation so that leaching occurs, a condition not easily found in arid and semi-arid regions [7]. In this investigation, we evaluated the capacity of three halophytes to desalinize salt-affected soils under low irrigation conditions with the aim of selecting species that are efficient and could be used for this purpose in arid and semi-arid regions.

## Material and methods

In this study *Sesuvium portulacastrum* L. (Aizoaceae) and *Arthrocnemum indicum* (Willd) Moq. and *Suaeda fruticosa* L. (Chenopodiaceae) were used. Seedlings were cultivated in 8-kg pots filled with a saline soil taken from the borders of the sabkha of Soliman (35 km northeast of Tunis). For each species, 30 plants were grown in six pots (5 plants per pot) and irrigated with tap water at almost 50% of the soil field capacity. Six additional pots were also filled with the same quantity of soil and irrigated in the same way but without plantation. After 170 days, shoots were cut, dried, weighed and ground. The nutrients were extracted from the powder using 0.5% $HNO_3$. $Na^+$ and $K^+$ concentrations were determined by flame spectrophotometry. Soil samples were also taken from the upper 15 cm from each pot, dried, ground, and analyzed for electrical conductivity (EC) of the aqueous extract (1 g soil in 10 ml distilled water) and soluble sodium content.

## Results and discussion

Our results revealed that the three halophytes used here considerably decreased soil salinity by absorbing high amounts of the sodium from soil (Fig. 1). EC, which was 1.4 dS $m^{-1}$ in the soil at the start of the experiment significantly diminished after 170 days. Similar results were obtained by Zahran et al. [8] who attempted to reclaim poorly drained soils in Egypt using two *Juncus* species (*J. rigidus* and *J. acutus*). They also reported that the EC of soil that had a 50% saturation decreased from 33 to 22 dS $m^{-1}$ in a single growth cycle of *J. rigidus*. Ravindran et al. [9], while evaluating the capacity of six halophytes to desalinize saline soils, observed that *Suaeda maritima* and *Sesuvium portulacastrum* caused very high decreases of EC in the soil. Our findings show that the variations in concentration of soluble sodium in the soil corresponded to changes in EC (Fig. 1), indicating the high capacity of the three halophytes to decrease the soil sodium content. These results also agree with those of Zhao [10], who found that *Suaeda salsa* plants reduced soil $Na^+$ content at a depth of 0–10 cm by 2.4% with a density of 15 plants $m^{-2}$ and by 3.8% with a density of 30 plants $m^{-2}$ . He also reported that soil $Na^+$ content

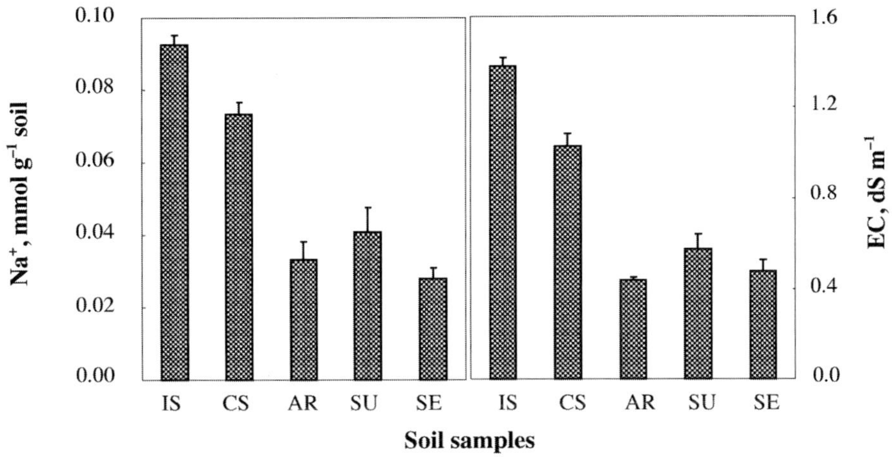


**Figure 1.** Soluble sodium content and electrical conductivity (EC) of aqueous extracts (1 g soil against 10 g distilled water) of saline soil samples taken from the upper 15 cm of 8-kg pots in which three halophytes [*A. indicum* (AR), *S. fruticosa* (SU), and *S. portulacastrum* L. (SE)] were grown over 170 days and constantly irrigated with tap water. IS and CS represent initial and control (irrigated without plantation) soils. Means of six replicates ± standard error at $p \leq 5\%$.

was more reduced at a depth of 20–30 cm. Our results also support the findings of Iwasaki [11], who found that the salt content of the upper 10 cm was reduced to less than one fifth of the original salt content after a single year of rice cultivation. Qadir and co-workers [3, 5, 12] has attributed such a decrease in soil salinity as well as in its sodicity to root effects on the soil in facilitating a $Na^+/Ca^{2+}$ substitution at the cation exchange sites and in ameliorating the soil physical properties, which facilitate drainage and consequently sodium leaching. However, he did not take into account a shoot contribution to bioremediation of salt-affected soils in addition to that of roots. Several authors [8–10, 13] have reported that plants that are able to accumulate sodium salts in their shoots could be used successfully to remove sodium from their substrates. Sodium ions are easily absorbed by the roots of the halophytes and transported to their shoots. High $Na^+$ concentrations, compared to potassium, were obtained with zero leaching in shoots of the three species (Fig. 2), indicating the ability of these plants to use sodium ions in the osmotic adjustment of their shoot cells; this was proved by the high water content of their aerial organs (Fig. 3). This result agrees with that of Messedi et al. [14], and Sleimi and Abdelly [15] who found a high capacity of sodium accumulation in *Sesuvium portulacastrum* and *Suaeda fruticosa*. Ravindran et al. [9] demonstrated that *Suaeda maritima* and *Sesuvium portulacastrum* accumulated 183.76 and 147 mg NaCl $g^{-1}$ dry weight (DW), respectively, in their shoots. In our study, values of shoot DW in each pot indicated that, among the three studied species, *Sesuvium portulacastrum* was the most productive and pro-

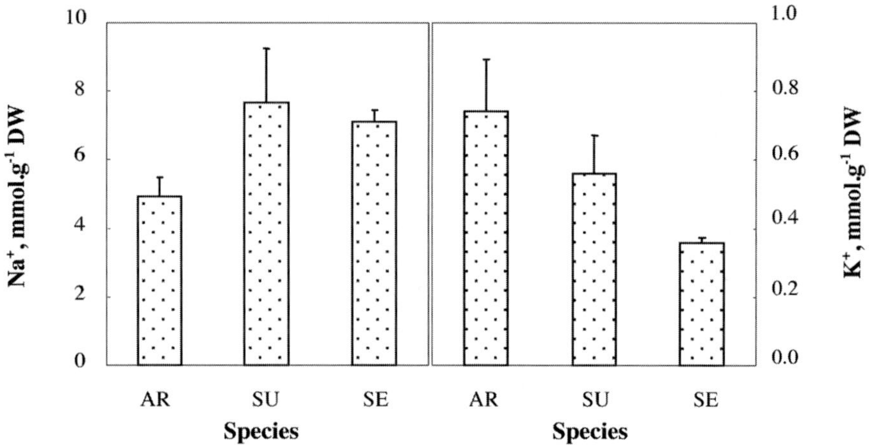


**Figure 2.** Sodium and potassium concentrations in shoots of three halophytes [*A. indicum* (AR), *S. fruticosa* (SU), and *S. portulacastrum* L. (SE)] grown over 170 days on a saline soil and constantly irrigated with tap water. Means of six replicates ± standard error at $p \leq 5\%$.

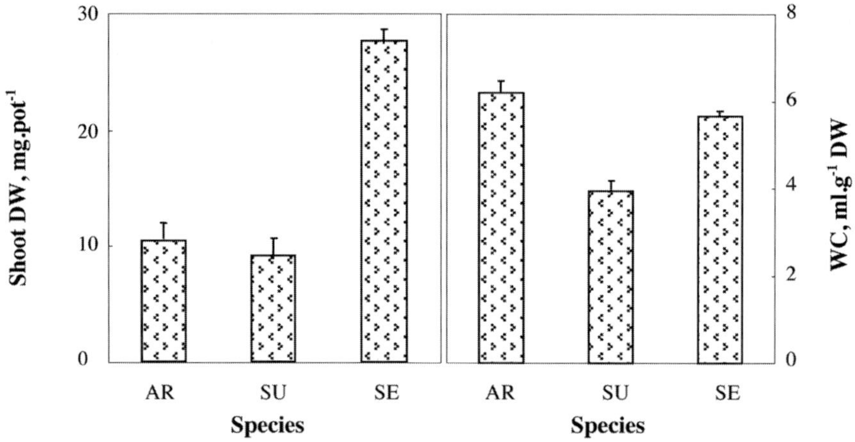


**Figure 3.** Shoot dry weight and their water content water in three halophytes [*A. indicum* (AR), *S. fruticosa* (SU), and *S. portulacastrum* L. (SE)] grown over 170 days on a saline soil and constantly irrigated with tap water. Means of six replicates ± standard error at $p \leq 5\%$.

duced 27 g DW pot$^{-1}$ over the 170 days, a value exceeding those obtained in *Arthrocnemum indicum* and *Suaeda fruticosa*. Thus, *Sesuvium portulacastrum* plants revealed the highest capacity to extract sodium from their root zones.

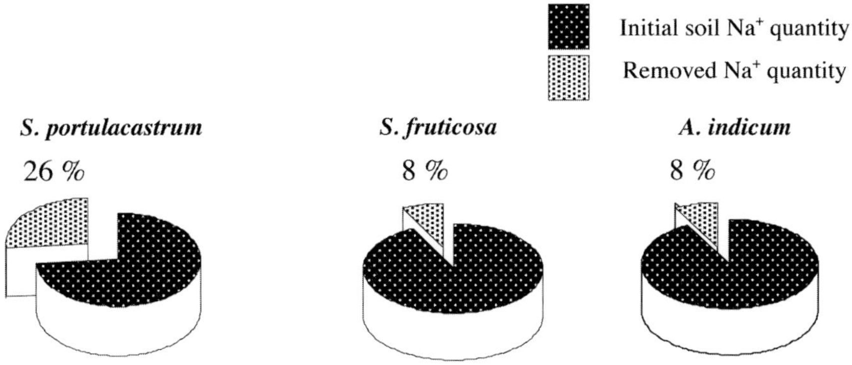


**Figure 4.** A comparison between desalination capacities in three halophytes. Results are percentages of quantities of extracted $Na^+$ from those initially present in the soil.

**Table 1.** Estimation of soil desalination capacity in three perennial halophytes grown over 170 days and constantly irrigated with tap water using the method of Tester and Davenport [16]

| **Species** | ***S. portulacastrum*** | ***A. indicum*** | ***S. fruticosa*** |
|---|---|---|---|
| DW per pot | 30 g | 10.50 g | 9.25 g |
| Exported Na quantity | 192 mmol $pot^{-1}$ | 55 mmol $pot^{-1}$ | 62 mmol $pot^{-1}$ |
| Per pot | 11.32 g $pot^{-1}$ | 3.217 g $pot^{-1}$ | 3.627 g $pot^{-1}$ |
| Exported Na quantity per hectare | 2504 kg $ha^{-1}$ | 711 kg $ha^{-1}$ | 802 kg $ha^{-1}$ |

It accumulated 26% of the original sodium of the soil over 170 days compared to 8% by *Arthrocnemum indicum* and *Suaeda fruticosa* (Fig. 4). These results indicate that *Sesuvium portulacastrum* is the most appropriate species for use on salt-affected soils in arid and semi-arid regions. With regard to biomass production, shoot sodium concentration, and the pot area, we estimated that *Sesuvium portulacastrum* was able to accumulate 2507 kg Na $ha^{-1}$ in 170 days (Tab. 1). These estimations are in agreement with those of Zhao [10] who found that *Suaeda salsa* produces about 20 tons DW $ha^{-1}$ containing 3–4 tons of salt. According to these estimations, *Sesuvium portulacastrum* can extract 14% of the salt that exists in the horizon 0–1 m of a soil having 10% water and a salt concentration that exceeds 200 mM NaCl [16].

In conclusion, *Sesuvium portulacastrum* seemed to be the most efficient plant to decrease soil salinity under laboratory conditions; however, field experiments should be carried out to elucidate the behavior of this plant on a salt-affected soil and its interactions with other species of the biotope.

## References


1 Summer ME (1993) Sodic soils: new perspectives. *Aust J Soil Res* 31: 683–750
2 Qadir M, Steffens D, Yan F, Schubert S (2003) Proton release by $N_2$–fixing plant roots: A possible contribution to phytoremediation of calcareous sodic soils. *J Plant Nutr Soil Sci* 166: 14–22
3 Qadir M, Oster JD (2004) Crop and irrigation management strategies for saline-sodic soils and waters aimed at environmentally sustainable agriculture. *Sci Total Environ* 323: 1–19
4 Shainberg I, Letey J (1984) Response of soils to sodic and saline conditions. *Hilgardia* 60: 76–120
5 Qadir M, Qureshi RH, Ahmad N (2002) Amelioration of calcareous saline-sodic soils through phytoremediation and chemical strategies. *Soil Use Manag* 18: 381–385
6 Qadir M, Ghafour A, Murtaza G (2000) Amelioration strategies for saline soils: A review. *Land Degrad Dev* 11: 501–521
7 Ponnamperuma FN (1972) The chemistry of submerged soils. *Adv Agron* 24: 29–96
8 Zahran MA, Abdel Wahid AA (1982) Contributions to the Ecology of Halophytes. *Tasks Veg Sci* 2: 235–257
9 Ravindran KC, Venkatesan K, Balakrishnan V, Chellappan KP, Balasubramanian T, Restoration of saline land by halophytes for Indian soils. *Soil Biol Biochem* 39: 2661–2664
10 Zhao KF (1991) Desalinization of saline soils by *Suaeda salsa*. *Plant Soil* 135: 303–305
11 Iwasaki K (1987) The effectiveness of salt-accumulating plants in reclaiming salinized soils. *Jpn J Trop Agric* 31: 255
12 Qadir M, Qureshi RH, Ahmad N (1996) Reclamation of a saline-sodic soil by gypsum and *Leptochloa fusca*. *Geoderma* 74: 207–217
13 Helalia AM, El-Amir S, Abou-Zeid ST, Zaghloul KF (1990) Bioremediation of saline-sodic soil by amshot grass in northern Egypt. *Soil Till Res* 22: 109–116
14 Messedi D, Labidi N, Grignon C, Abdelly C (2004) Limits imposed by salinity to the growth of the halophyte *Sesuvium portulacastrum*. *J Plant Nutr Soil Sci* 167: 720–725
15 Sleimi N, Abdelly C (2002) Growth and mineral nutrition of some halophytes under seawater irrigation. In: R Ahmad, KA Malik (eds): *Prospects for saline agriculture*. Academic Press, Dordrecht, 403–410
16 Tester M, Davenport R (2003) $Na^+$ tolerance and $Na^+$ transport in higher plants. *Ann Bot* 91: 503–5027

Biosaline Agriculture and High Salinity Tolerance
Edited by Chedly Abdelly, Münir Öztürk, Muhammad Ashraf and Claude Grignon


# New approaches for biosaline agriculture development, management and conservation of sandy desert ecosystems

Kristina N. Toderich[1], Shoaib Ismail[2], Ekaterina A. Juylova[1], Abdullo A. Rabbimov[3], Batyr B. Bekchanov[3], Elena V. Shyuskaya[4], Lilya G. Gismatullina[1], Kozan Osamu[5] and Toshpulat B. Radjabov[1]

[1] *Department of Desert Ecology and Water Resources Research, Samarkand Division of the Academy of Sciences of Uzbekistan. 3, Timur Malik str., 703000 Samarkand, Uzbekistan*
[2] *International Center for Biosaline Agriculture (ICBA), PO Box 14660, Dubai, UAE*
[3] *Uzbek Institute of Karakul Sheep Breeding and Desert Ecology Research, 47, Mirzo Ulugbek street, 703054, Samarkand, Uzbekistan*
[4] *K.A. Timiriazeva Plant Physiology Institute of Russian Academy of Sciences. Botanicheskaya st. 35, Moscow, 127276, Russia*
[5] *Disaster Prevention Research Institute, Kyoto University, Japan*

## Abstract

The major factors and current trends of soil salinization and rangelands degradation in the arid/semiarid zones of Aral Sea Basin were discussed. The bioremediation of abandoned saline lands using natural marginal resources could be considered as one of a number of strategies that can be employed to bring these lands back to their full production potential. Different ecological groups of halophytes were characterized according to taxonomy, mineral composition and salt tolerance. A positive relationship between mineral content of wild halophytes and biomass has been identified. Ion contents of evaluated wild native halophytes were relatively low and hence these species could be recognized as alternative forages, both in pure halophytic pastures and/or in mixture grass stands. Most of evaluated halophytes being late flowering and seed maturating species are recommended as a fattening feed during autumn and winter seasons, when there is a deficit of forages on the pasture lands. Introduction of strip cropping system represent an alternative for private farms in the livestock-based farming system, as well as a way to diversify feed resources under unfavorable environments. It also leads to the uniform distribution of good quality feed resources throughout the year and during difficult periods while preserving soils, water and phytogenetic resources. Another technique used in the salt-affected sandy desert environments is to plant shrubs as windbreaks to spare the land for other crops and help protect the soils from wind erosion and sand encroachment.

## Introduction

Degradation of desert rangelands throughout the whole Central Asian region has reached an alarming degree, calling for prompt action. The increasing human population and expanding agricultural areas have resulted in heavier grazing pressures in rangelands in spite of the increasing role of crop residues and grains in livestock production [1, 2]. As a result of erratic cropping in low rainfall zones, overgrazing of the good rangelands, cutting of shrubs by the local population for firewood, the natural vegetation of these desert areas are under pressure from anthropogenic degradation factors. This leads to the eradication of useful, endemic or rare desert plant species and to the reduction of rangelands productivity. All the above have induced the disintegration of rural infrastructure in arid zones of Zarafshan River Basin and Kyzylkum Desert that has led to the migration of local population from the native areas to neighboring cities or countries.

Salinization is one of the major ecological and production problems currently facing the agricultural sector in arid and semi-arid areas of the Central Asian countries. Continuous use by the former USSR of the major rivers of Central Asia (Amu Darya, Zarafshan, Syr Darya) for production and exports of cotton, oil, and minerals, has resulted in rising water tables, waterlogging, and the saline lands around the Aral Sea. Secondary salinization (human induced) is increasing rapidly and crop production under these conditions is becoming less sustainable. Drought and salinity can have a far greater effect on food security in Central Asian than in other areas. Besides salinization, contamination by heavy metals and chemical compounds released by agriculture, uranium, oil and gas industries has been frequently reported for Kyzylkum sandy Desert [3–10]. As the result, the formerly high productive livestock system has deteriorated and livelihoods of the people have dramatically declined. In Uzbekistan, natural desert pastures valuable for livestock development occupy 17.5 million ha; however, only 485.1 thousand ha are presently under irrigated agriculture. Saline pastures cover about 2 million ha. The main vegetation distributed here are halophytes, which have importance as feed, fodder, technical, medicinal plants, etc. These phyto-genetique resources might play very important role for soil desalinization, water table control, and valorization of non-conventional water resources, landscaping purpose and sand-dune fixation. Several options are available for improving the livelihoods of the rural population in the Central Kyzylkums. The hot water from many of the artesian wells could be used not only for development of arid fodder production, but also for recreation, vegetable production, and other purposes. The creation of highly productive systems for arid livestock fodder farming will entail the safety of natural habitat and increase the income levels of rural communities.

This paper is an attempt to contribute to a better understanding of the very difficult arid environments in Uzbekistan, and of the conservation and rational use of halophytic rangelands resources to encourage the sustainable development of saline/sandy deserts soils by mobilization of phytogenetic resources, involving both native and introduced salt-tolerant plants. Studies on their ecological, morphological and structural-functional properties, seed and biomass production, and testing of alternative low-cost technologies to optimize the selection and domestication of halophytic arid plants are different aspects for model studies for prediction of rangelands productivity. Another significant goal if that of salinity control, remediation and economic development of arid/saline lands. One of the most promising uses of halophytes will be the production and conservation of important seeds germplasm. The demand for seeds of salt-tolerant species has increased and a number of farmers have become interested in applying biosaline agriculture techniques as a feasible option for their marginalized farms. An innovative selection programs and development of suitable modern agro-technologies are needed to multiply seeds and/or salt-tolerant plant material, establish them within natural plant communities and introduce them into different ecosystems in which they are suitable. The outline of the general strategies for natural rangelands regeneration and management in new changing desert/semi-desert environments are proposed and modeled.

The main objectives for implementation of such studies are:

- To select appropriate plant materials from natural halophytes with high production value (both green biomass and seeds) with high nutritional value
- To evaluate optimal agronomic practices of different plant groups based on their reproductive biology, biochemistry, and physiology
- Study mixed cropping system of $C_3/C_4$ plants in different planting ratios and their effect on seasonal biomass production under arid saline conditions from biosaline agriculture perspectives
- Efficient use of non-conventional water resources (hot artesian/ground mineralized water), and forage alternatives, while protecting natural resources (soil and biodiversity of natural rangeland plants).

## Materials and methods

The study area covers waste marginal lands and natural rangelands affected by aridity and salinity mostly located in the lower reaches of Zarafshan River Valley and Kyzylkum Deserts, including Karakata saline depression. The target area (Fig. 1) constitutes a salt depression formed by freely flowing saline hot artesian water (vertical drainage water), which is the only water source available for cultivation under the sandy Kyzylkum desert conditions. Unless the water is used in halophyte production, it would induce further salt accumulation and soil deterioration.

Soils collected from different points (0–100 cm) of surveyed area were submitted for detailed physical and chemical analysis using atomic adsorption

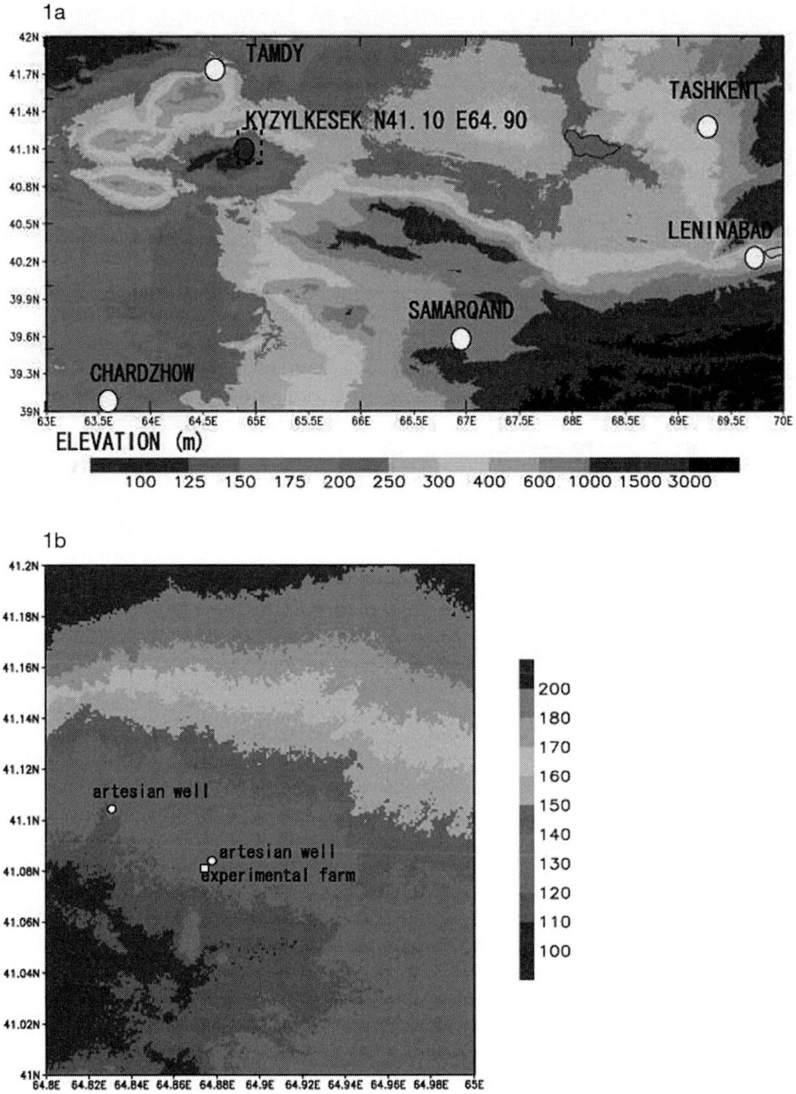


**Figure 1.** (a, b) Altitude maps in large and small scale. The dotted square in large scale map indicates the left figure area.

spectrophotometry. Total concentrations of trace elements were analyzed by an ICP-MS (Hewlett Packard HP 4500 system) in the Radioisotope Research Center, Kyoto University. The electric conductivity (EC) as well as the soluble cations and anions were determined in aqueous extracts (soil:water 1:10 w/w). All the data were statistically analyzed using "Statistics 9.0", "Surfer v.7.04" and ArcGIS 9.1.

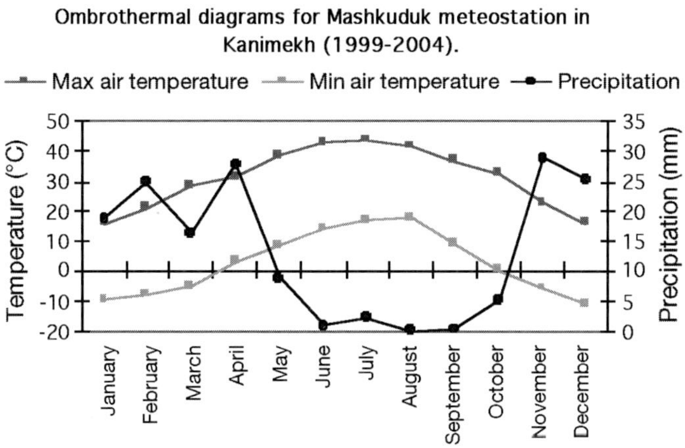


**Figure 2.** Climatic conditions for the Kanimekh agroecological zone (Central Kyzylkums, Uzbekistan).

A trial for a strip livestock-farming system was established at the Kysylkesek Experimental Site, Kanimekh district Central Kyzylkum), where the average annual rainfall varies between 80 and 110 ml. Strips of different forage halophytes were cultivated, each separated by a row of fodder shrubs. Each strip was 5–10 m wide.

## Agro-climatic conditions and chemistry of soils and underground water

The agro-climatic environments and availability of water sources for cultivation of crops on the desert salt-affected soils are radically different from the agricultural irrigated areas. Climatic conditions for the Kanimekh agro-ecological zone, the Navoi region (Central Kyzylkum), which was chosen as a biomonitoring site of salinity trends and halophytic botanic diversity changes, are shown in Figure 2.

Spatial distribution (zonation) of natural vegetation in relation to soil salinity was studied in an area of about 10 ha, located between two artesian hot springs (Fig. 3). The most important factor for the zonation was soil salinity and the salt-tolerance limits of species. In general, the width of each zone, as is seen in the Figure 3, differs according to the relief, floristic composition, and the salt concentration.

The soil type of the surveyed area is silt-sandy loam throughout the profile up to a depth of 60 cm. Representative soils samples were collected at different depths from 0–100 cm. The soil was highly saline in the topsoil and in the lower layers. The predominant salinity type is sulfate-chloride. Ground water salinity varies from 2.0 to 8.2 g/l. Sodium and magnesium are the dominating cations. The ground water table fluctuates from 0.8 to 2.5 m during May-July at the experimental agricultural plot and up to 5–8 m in the desert pasture

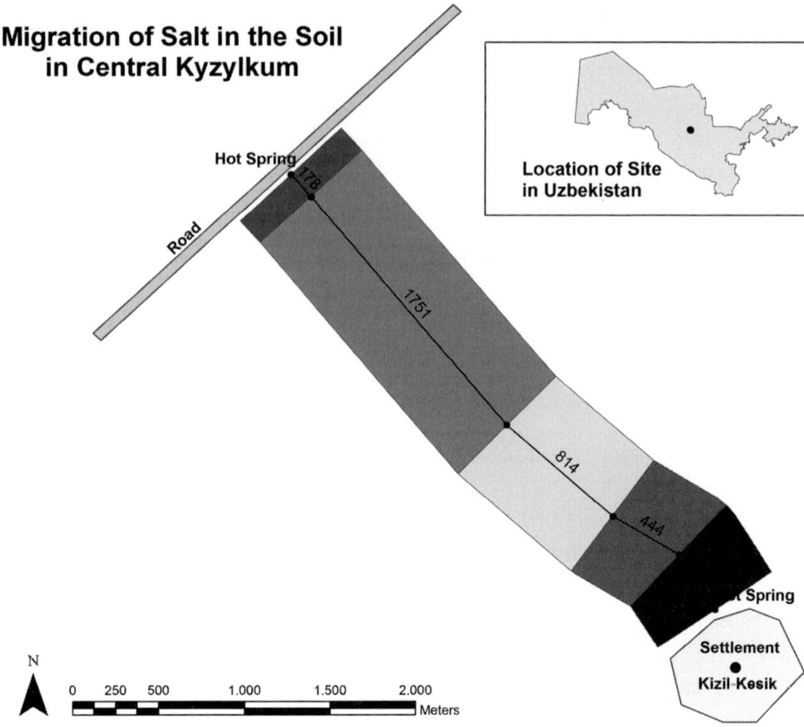


**Figure 3.** Spatial distribution of different desert landscapes according salinity gradient: Low salinity in the virgin desert *Artemisia* rangelands (slight grey color); moderate salinity under artificial *Haloxylon aphyllum* forest (white) and high salinity (salt marshes solonchaks with *Tamarix*, *Halostachys* and *Aeluropus* plant community (black color).

area. The dynamics of soil moisture under different plant communities, which are distributed in the degraded pastures, are shown in Figure 4.

Due to low transpiration capacity, *Alhagi* plant communities promote retention of soil moisture in the topsoil. The same situation is also observed on saline areas near agricultural plots, where *Climacoptera* and others annual *Salsola*'s species dominate. Additionally, the high evaporation rates dries the ponds in summer. The poor natural drainage system of marginal cropping irrigated lands has caused an increase in the salt content at the superficial crust and groundwater that induces secondary salinization of the soils. Very intensive processes of soil salinization occur around the artesian wells. The introduction of salt-tolerant wild halophytes has supported studies of groundwater with reference to salinity sources and irrigation management.

Average EC of the irrigation water (artesian hot spring) varies between 8.30 and 18.1 dS/m, pH between 7.3 to 8.1 (Tab. 1) The dominant cation is $Na^+$ and the dominant anion is $SO_4^{2-}$.

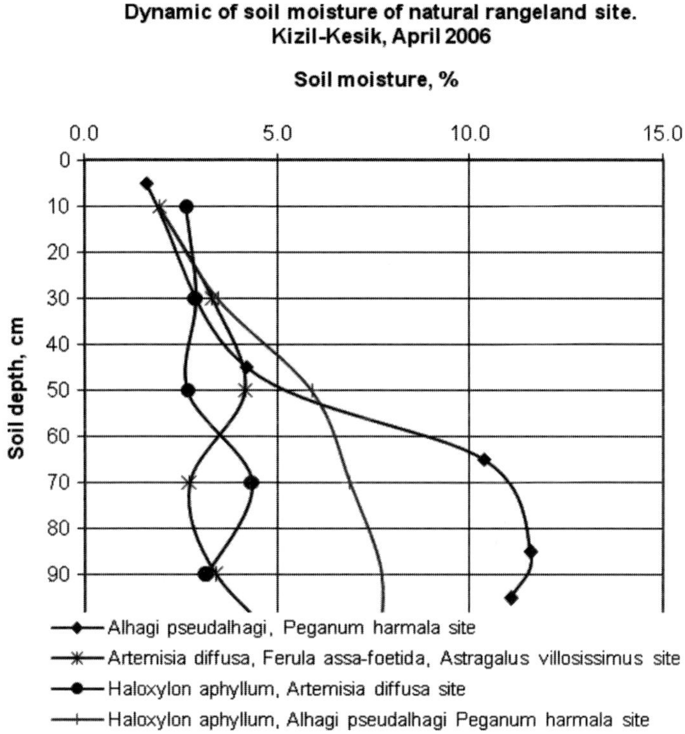


**Figure 4.** Dynamics of soil moisture under different desert plant communities.

**Table 1.** Chemical properties of the drainage and ground water of Uzbekistan (Observations are for August 2006 period).

| Site ID | Chemical properties | | | | | | | | | |
|---|---|---|---|---|---|---|---|---|---|---|
| | pH | EC | TDS | Soluble cations | | | | Soluble anions | | |
| | | | | $Ca^{2+}$ | $Mg^{2+}$ | $Na^{+}$ | $K^{+}$ | $HCO_3^-$ | $Cl^-$ | $SO_4^{2-}$ |
| | | dS/m | mg/l | ppm | ppm | ppm | Ppm | ppm | ppm | ppm |
| Drainage water | 7.93 | 18.1 | 8800 | 584.3 | 364.80 | 1778.40 | 7.00 | 268.5 | 1996.9 | 3695.7 |
| Drainage water | 7.85 | 17.8 | 8750 | 574.3 | 352.64 | 1788.94 | 7.00 | 262.4 | 1996.9 | 3607.3 |
| Drainage water | 8.10 | 12.5 | 9000 | 592.3 | 370.88 | 1776.98 | 7.00 | 262.4 | 2196.7 | 3697.4 |
| Artesian irrigation water | | | | | | | | | | |
| Well 1 | 7.40 | 10.9 | 2654 | 184.1 | 91.20 | 565.11 | 8.00 | 176.9 | 621.2 | 1014.6 |
| Well 2 | 7.30 | 9.86 | 2716 | 184.1 | 85.12 | 574.31 | 8.00 | 164.8 | 621.2 | 1018.3 |
| Well 3 | 7.40 | 8.30 | 2826 | 190.1 | 93.69 | 585.58 | 8.00 | 176.9 | 621.2 | 1078.7 |

TDS: Total dissolved salts

The drainage water is ~3.5 times more saline than irrigation water. The micro element composition for the Kyzylkesek water is: Sr > Ba > Ti > Mn > Cr > Ni > Cu > Mo > Pb.

## Floral composition of vegetation of salt-affected lands and evaluation of halophytic germplasm

Studies of the region undertaken by different authors have shown presence of different ecological habitats and high plant diversity along the Zarafshan River Valley (old agricultural zone) and Kyzylkum Desert [4, 6, 11, 12]. During the present survey more than 380 species of different groups of salt-loving plants (wild halophytes representing 19 taxonomical families) were described. The study areas show a high endemism in plants (about 3.4% of the total species). Most noticeable is the relative richness of the Chenopodiaceae, comprising nearly 33%, equivalent only with Australia chenopods. It is also quite rich in Asteraceae (20%), Poaceae (11%); Fabaceae and Brassicaceae (about 11%). Species belonging to Polygonaceae, Plumbaginaceae, Zygophyllaceae, Cyperaceae account for a smaller share (3–5%), whereas Eleagnaceae, Plantagainaceae and Frankeniaceae make up an even smaller part (<1.0%) of rangelands halophytic pastures. Among cited plant resources, there are a number of native and exotic halophytes, both $C_3$ and $C_4$ plants, suitable for reclamation of arid and semi-arid, salt-affected and waterlogging areas, which have been proven very useful in demonstration trials. These areas have not yet been widely used as part of the arid production system of Uzbekistan by the pastoralists and farmers.

These species grow well in association with a variety of species but often provide severe competition to perennial species, both in natural and introduced pastures on saline and disturbed mine sites. However, many of the germplasm are on the verge of disappearing due to overgrazing, which may result in an irreversible loss of biodiversity resources. Most populations of the desert halophytes within the flora of the Uzbekistan are local and small, sometimes fragmented. They frequently have incomplete life cycles with little ability to reproduce, and low indices of renewal and replacement.

Each of zones consists of different dominant species. In the first zone, where there is a high mineral content, species of genus *Salicornia*, *Halostahys*, *Halimocnemis*, *Climacoptera* are largely distributed. The vegetation period begins fairly late because the marshes are under water for a large part of the year. The conventional salt-tolerant crops (sorghum, pearl millet, fodder beet, safflower etc.) as shown in Figure 5 occupy an intermediate place between true halophytes and xero-halophytes.

Based on soil characteristics, water table level, mineral composition of above-ground biomass, morphological and reproductive traits, and carbon

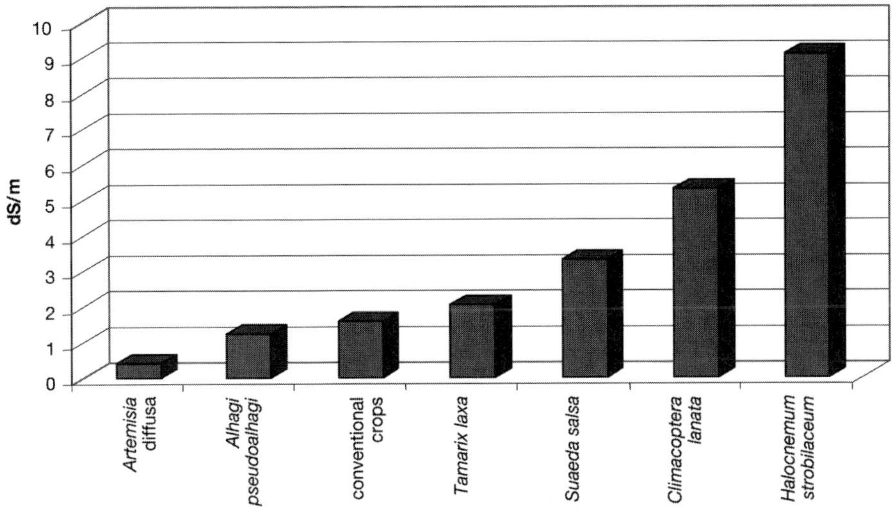


**Figure 5.** Ecological row of plant species distribution including conventional crops by increasing salinity in the irrigation water.

discrimination values, a new concept for classification of halophytes was developed. Based on these, different ecological groups of halophytes within the flora of Uzbekistan could be differentiated.

*Group I: Hyper-halophytes*

These group of plants are characterized by their presence with highly saline water (full-strength seawater; up to 100 dS/m), where the water table varies from 0.5 to 1.5 m depth, with solonchak-alkaline and solonetz soils. Plants are succulent and have both $C_3$ and $C_4$ types of photosynthesis.

The main species of this group include *C. longistylosa*, *C. kasakaroum*, *C. bucharica*, *C. crassa*, *C. subcrassa*, *S. transoxana*, *C. ferganica*, *C. aralensis*, *C. turcomanica*, *C. turgaica*, *C. itricata*, *C. turkestanica*, *Plantago coronopus* L. ($C_3$ photosynthesis), *Salicornia* species ($C_3$ photosynthesis), *Halostachys belangerana* (Moq.) Botsch, *Halocnemum strobilaceum* (Pall.) M.B. ($C_3$ photosynthesis); *Petrosimonia crassifolia* (Pall.) Bunge, *Petrosimonia litwinowi* Korsh, *Halocnemis varia*, and *Halogeton glomeratus*.

*Group II: Hydro-halophytes*

Species within this group that has $C_3$ photosynthesis occur in wet to standing water, varying from freshwater to brackish water marshes, and ditches as well as around seeps and springs. It can tolerate up to 10 000 ppm, and beyond that. The main species include *Phragmites australis* (Cav.) Trin, *Phragmites communis* Trin, *Arundo* spp., *Typha* ssp. *Hippophae rhamnoides*, *Populus diversifolia*, and *Elaeagnus angustifolia*.

### *Group III: Euhalophytes*

This group of plants is a good indicator of superficial distribution (1–2 m) of underground water; and mainly consists of salt-accumulator and salt-excluder plants. They occur in wet sandy areas at the edge of salt flats, marshes and salt deserts; and in wet-marsh salted soils on the margin of lakes, tugai, salted desert depression, and takyrs. The vegetation mainly comprises of *Suaeda*, annual *Salsola*; *Aeluropus repens* (Desf.) Parl., *Aeluropus littoralis* (Gouan) Parl, *Aeluropus villosus* Hook, *Poa littoralis* Gouan; *Tamarix*; *Dactylis littoralis* (Gouan) Willd., and *Kochia scoparia*, which tolerates full-strength seawater. The plants mostly have $C_4$ photosynthesis.

### *Group IV: Halo-xerophytes*

These plants mainly occur where the water table is at >4 m in depth on areas of different types of soils (from sand to clay): gray with gypsum content, alkaline meadow salt marshes, and sandy desert soils. *Haloxylon aphyllum*, perennial *Salsola* spp., *Ephedra strobilaceae, Halothamnus subaphylla, Campharosma lessingii, Kochia scoparia, Zygophyllum* sp., *Alhagi pseudoalhagi, Lycium turcomanicum, Lycium ruthenicum*, some *Calligonum* sp.; *Ceratoides ewersmanniana, Anabasis annua, A. salsa, Anabasis aphylla* L., *Anabasis eriopoda* (Schrenk.) Benth, and *Anabasis ferganica* (Drob.) are the major contributors of this group.

### *Group V: Halogemimezophytes*

This group of plants often occur where the water table varies between 1.5 and 2.5 m in depth, and inhabits steppes, semi-desert and desert zones mostly on solonetz-alkaline soils, lake shores and river banks. *Cynadon dactylon, Limonium gmelinii, Salsola arbuscula, Karelinia caspica, Frenkenia* spp., *Zygophyllum fabago, Halimodendron halimodedron, Agropyron desertorum, Eremopyrum orientalis, Psylliostachys suvorovii, Atriplex tatarica, Bassia hyssopifolia, Glycyrrhiza glabra*, and *Limonium otolepis* are the major dominants of the group. It includes both $C_3$ and $C_4$ plants.

*Group VI: Halo-gemipetrophytes*

These are found where the water table varies between 1.5 and 4.0 m where the plants grow on a stony skeletal saline substrate. This is a small group of plants mostly shrubs, semishrubs and semishrublets and is represented by: *Haloxylon ammodendron, Salsola arbusculiformis*, and some species of *Atraphaxis, Nanophyton*, and *Anabasis*.

*Group VII: Metallo-halophytes*

Most hyper-accumulators of heavy metals and ions grow slowly and have a smaller biomass, delayed flowering and low seed quality [8, 9]. Most distinguishable species within the flora of Uzbekistan are *Artemisia diffusa* H. Krasch, *A. halophyla, Karellinia caspica, Allysum desertorum, Tamarix hispida*, and *Frankenia translocate*; they accumulate toxic metals from soils into the harvestable portions of roots and surface biomass (shoots, leaves, etc.).

**Mineral value of halophytic edificators**

Studies have provided some understanding of processes involved in the accumulation of essential elements, but little is known about the mechanism(s) of mobilization, uptake and transport of environmentally hazardous heavy metals [7, 9, 13–15]. Many reports indicate that the plants containing higher concentrations of salts are toxic for livestock and responsible for different types of diseases and physiological disorders [16].

Our investigation on chemical composition of desert plants for ions such as $Cl^-$, $SO_4^{2-}$, $HCO_3^-$, $Na^+$, $K^+$, $Ca^{2+}$, $Mg^{2+}$, as well as phosphorus and iron, have shown significant changes within different halophytic forage species. The naturally growing plants, e.g., *Halocnemum strobilaceum, Tamarix hispida, Climacoptera*, and *Halothamnus subaphylla* contains higher $Na^+$ concentrations near the critical limit for livestock, while legumes (*Alhagi pseudoalhagi*) and some graminous fodder grass mostly accumulate $K^+$ (Fig. 6).

Concentration of salt ions, phosphorus and general iron in the dry matter biomass of all species varies depending on physical and chemical properties of soils and irrigation water, as shown in Figure 7a–d for *Alhagi pseudoalhagi, Tamarix laxa, Poaceae* spp. and *Climacoptera lanata* inhabiting the bio-monitoring points along Zarafshan River and Kyzylkum Desert.

Highest concentration of $K^+$ was found in *Kochia scoparia*, closely followed by *Atriplex nitens* and *Suaeda arcuata*. Figure 6 shows that the total mineral ions were maximum in *Halocnemum strobilaceum*, followed by *Climacoptera lanata, Suaeda salsa, Salsolas* spp., *Zygophyllum* spp., *Tamarix laxa, T. hispida* and *Haloxylon aphyllum*. On the basis of salt concentration of the main cations ($Ca^{2+}$, $Na^+$, $K^+$, $Mg^{2+}$) and anions ($SO_4^{2-}$, $Cl^-$, $HCO_3^-$), *Alhagi pseudoalhagi,*

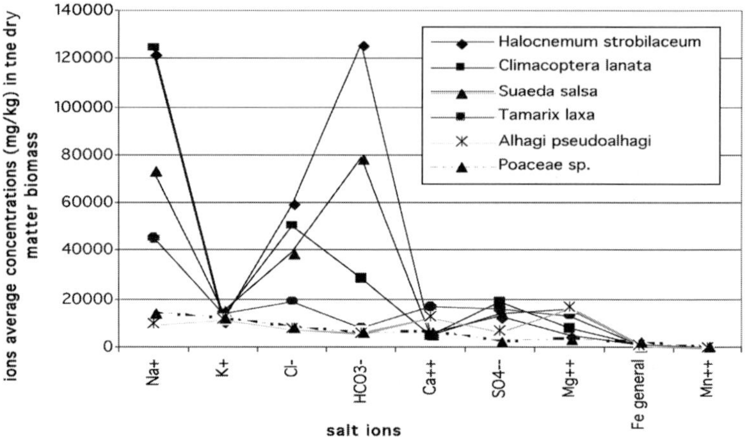


**Figure 6.** Changes in chemical composition of dry matter biomass of various halophytic forage species growing under sandy saline desert environments.

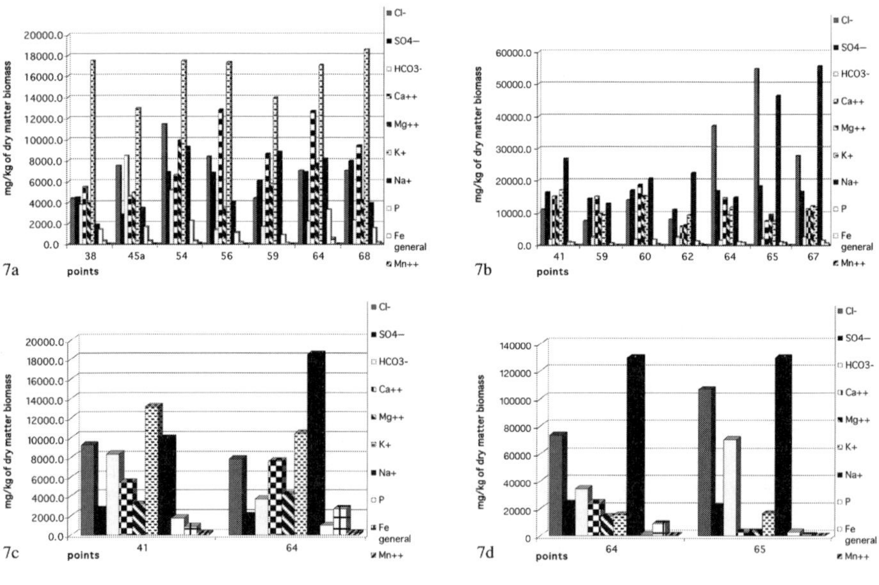


**Figure 7.** Concentration of salt ions, phosphorus and general iron in the dry matter biomass of (a) *Alhagi pseudoalhagi*, (b) *Tamarix laxa*, (c) *Poaceae* spp. and (d) *Climacoptera lanata*.

*Poaceae* spp., and *Artemisia diffusa* have a minimum concentration of mineral ions and could be categorized as relatively more palatable as forage plants. However, the fresh biomass of *Alhagi pseudoalhagi* from highly saline habitats sharply decreases with the increasing of gradient of salinity (Fig. 8a, b).

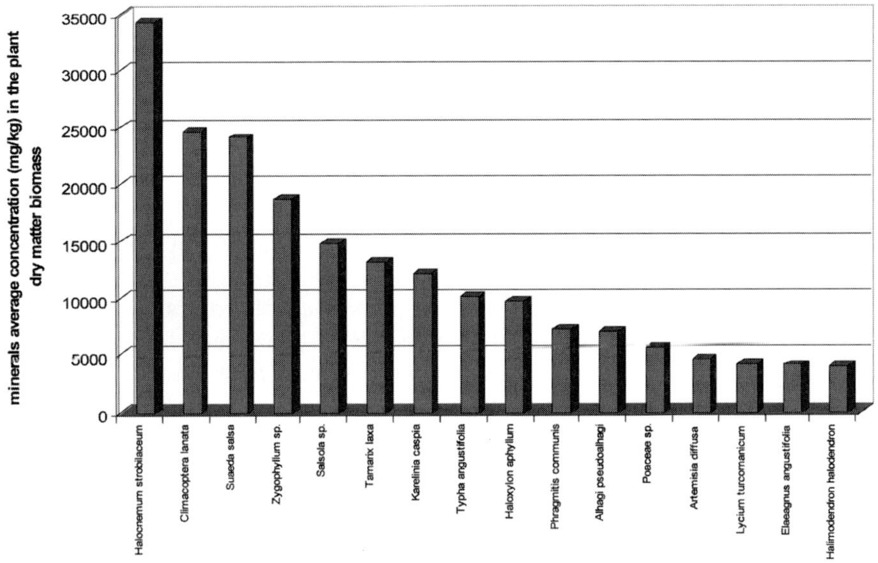


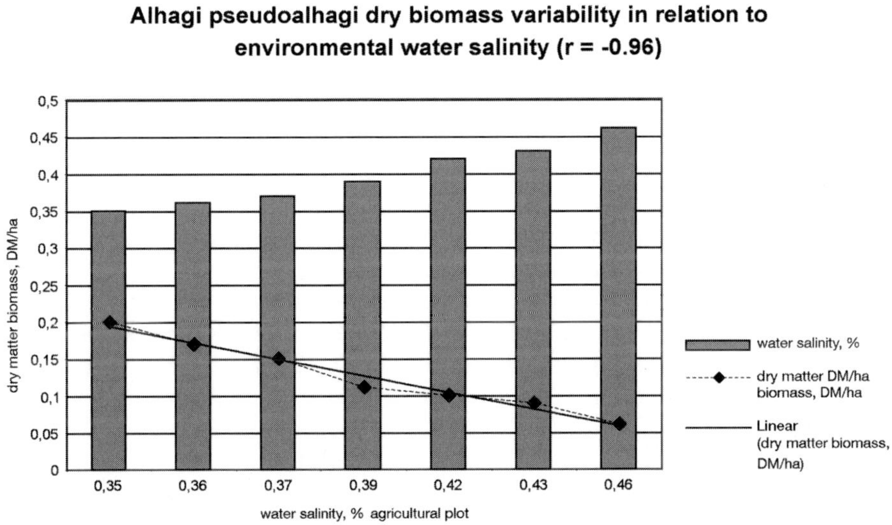


**Figure 8.** (a, b) Zarafshan River Valley and Kyzylkum Desert inhabit plant species distribution row by mineral accumulating ability (a); changes of dry biomass of *Alhagi pseudoalhagi* in relation to soil salinity (b).

*Alhagi pseudoalhagi, Poaceae* spp. and *Artemisia diffusa* are more suitable as animal fodder than other salt-tolerant plants. These species can be recommended for direct grazing or feeding. Salt bushes also maintain relatively higher amounts of Ca and Mg. Potassium concentration was found at higher levels in *Kochia scoparia* and *Agropyron desertorum*, closely followed by *Atriplex nitens* and *Suaeda salsa*, while annual *Salsola* and grasses like *Bromus*, *Aeluropus*, and *Eremopyrum* contain low amount of mineral ions because excess salts are excluded through salt glands present abundantly on the surface of epidermis, especially of the latter-mentioned species.

## Livestock farming-cropping system in sandy saline desert environments

Planting herbaceous fodder crops in strips between fodder shrubs on intensive agro-forestry plantations could solve the karakul sheep feeding problem in degraded (both by overgrazing and salinity) sandy desert areas. In addition, wild halophytes species planted in widely spaced patterns (15–25 m) allows for easy mechanical cultivation and harvesting of grass and cereals. Salt-tolerant trees and/or shrubs species, e.g., *Populus, Salix, Elaeagnus, Morus,* and *Hippophae rhamnoides* established on good deep soils have good potential as part of the production system. In the present case, fodder shrubs were associated with a cereal farming system, including rangeland species alone, or mixed with different salt-tolerant fodder crops. As part of the desert land revegetation, saltbushes *Atriplex canescens, A. undulata, A. nummularia* and *A. amnicola* were recently introduced at the saline sandy desert zones of Kyzylkum. Seedlings were produced at the nursery of the Plant Industry Institute, Uzbekistan, and were transplanted to the site under appropriate management in the fall period.

Tables 2 and 3 indicate the mean yield of native halophytes in mixture with different promising salt-tolerant crops during 2005–2006. Better plant growth, accumulation of green biomass and consequently yield of both fresh and dry matter were significant for *Kochia scoparia, Climacoptera lanata*, and *Atriplex nitens* grown in a mixture with salt tolerant crops.

There were significant differences in the yield of fodder mass and seeds between annual and perennial species of halophytes. Yield of all tested species decreased with increased planting density. Use of fertilizers also affected the biomass yield.

## Conclusions

The *Poaceae* species, *Artemisia diffusa* and *Alhagi pseudoalhagi* are more suitable as animal fodder than other salt-tolerant plants. Grass species can be recommended for direct grazing or feeding. For instance, *Agropyron desertorum* in mixture with *Kochia scoparia, Climacoptera lanata*, and *A. nitens* in pure stands and in mixture with many forage perennial halophytes were introduced under irrigation (with artesian mineralized water), and can be recommended to farmer's for cultivation. This would create a livestock grazing system and also form a fodder reserve on salt-affected wastelands. It was

**Table 2.** Plant growth and productivity of annual halophytic pastures irrigated with artesian mineralized water at Kyzylkum site. Plantation was done in 2006 .

| **Species/ Acc. No.** | **Date of seed bedding** | **Number of plants (1000/ ha)** | **Height of plants (cm)** | **Yield of fodder mass (T/ha)** | | **Yield of seeds (T/ha)** | **Vegetative period (Days)** |
|---|---|---|---|---|---|---|---|
| | | | | **Wet** | **Dry** | | |
| *Kochia scoparia* | | | | | | | |
| K-599 | 14.03.06 | 262.9 | 160.7–192.0 | 29.5 | 14.6 | 1.19–4.60 | 82–96 |
| | 14.03.06 | 231.3 | 101.0–169.0 | 21.0 | 11.0 | 1.37–2.40 | 80–85 |
| K -598 | 14.03.06 | 214.6 | 148.0–188.0 | 28.1 | 9.5 | 2.80–2.92 | 78–80 |
| Control/ Kyzylkum population | Self-reproduction | 301.4 | 121.0–160.0 | 15.6 | 7.8 | 0.75–1.34 | 80–101 |
| *Atriplex nitens* | | | | | | | |
| K-550 | 14.03.06 | no data | 121.0±3.9 | 21.0± 5.8 | 9.6± 3.8 | 2.90–3.60 | 125–135 |
| K-632 | Self-reproduction | no data | 140.0±1.2 | 190 ± 4.1 | 9.1 ± 3.2 | 2.08–4.30 | 136–148 |
| K-620 | Self-reproduction | no data | 180.0±5.4 | 21.0±7.1 | 9.69±5.8 | 2.96–5.60 | 140–150 |
| *Climacoptera lanata* | | | | | | | |
| K-621 | 25.02.06 | 173.3 | 47.3±0.7 | 58.8 | 17.9 | 0.09–0.45 | 180–220 |
| K-602 | 25.02.06 | 113.3 | 75.3±0.9 | 114.6±2.3 | 21.1 | 1.43–1.56 | 190–232 |
| *Climacoptera* + Sorghum | 14.04.06. | 146.0<br>620.0* | 85.0±3.4<br>205.0±4.9 | 175±6.9<br>22.33±2.6 | 8.85±4.6<br>13.4±4.9 | 3.90–8.48<br>5.10–6.85 | 190–235<br>125–136 |
| *Climacoptera* + Sorghum (+ Fertilizer) | 10.04.04 | 104.5<br>138.0 | 73.3±2.2<br>210.0±7.8 | 96.8± 2.9<br>68.0±10.1 | 6.17± 3.4<br>23.8±9.3 | 1.69–4.2<br>8.30–12.47 | 180–210<br>110–125 |
| *Climacoptera* + *Zea maize* | 10.04.06 | 135.0<br>212.0 | 67.7± 0.6<br>250.0±3.9 | 61.6 ± 2.1<br>13.5±4.6 | 7.04± 3.6<br>25.5±7.1 | 1.92–3.6<br>6.40–7.22 | 180–210<br>95–130 |
| *Climacoptera* + *Zea maize* (+ Fertilizer) | 10.04.06 | 153.3<br>234.0 | 89.0±6.5<br>280.0±6.2 | 45.0±6.9<br>42.0±8.1 | 22.0±5.6<br>19.8±4.9 | 4.90–13.2<br>5.40–10.20 | 195–205<br>85–110 |
| Salt depression (control) *Climacoptera* pure stands | Self-reproduction | 51.1 | 46.7±1.7 | 62.2 | 11.5 | 0.80–1.00 | 210–235 |
| *Suaeda arcuata* | | | | | | | |
| Salt depression | Self-reproduction | 15.3 | 113.6 | 54.4 | 13.6 | 0.57–1.09 | 170–190 |

* Values mentioned for graminous crops; Sorghum or Zea mays

**Table 3.** Plant growth and productivity of perennial halophytic pastures irrigated with artesian mineralized water at Kyzylkum site. Plantation was done in 2006.

| Name of plant species | Date of seed bedding | Number of plants (1000/ha) | Height of plants (cm) | Yield of fodder mass (T/ha) | | Yield of seeds (T/ha) | Veg-etative period (Days) |
|---|---|---|---|---|---|---|---|
| | | | | Fresh | Dry | | |
| *Salsola orientalis* | Self-reproduction (II–III) | 9.29–12.3 | 75–155 | 2.0–2.8 | 1.0–2.2 | 0.05–0.15 | 250–254 |
| *Halothamnus subaphylla* | 20.10.05 | 9.25–10.28 | 109–150 | 1.9–2.06 | 1.24–2.1 | 0.90–0.93 | 235–250 |
| *Kochia prostrata* | Self-reproduction | 17.22 | 94–103 | 2.6–3.09 | 2.06–2.16 | 0.30–0.67 | 250–260 |
| *Camphorosma Lesingii* | Self-reproduction (II-III) | 14.98 | 88–92 | 1.6–2.07 | 1.2–1.6 | 0.02–0.21 | 230–235 |
| *Atriplex canescens* K-4773 | 14.03.06 (seedlings) | 104 | 35.1–35.3 | 6.19±5.3 | 2.05–2.25 | | 220–238 |
| *Agropyron desertorum* | 14.03.06 | 812 | 125–157 | 0.8–1.5 | 0.40–0.60 | | 105–108 |
| *Alhagi pseudoalhagi* | Self-reproduction (II-IY) | 960/1004 | 58–161 | 0.8 5–2.40 | 1.24–1.60 | 0.10–0.19 | 218–225 |
| *Glychyrriza glabra** | | | | | | | |
| K-609 Karakalpakstan ecotype | 25.02.06 | 231–345 | 45–110 | 4.10–7.90 | 1.27–3.10** | 0.79–0.9 | 216–220 |
| K-603-Mirzachuli steppe | 25.02.06 | 380–416.0 | 60–150 | 8.40 | 3.46 | 0.46 | 190–205 |
| *Hippophae rhamnoides* K-656 | 25.02.06 | 280–310 | 68–105 | – | – | – | 205–220 |

*Expected yield of dry roots 2.5 t/ha.

**Yield of fodder mass at the first year of vegetation.

found that seed sowing mixture of *Kochia scoparia* and *Agropyron desertorum* in association with wild growing *Alhagi pseudoalhagi* could produce up to 1.3 t dry matter/ha.

Salt bushes and perennial species of *Salsola, Halothamnus, Ceratoides*, and *Camphorosma* that have high potential to tolerate strongly saline environments also maintain relatively higher Ca and Mg ions. The annuals and grasses like species of genus *Bromus, Aeluropus, Eremopyrum* and *Carex* contain low amounts of mineral ions. This could be due to the excretion of salts through salt glands, which are absent in typical for Kyzylkum desert halophytes such as, *Atriplex canescens, A. nitens* and *Climacoptera lanata*, and can be utilized in the animal ration by mixing feed (with conventional forages).

The ion contents of the evaluated wild native halophytes were relatively low and hence these species could be recognized as alternative forages, both in pure halophytic pastures and/or in mixture grass stands. As most of evaluated halophytes are late flowering and seed maturating species, they can be recommended as a fattening feed during autumn and winter seasons when there is a deficit of forages on the pasture lands.

Introduction of a strip cropping system represents an alternative for private farms in the livestock-based farming system, as well as a way to diversify feed resources under unfavorable environments. It also leads to the uniform distribution of good quality feed resources throughout the year and during difficult periods, while preserving soils, water and phytogenetic resources. Another technique used in the salt-affected sandy desert environments is to plant shrubs as windbreaks to spare the land for other crops and help protect the soils from wind erosion and sand encroachment. At present efforts are being made to provide plants materials (seeds, seedlings, younger plants) to the farmers and train them on how to manage the improved rangelands.

## References

1 Nordblom T, Shomo F, Gintzburger G (1997) Food and feed prospects for resources in Central Asia. In: M Demment (ed): *Central Asia regional livestock assessment workshop*. Tashkent , Uzbekistan. Management Entity, Small Ruminant CRSP, University of California, USA, 226

2 Gintzburger G, Toderich KN, Mardonov BK, Makhmudov MM (2003) *Rangelands of the arid and semi-arid zones in Uzbekistan*. Centre de Cooperation Internationale en Resherche Agronomique pour le Development (CIRAD), Montpellier

3 Goldstein RI (1997) Ecological situation of Kyzylkum in connection with its industrial development. In: *Bulletin of SCST of the Republic of Uzbekistan* 3–4: 70–75

4 Goldshtein RI, Tsukatani T, Toderich KN, Ashurmertov OA, Aparin VB (2000) *Ecological state and conservation of arid salt/affected lands using Asiatic Salsola species*. Abstracts Book of International Seminars for Saline Agriculture, Islamabad, 10–12

5 Tsukatani T, Katayama Y (2001) *Water quality of Zarafshan River basins*. Kier Discussion paper no. 527. Kyoto University Publisher, Kyoto

6 Toderich KN, Goldshtein RI, Aparin VB, Idzikowska K, Rashidova GS (2001) Environmental state and an analysis of phytogenetic resources of halophytic plants for rehabilitation and livestock feeding in arid and sandy deserts of Uzbekistan. In: SW Breckle, W Wuhrer (eds): *Sustainable land use in deserts*. Springer, Berlin, 154–165
7 Toderich KN, Tsukatani T, Black CC, Takabe K, Katayama Y (2002) *Adaptations of plants to metal/salt contained environments: glandular structure and salt excretion*. Kier Discussion paper, no. 552, Kyoto University, Kyoto
8 Toderich KN, Tsukatani T, Petukhov OF, Grutsinov VA, Khujanazarov T, Juylova EA (2004) Risk assessment of environmental contaminants of Asiatic deserts ecosystems in relation to plant distribution and structure. *J Arid Land Stud* 14S: 33–36
9 Toderich KN, Yensen NP, Katayama Y, Kawabata Y, Grutsinov VA, Mardanova GK, Gismatullina LG (2006) Phytoremediation technologies: Using plants to clean up the metal/salts contaminated desert environments. *J Arid Land Stud* 15S: 183–186
10 Aparin V, Kawabata Y, Ko S, Shiraishi K, Nagai M, Yamamoto M, Katayama Y (2006) Evaluation of geoecological status and anthropogenic impact on the central Kyzylkum desert (Uzbekistan). *J Arid Land Stud* 15–4: 129–133
11 Akjigitova NI (1982) *Halophytic vegetation of Middle Asia and its indicator properties*. "FAN" Publisher, Tashkent
12 Shamsutdinov ZS, Savchenko IV, Shamsutdinov NZ (2000) *Holophytes of Russia and their ecological characteristics and use*. Academy of Agricultural Sciences, Russian Fodder Institute, Moscow
13 Toderich KN, Tsukatani T, Shuyskaya EV, Khujanazarov T, Azizov AA (2005) Water quality and livestock waste management in the arid and semiarid zones of Uzbekistan. In: *Proceedings of the University of Obihiro*, Obihiro, Japan, 574–583
14 Toderich KN, Li VV, Black CC, Yunusov TR, Shuiskaya EV, Mardanova GK, Gismatullina LG (2005) Linkage studies of structure, isoenzymatic diversity and some biotechnological procedures for *Salsola* species under desert saline environments. In: M Öztürk, Y Waisel, MA Khan, G Görk (eds): *Biosaline agriculture and salinity tolerance in plants.* Birkhäuser, Basel, 73–82
15 Toderich KN, Black CC, Ashurmetov AA, Japakova U, Gismatullina LG, Yunusov T (2006) *Salsola* spp. as a model for understanding plants reproductive system. In: B Sapaeva (ed): *Biochemistry and Environmental Physiology in Desert Ecosystems*. Baytenovskie Chteniya, Almata, 204–206
16 Escarre J, Lefebvre C, Gruber W, Leblane M, Lepart J, Riviere Y, Delay B (2000) Zinc and cadmium hyper accumulation by *Thlaspi caerulescens* from metalliferous and non metalliferous sites in the Mediterranean area: Implications for phytoremediation. *New Phytologists* 145: 429–437

Biosaline Agriculture and High Salinity Tolerance
Edited by Chedly Abdelly, Münir Öztürk, Muhammad Ashraf and Claude Grignon


# Computer supported system for the risk assessment and action recommendation for the water objects in Uzbekistan based on the already developed databank

Aydar B. Nasrulin[1], Faruh S. Shaazizov[1], Helmut Lieth[2]

[1] *Institute of Water Problems of the Uzbek Academy of Sciences Uzbekistan, 100041, Toshkent sh. F. Khojayev kochasi, 25"A", Tashkent, Uzbekistan*
[2] *Institute of Environmental Systems Research (USF), 49069 Osnabrück, Germany*

## Abstract

The water reservoirs in Uzbekistan were built about 50 years ago to manage agricultural irrigation for rice and cotton. Since the separation from Russia the country's needs for crops and water allocation are drastically changing. Due to world market conditions, the country now needs more food crops with less water requirement and the development of husbandry. The reservoir management had inflicted problems upon the Aral Sea with large impacts upon fisheries, natural forests and wildlife in the delta regions of the two rivers Amudarja and Syrdarja. For the future management of the large reservoirs upstream of both rivers new geographical information system-based decision support systems are under construction. This is needed to plan the water distribution for irrigation and urban usage as well as for risk assessments in case of catastrophic events and dam breaking. Here we present monitoring and modeling attempts for the management of all dam, reservoirs and pumping facilities in Uzbekistan.

## Introduction

The Republic of Uzbekistan needs a sustainable management concept for development for its water resources, because the current social economic situation cannot be improved without a realistic management of water resources. The past century has left a lot of problems for Uzbekistan, which require solutions in this century. First of all, it will be impossible to reform the agriculture and water economy without the knowledge of the exact demand of quality and quantity for the projected plans and reforms. The political and economical stability of Uzbekistan depends on an effective supply of water resources and a sustainable environmental policy. For this we need to develop hydroecological monitoring of the quality and quantity of water resources, as they are required for specific regions and for growing specific agricultural products. It does not make sense to grow plants such as cotton and rice in regions that do not have sufficient resources of water. It may be better to promote cattle breeding on pastures in such areas, in particular now, when Uzbekistan can loose part of its water supplies owing to the USA's projects of constructing channels in Afghanistan and the project of constructing a water reservoir in Turkmenistan.

To save water resources, Uzbekistan will have to introduce charges for using water in agriculture, as it is already done in communal services where water meters were installed to differentiate the payments for drinking water. Tto survive and be profitable, farmers will have to reduce water consumption and to grow plants that need less water. This requires new scientific approaches, i.e., a geographical information system (GIS) and a decision-support system (DSS) parallel to implementing hydroecological monitoring. In our project, special attention is given to the elaboration of criteria of safety, exploitation of large water objects and hydrotechnical construction of Uzbekistan.

More than 95% of agricultural products are produced using artificial irrigation. To provide the needs of the national economy of the Republic of Uzbekistan, particularly for irrigation agriculture, 55 reservoirs with total water volume more than 15 $km^3$ have been built, and more than 40 big water supply systems have been established. About 15 thousand pumping stations have been installed, and canals with the total length of 2.7 thousand km and 25 main collectors have been constructed. The majority of them were built in 1970s–1980s and have now completely or partially exhausted their resources. Together with the decreasing lifetime of the water reservoirs and technical constructions, the risk of accidents and aging constructions needs attention.

Natural calamities, like the severe flooding that occurred in southern regions of Russia, USA, Czech Republic, Germany and France, were aggravated basically by accidents that had happened on water objects. The water objects in above-mentioned countries were constructed to cope with a normal volume of water flows that amounts to about 1.0% of the catastrophic events. The water reservoirs of Uzbekistan are in very complicated conditions. Several big reservoirs are situated up stream of Central Asian rivers, such as: Toktogul reservoir with a water volume of 19.5 $km^3$, Sarez lake with a volume of 18 $km^3$, Nurek reservoir with a volume of 10.5 $km^3$, and Charvak reservoir with a volume of 2.0 $km^3$. The age of the hydrotechnical constructions and equipment in these water objects do not allow the full utilization of their functional capacity. In some cases the objects are situated in accident risk regions. This situation creates serious threats to the economical and ecological safety of our country. The conditions of water intakes in big water objects, e.g. Amu-Bukhara, Karshi and Jisakh systems of main canals and dams of Charvak and Pachkamar reservoirs, Ravatkhoji and Takhiatash distribution schemes, are very problematical and are at risk.

## Methods

In the last 5 years we have developed the methodology of hydroecological monitoring for the construction of the GIS-based DSS, using the Amudarya

Databank and the main part of the methodology for hydroecological monitoring developed by the Osnabrueck Institute of Environment Systems Research (1999–2001). Materials of field research, GIS maps obtained in the INTAS Aral Sea Project 00-1039 "Restoration and Management Options for Aquatic and Tugai Ecosystems in the Northern Amudarya Delta Region" were accumulated [1, 2]. The maps and databank on the quality of the water in the Amudarya river, prepared with the use of GIS technology, can be found on http://www.usf.uos.de/projects/aral/. They show an example of GIS usage (system ArcView. GIS 3.2) with the purposes of hydroecological monitoring of the Aral Sea Basin [3]. The results of the cooperative project of 2003–2005 have been saved in tabular format DBF4 (dBASE IV).

Using similar methods, a DSS for sustainable development in the Amudarya Delta Region of Uzbekistan [4] is in preparation. The work is being done at the Institute of Water Problems of the Academy of Sciences of the Republic of Uzbekistan beginning with the introduction of the methodology of hydroecological monitoring on specific territories. This will help us in elaborating the criteria of the safety problems with water facilities of the Republic of Uzbekistan on the basis of our hydroecological monitoring and our own GIS techniques.

A large database of water resources quality has been created. We have constructed a series of digital hydroecological maps. At present we still need to use previously obtained materials from other important fields of science for different regions of Uzbekistan.

The hydrotechnical constructions, such as reservoirs, dams and pumping stations have become technically outdated over their long period of exploitation and due to the changes of the hydrological and hydrogeological situation in the regions. These constructions now present a serious problem. The situation has become aggravated by the need for scientifically based recommendations to improve the safety of the constructions. Because of the changing crop rotations in Uzbekistan the reservoirs and pumping facilities work all year round. We need now relevant software and GIS assistance on existing and potential problems in several regions and for various constructions. The hydrotechnical constructions of Uzbekistan were intended to provide irrigation water for agriculture (4.3 million hectares), drinking water and industry. The consequences of the drastic changes and material damage to the constructions may create situations comparable with the impacts of natural catastrophes. Big hydrotechnical constructions of our Republic were built 25–30 years ago. In that period was no attention paid to safety problems. The ignoring of international norms and rules for hydrotechnical constructions has fatally affected some scientific technical problems in this field. In Uzbekistan more than 53 high dams exist. Of these 29 are ground-homogeneous; 17 stone-ground with core; 6 ground with shield, and 1 concrete. It is necessary to note that the presently available controlling system for dams and safety other hydrotechnical constructions are not adequate. We need urgently information on risk priorities and different usages to achieve our goals. We plan to create a series of GIS maps, which will include all water objects together with risk parameters, determination of flood areas, prognosis of different scenarios for floods, effects of destroying the dams and estimations of the destructed area.

In detail we are planning to perform the following:

- Creation of the database of big water objects, which include data about their state and their basic parameters
- Elaboration of standard method for the estimation of criteria for critical states
- Creation of GIS maps of investigated regions and elaboration of digital maps especially for the most dangerous areas
- Assessment of the influence of the hydrochemical state of water to the hydrotechnical constructions
- Elaboration of a scientifically established base for predicting the safety of water objects in order to protect human life and their interests, environment and economic objects.

For our project we need further consultations with experts from the Institute of Environmental Systems Research (Osnabrück) or from other international institutions on GIS and modeling.

Here we apply this method in a form published by the ESRI Corporation. We have generated all the computer models required apart from the geographical data handling so that their output can be handled with the specifications for commonly used GIS maps.

A variety of patterns had to be generated to satisfy the demands of the specialists among the decision makers. From our side we have produced analytical models for 20 hydrochemical components divided into four groups: mineral content, metals (especially heavy metals), organic compounds and biological partials.

The combination of all methods enabled us to use our quality monitoring system, particularly for distribution of surface waters, to solve a number of national socio-economical issues. These problems include the water quality assessment for drinking and technical needs; planning public and industrial water-supply management, industrial complexes and construction locations; making decisions for regulation and redistribution of the river run-offs; managing agricultural irrigation systems exploitation; managing the water reservoirs used for fishery; assessing ecological situation for public health, etc. We have evaluated the period 1978–2004. Now, with the obtained material, we can further improve chosen hydrotechnical constructions by adding calculated data for different scenarios. In preparing this work we are using the programs and models prepared by German scientists to obtain good results.

As the basis of this work, we can use the principle of hydroecological monitoring, as shown in Figures 1 and 2, for constructing the DSSs. The location of a water reservoir and its hydrotechnical constructions is shown on a digital map. Data on the hydrochemical compounds are entered into an interactive map using tabular formats and computer subroutines.

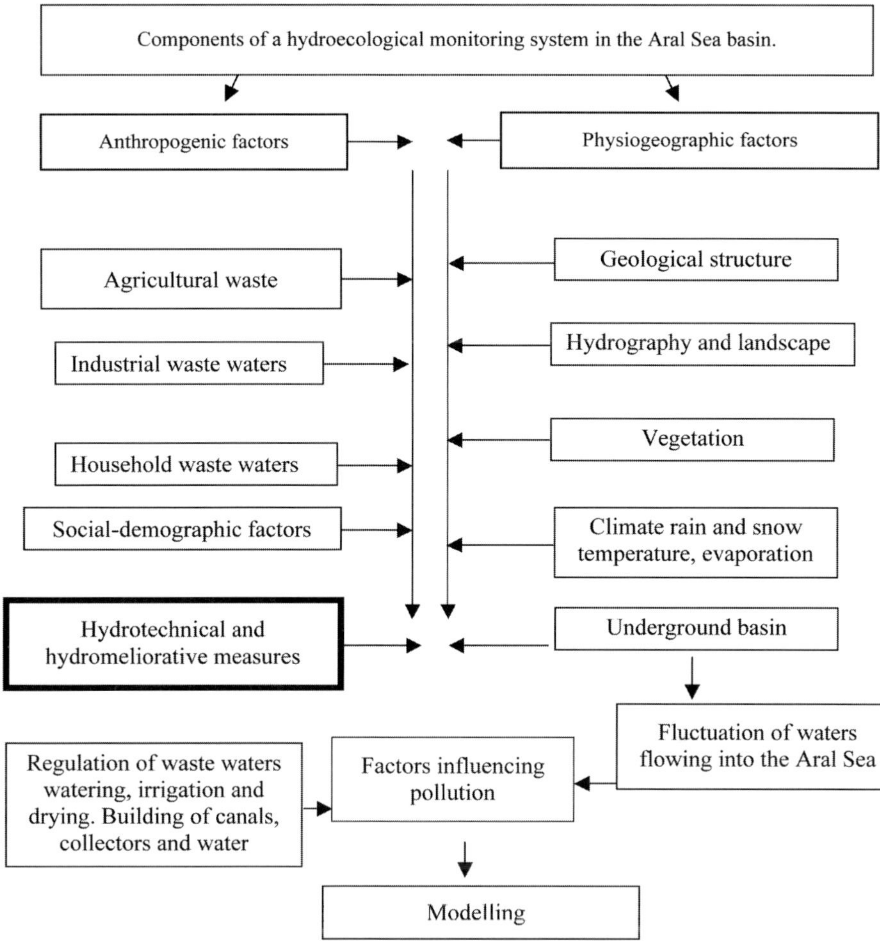


**Figure 1.** Major monitoring categories in the Aral Sea basin monitoring systems (ASBMS). This flow chart was elaborated in 2000 [5] and has since been filled with relevant data of hydrotechnical construction as the problem of water quality and quantity depends to a large part on their risk criteria.

With respect to the project, the parameters of the more than 56 large water reservoirs of Uzbekistan have been put on a digital map. It is planned to input for additional 120 water-related objects the following data: water reservoir, hydrounits, mudflow storehouses and pump stations important for the management of hydrotechnical constructions.

The use of GIS technology (especially ArcView GIS 3.2) will help to solve a lot of scientific problems in creating ecological management maps using data collected by different scientists. GIS provides the opportunity to convert the cartographical image into a dynamic ecological model. ArcView GIS can

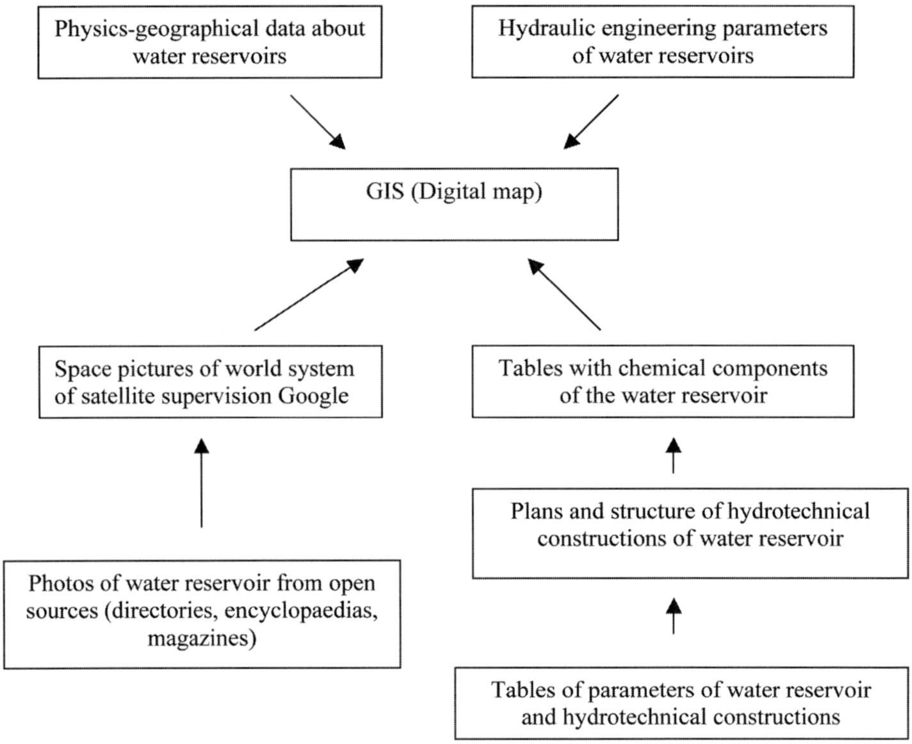


**Figure 2.** The block diagram for monitoring hydraulic engineering.

use the same base map to create hundreds of similar maps with different contents. The data can be accessed in a format of a ship-file (format ArcView), and may also be used in many other formats. ArcView can also be used to create new geographical data.

Using the "Hydrochemical databank" (which includes data of river water quality since 1938), we have conducted an estimate of modern contamination levels of river waters, distinguishing five classes of quality: good, satisfactory, dangerous, bad and high dangerous. The Microsoft Excel database of the delta of Amudarya River was created using the data from the Hydrometeocentre of Middle Asia Republics Hydrometeorological Central Directorate, the Ministry of Water Resources and the Institute of Water Problems of Uzbekistan Academy of Science [4].

## Results

The main objective of this project was to develop a scientific base for a technique to operate the hydrotechnical constructions safely and to protect the life of residents, the environment and the economic objects located near to these constructions [6, 7].

We have provided data for the following parameters in our software systems:

1) System of coordinates
2) System to fill some spatial layers of digital maps, raster and vector objects
3) A databank with several files
4) A graphical interface of investigated objects on GIS technology
5) An output to help in the realization of experimental work

Structure and a format of the information used for the DSS on hydrotechnical construction are:

1) The text information (methodical messages, instructions for use)
2) Digital information (statistics, tables)
3) Graphic information (diagrams, photos, maps)
4) Vector information layers for the user of thematic and topographical maps
5) References to the files, containing information bases (Microsoft Excel-databases)
6) Digital map of Uzbekistan on the basis of GIS technology

Components of the development of computer support system:

1) Choice of a research direction
   - The estimation of a modern condition
   - The substantiation of the project
2) Theoretical and experimental researches
   - Base of used materials
   - Methods of processing
   - Algorithms of processing
3) Technical and economic parameters
4) Practical use

The digital map for computer DSSs on water reservoir hydrotechnical constructions contains:

1) Images, schemes (plans) and photos
2) Diagrams
3) Texts.

*Acknowledgement*

We thank Brigitte Herzog for writing a revised version of this manuscript.

## References

1 Schlüter M (2003) *Development of integrated GIS-based simulation tool to support ecologically sound water management in the Amudarja river delta.* Ph.D Thesis, USF University of Osnabrück, Faculty of Mathematics/Informatics

2 Nasrulin AB (2000) Hydroecological monitoring of the Aral Sea Basin for the purpose of Ecological safety. *Water Resource* 1: 109–113 (in Russian and English)

3 Nasrulin AB (2000) The socio-economic development of Uzbek Republic and the water quality hydroecological monitoring of Aral Sea Basin. *Proceedings of the 1st World Water Congress of the International Water*, 3–7 July 2000, Paris

4 Savitsky A, Nasrulin AB (2004) Sharing GIS-technologies of know-how and optimisation models for the development of ecologically applicable modes of exploitation of hydraulic engineering buildings. / Use of geographic information systems and simulation models for research and Decision support in Central Asian river basin. Humboldt-Kolleg, International Conference, 6–10 July 2004, Tashkent, Uzbekistan, 156–161 (in Russian)

5 Nasrulin AB, Lieth H (2000) Halophytic vegetation of Uzbekistan (natural and man induced) / EU Concerted Action IC18CT96–0055 "Sustainable Halophyte Utilisation in the Mediterranean and Subtropical Dry Regions" Final Meeting "Cashcrop Halophytes": Potentials, Pilot Projects, Basic and Applied Research on Halophytes and Saline Irrigation, 29 May–2 June 2000, Osnabrück, Germany

6 Nasrulin AB, Lieth H (2001) Elaboration of systems hydroecological monitoring of Aral Sea Basin. In: M Matthies, H Malchow, J Kriz (eds): *Integrative systems approaches to natural and social dynamics*. Springer, Berlin, 249–261

7 Nasrulin AB, Lieth H (2004) Influence of Alexander von Humboldt ideas on a geographic complex method. Fundamentals of modern GIS technology and Decision support system in Central Asia. / Materials 2 International Conference "Historical role of Alexander von Humboldt and his expedition in the development of world, regional and national sciences", 14–16 October Almaty, Kazakhstan, 128–133 (in Russian)

Biosaline Agriculture and High Salinity Tolerance
Edited by Chedly Abdelly, Münir Öztürk, Muhammad Ashraf and Claude Grignon


# Functional genomics to discover genes for salt tolerance in annual and perennial plants

Isacco Beritognolo[1], Maurizio Sabatti[1], Mikael Brosché[2] and Giuseppe Scarascia Mugnozza[1]

[1] *DI.S.A.F.RI – Università della Tuscia, via S. Camillo de Lellis, snc, 01100 Viterbo, Italy*
[2] *Department of Biological and Environmental Sciences, University of Helsinki, 00014 Helsinki, Finland*

## Abstract

With the progress in plant genomics, more and more information is being gained about genes that respond to different stresses. Microarray analyses of transcriptome regulation under salt stress have uncovered the complex gene networks involved in mechanisms of sensing, signaling, and short-term response. Most of this knowledge has been derived from shock-stress experiments conducted on one genotype under laboratory conditions, but the long-term acclimation to salt stress has been addressed by only few studies. The genes responsible for the variability of salt tolerance could be valuable resources in breeding programs but they are difficult to identify in typical microarray experiments. The genes revealed by transcriptome analyses of salt-stressed plants are often common to other stresses and other species and do not explain the heritable variation. Comparative genomics is based on the comparison of genotypes differing in phenotypical behavior and is a promising approach to identify genes that control the heritable genetic variation of salt tolerance.

## Physiology of salt stress

Soil salinity strongly limits plant growth and productivity of agricultural and forest lands. Salinity affects about 7% of the world's land and this area is continuously increasing because of secondary salinization [1, 2]. The stress induced by salinity on plants is the result of two mechanisms: hyperosmotic stress, due to decreased soil water potential, and ion imbalance, leading to the accumulation of toxic ions (sodium, chloride, boron). Two phases can be distinguished in the response of plants to salt stress [3]. In a first phase, the effects of salt stress are exerted by salt outside the plants. The salt in the soil decreases the availability of water, which limits the photosynthetic activity and plant growth. This phase is common to drought stress. Within the leaves, under high light intensities, water limitation can lead to production of reactive oxygen species (ROS), which are responsible of cellular oxidative stress [4].

In a second phase, the toxic effect is due to the salt entering the plant. Sodium plays a central role when salinity is caused by the excess of NaCl. Sodium can enter the cytoplasm by non-selective cation channels (at low salinity) and potassium channels (at high salinity) and reach concentrations that are toxic to cytosol enzymes.

The ability of plants to grow in saline soils is associated with the capacity of plant cells to maintain low cytosolic sodium concentrations and high $K^+/Na^+$ ratio. The strategies for the maintenance of a high $K^+/Na^+$ ratio are sodium extrusion and/or intracellular compartmentalization of sodium into a vacuole. To maintain homeostasis, the excess of sodium is actively transported out of the cytoplasm mainly by $Na^+/H^+$ antiporters. The energy for this active transport of sodium is provided by plasma membrane $H^+$-ATPases (P-type $H^+$-ATPases) and tonoplast $H^+$-ATPases (V-type $H^+$-ATPases), which maintain the proton gradient by consumption of ATP energy [5, 6]. The synthesis of osmolytes and compatible solutes is an additional adaptive mechanism for tolerance to salt stress. Compatible solutes found in plant species are proline, glycine betaine, sucrose, raffinose, mannitol, sorbitol, and cyclic polyalcohols. These substances decrease the osmotic potential of the cytoplasm and enable the cells to maintain water homeostasis and turgor against osmotic stresses (osmotic adjustments). Compatible solutes are important in counterbalancing the osmotic potential of salt actively compartmentalized in the vacuole. Certain compatible solutes have an additional protective role against ROS, but the biochemical or chemical mechanisms are not yet clear. An important mechanism of protection against oxidative stress is the synthesis and activation of enzymes that detoxify ROS, such as superoxide dismutase, catalase, and enzymes of the ascorbate-glutathione cycle [7, 8].

## Genomics of salt tolerance

Most genes involved in tolerance to salinity and other abiotic stresses have been identified by transcriptome studies based on microarray technology. The term transcriptome indicates the full complement of activated genes (transcripts) in a particular tissue at a particular time. Microarray analyses enable scientists to detect the transcripts of many genes at the same time. Thousands of gene-specific DNA probes are immobilized in a geometrically ordered array onto the surface of a small glass slides (roughly the size of a microscope slide). The probes can be cDNA clones, polynucleotides, or *in situ*-synthesized oligonucleotides. The messenger RNAs (mRNA) extracted from a particular sample are retro-transcribed in cDNA, then labeled by fluorescent dyes, and finally hybridized onto the array. The fluorescent transcripts hybridize against the complementary immobilized probes and emit a spot fluorescence signal. The intensity of fluorescence emitted by a particular probe is proportional to the concentration (transcript level) of that specific transcript in the total mRNA population. Typical microarray experiments aim at comparing two samples to get information about the differential gene expression. In brief, mRNA from two alternative samples (for example, treatment *versus* control) are labeled by different fluorescent dyes (i.e., red and green) and hybridized onto the same microarray. The two color fluorescence signals are then detected separately by a scanner. The probe-specific signal intensity is extracted by image analysis, followed by background subtraction and data normalization. The log ratio between the two fluorescence signals expresses the ratio of transcript level between the samples analyzed. Microarray experiment data can be statistically treated similarly to classical experiments even though specific algorithms for normalization and statistical test have been developed. The development of microarray platforms in plant species relies on the availability of sequence information from subtracted cDNA libraries, expressed sequence tags (EST), and genome sequence. This technology is widely applied in model plant species, such as *Arabidopsis thaliana*, and *Oryza sativa* (rice), among herbaceous species and *Populus* spp. among perennial species. In these species, microarray analyses have been applied to study the adaptation to salinity and other abiotic stresses.

A compilation of transcriptome analyses of plant response to salt stress is reported in Table 1. Many experiments addressed the early response to shock salinity stress in short-term experiments. Great progress has been achieved by this strategy, which has uncovered complex gene networks and provided a first insight of the transcriptome regulation induced by salt stress, especially in very early phases. However, the severe treatment imposed by shock salt stress is subject to criticism, because these artificial conditions are very different from salinity that occurs naturally in the field [3]. In addition, the very early sampling times are representative of stress sensing, signaling, and rapid response to a severe shock but the acclimation/recovery phase is not represented in most experiments. Analyses at very early phase (within hours) of shock stress revealed a wide transcriptome regulation, but most genes involved in this response were not specific of salt stress. Microarray experiments conducted with multiple abiotic stresses helped in discriminating the nonspecific response from the specific response to a particular stress. This approach has revealed a wide cross-talk among the gene networks induced by several abiotic and biotic stresses. At early phases, many genes are commonly induced by different stresses and only a few genes are specifically regulated by a particular stress. However, the comparison of gene complements induced by different stresses allowed the subtraction of common genes and the identification of a reduced number of stress-specific genes.

Classical transcriptome analyses of salt-stress response discovered many genes involved in physiological adaptive response, non-adaptive response, cellular homeostasis of osmotic and ionic stress, and heritable responses. However, the genetic control of stress tolerance and the heritable variation are difficult to extract from this information. Comparative genomics can solve this problem [9]. The rationale of this approach is to compare genotypes or related species that differ in salt tolerance. The comparative

**Table 1.** Studies of transcriptome regulation under salt stress

| Reference | Species | Plant material tissues/organs | Treatment | Time points | Technique |
|---|---|---|---|---|---|
| Gong et al. [32] | *Arabidopsis thaliana* genotypes: *gl1* (wild type) *sos1*, *sos2*, *sos3* (mutants) | Seedlings, whole plant | Salinity | 4 h | ESTs |
| Chen et al. [22] | *Arabidopsis thaliana* | 4-week-old plants, roots, shoots | Salinity, osmotic (mannitol), cold, bacterial infection | 3 h, 27 h | Microarray |
| Kreps et al. [23] | *Arabidopsis thaliana* | 4-week-old plants, roots, leaves | Salinity, osmotic (mannitol), cold | 3 h, 27 h | Microarray |
| Seki et al. [21] | *Arabidopsis thaliana* | 3-week-old plants, whole plant | Salinity, drought, cold | 1 h, 2 h, 5 h, 10 h, 24 h | Microarray |
| Maathuis et al. [33] | *Arabidopsis thaliana* | Plants (growth stage 5.10), roots | Salinity, $K^+$ starvation, $Ca^{2+}$ starvation | 5 h, 10 h, 24 h, 96 h | Microarray |
| Gong et al. [14] | *Arabidopsis thaliana* *Thellungiella halophila* | 4-week-old plants 6-week-old plants, whole plant | Salinity | 3 h, 24 h | Microarray |
| Taji et al. [13] | *Arabidopsis thaliana* *Thellungiella halophila* | 3- or 4-week-old plants, whole plants | Salinity | 2 h, 5 h, 10 h, 24 h | Microarray |
| Wang et al. [12] | *Thellungiella halophila* | Seedlings, aerial part tissue | Salinity | 48 h | ESTs |
| Wong et al. [34] | *Thellungiella halophila* | 4-week-old plants, leaves | Salinity, cold, drought, | 24 h, 1 week, 3 week (cold), point of wilting (drought) 3 h, 24 h, 3 days (salinity) | ESTs |
| Kawasaki et al. [16] | *Oryza sativa*, 2 genotypes: Pokkali (tolerant), IR29 (sensitive) | Seedlings, roots | Salinity | 15 min, 1 h, 2 h, 6 h, 24 h, 7 days | Microarray |
| Rabbani et al. [19] | *Oryza sativa* | 2-week-old seedlings, roots, leaves | Cold, drought, salinity, ABA | 5 h, 10 h, 24 h | Microarray |
| Walia et al. [17] | *Oryza sativa*, 2 genotypes: FL478 (tolerant), IR29 (sensitive) | 1-month-old plants, shoots | Salinity (progressive stress) | 30 days | Microarray |

| | | | | | |
|---|---|---|---|---|---|
| Shiozaki et al. [35] | *Oryza sativa* | 2-week-old plants, whole plants | Salinity, dehydration, cold, heat, osmotic (mannitol), ABA, wounding | 0 h, 0.5 h, 1 h, 2 h, 5 h, 10 h and 24 h for temporal analysis of salinity stress 5 h, 10 h for the other stress | ESTs |
| Ueda et al. [18] | *Oryza sativa*<br>*Hordeum vulgare* | Seedlings, roots, leaves | Salinity | 1 h, 24 h | Microarray |
| Wang et al. [36] | *Zea mays* | 4-week-old, roots, leaves | Salinity | 1 h, 3 h, 6 h, 12 h. 24 h, and 72 h | Microarray |
| Buchanan et al. [37] | *Sorghum bicolor* | 1-month-old plants, roots, shoots | Salinity, osmotic (peg), ABA | 3 h, 27 h | Microarray |
| Kawaura et al. [38] | *Triticum aestivum* | 14-day-old seedlings, roots, shoots | Salinity | 6 h, 24 h | Microarray |
| Ozturk et al. [39] | *Hordeum vulgare* | 3-week-old, roots, leaves | Drought, salinity | 6 h, 10 h (drought stress) 24 h (salinity) | Microarray |
| Atienza et al. [40] | *Hordeum vulgare* | 7-day-old plants, leaves | Cold, drought, salinity, high light, copper | 2 h to 24 h | Microarray |
| Walia et al. [41] | *Hordeum vulgare* | 14-day-old plants, shoots | Salinity | 3 h, 8 h, 27 h | Microarray |
| Kore-eda et al. [42] | *Mesembryanthemum crystallinum* | 6-week-old plants, leaves | Salinity | 30 h, 48 h | EST |
| Rensink et al. [43] | *Solanum tuberosum* | 5-week-old plants, roots, leaves | Salinity, cold, heat | 3 h , 9 h, 27 h | Microarray |
| Sreenivasulu et al. [44] | *Setaria italica*, 2 genotypes: Prasad (tolerant), Lepakshi (sensitive) | 5-day-old seedlings, whole plant | Salinity | 5 days | Macroarray |
| Gu et al. [30] | *Populus euphratica* | 3-month-old, micropropagated, whole plant | Salinity recovering | Salt: 30 min, 3 h, 8 h, 24 h, 72 h recovering: 1 h, 8 h, 24 h, 48 h | Microarray |
| Brosché et al. [29] | *Populus euphratica* | Aged trees, leaves | Salinity (long-term adaptation) | - | Microarray |

analyses of transcriptome under different growth conditions can reveal the genes showing different expression profiles. These genes are likely to be responsible for the different phenotypic behavior of the genotypes studied and are good candidates to explain the genetic variability in tolerance. Here we present some studies of comparative genomics applied to investigate the tolerance to salt and other abiotic stresses in the model plants *Arabidopsis*, rice, and *Populus*.

## Comparative genomics of salt-stress response in annual plants

### *Arabidopsis*

A very informative study that explored the natural transcriptome variability in *A. thaliana* was carried out by Chen et al. [10]. That study does not include stress experiments but provides valuable information to understand the genomics of plant adaptation to environmental conditions. Five *A. thaliana* accessions were compared by analysis of transcriptome in different organs and at different developmental stages. Globally, the transcription profile was consistent across the five accessions. As expected, differences among tissues were more evident than differences among accessions but the interesting information derived from genes whose expression profile varied among the accessions. Of the whole transcriptome, relatively few genes were differentially expressed among accessions. The most plastic genes were involved in transcription regulation and stress/defense response. These biological functions are important in the adaptive response to environmental conditions and suggest that the differential expression of these genes could be a target of natural selection. The authors [10] emphasized the key role of transcription factors and regulatory regions as a major force for the evolutionary adaptation to environment. This study demonstrated the power of transcriptome analysis as a tool to study plant adaptation. Moreover, the wide variation observed in the transcription of stress-related genes opened perspectives for the investigation of tolerance to stresses in *A. thaliana*, as well as in other plant species.

*A. thaliana* was compared to its relative *Thellungiella halophila* (salt cress) to study the genomics of salinity tolerance. Salt cress is an annual crucifer species that is adapted to extreme habitats. It is very tolerant to cold, drought, and salinity (up to 500 mM NaCl) and has been adopted as a model plant for genomics of salt tolerance because it is closely related to *A. thaliana*. The two species share a high nucleotide identity (92–95%) of coding sequences and this enables an easy comparison of their genomes and transcriptome [11, 12]. Taji et al. [13] compared the two species by microarray analysis of transcriptome at early phases of shock salinity stress. A higher number of salt-induced genes was found in *A. thaliana* than in salt cress and only two genes were commonly regulated in both species. To better understand the gene networks underlying the higher salt tolerance of salt cress, the authors performed a direct on-array transcriptome comparison between *A. thaliana* and salt cress when grown under normal conditions (non-stressed). Interestingly, many stress-inducible genes were up-regulated in non-stressed salt cress relative to non-stressed *A. thaliana*. These results explained why a reduced number of genes were stress-induced in salt cress. Some salt cress-specific genes were directly related to specific mechanisms of tolerance. Salt cress is specifically resistant to oxidative stress (as tested by application of paraquat herbicide) and accumulates low amounts of sodium in leaves when salt-stressed. Consistently with this, oxidative-stress genes (FeSOD) and SOS1 genes ($Na^+/H^+$ antiporter) were over expressed in salt cress relative to *A. thaliana*.

A similar comparison between *A. thaliana* and salt cress was carried out by Gong et al. [14]. The authors applied two levels of salinity stress, taking into account the different sensitivity of the two species. Many genes were commonly regulated in the two species. Most of them were down-regulated and were functionally classified as related to repression of protein synthesis, photosynthesis, and cell division. The commonly up-regulated genes were involved in redox homeostasis, growth remodulation, and leaf senescence. The most interesting information derived from the genes that showed different regulation between the species. Consistent with results reported by Taji et al. [13], a greater number of transcripts were up-regulated in *A. thaliana* than in salt cress. The comparison of non-stressed transcriptome revealed a high number of genes significantly different between the species. These genes were related to stress signaling, redox control, protein folding, and cell wall synthesis. Interestingly, many salt cress-specific genes were classified as unknown, and this suggested novel mechanisms for salt tolerance not previously identified in *A. thaliana*.

The two studies reported above arrived at similar conclusions: under normal growth conditions the stress-inducible signaling pathways are constitutive and active in salt cress and the constitutive expression of stress-related genes is probably the key for the high salt tolerance of this species. Under stress conditions, *A. thaliana* responds to salinity with a wide reaction, which consumes energy in multiple nonspecific pathways. In contrast, the response of salt cress is more precise and involves a smaller set of genes. A similar comparative approach was applied to study a different abiotic stress, such as the adaptation to metals [15]. *A. thaliana* (sensitive to Zn) was compared to *Arabidopsis hallery*, which is a hyperaccumulator of Zn. The root transcriptome of non-treated plants was analyzed in the two species using microarrays. This strategy aimed at determining the differences in constitutive gene expression. Only 25 genes showed marked differences in regulation between *A. thaliana* and *A. hallery*. Some candidate genes for metal hyperaccumulation and metal homeostasis were highly up-regulated in *A. hallery*. Moreover, protein analyses supported the role of these genes in Zn hyperaccumulation.

*Rice*

*O. sativa* (rice) is a model plant for cereals, and extensive research has been conducted to study the genomics of stress tolerance in this species. Comparative genomics studies were performed among rice genotypes and between rice and other plant species.

Kawasaki et al. [16] performed a microarray analysis of the root transcriptome at the early phases of salt stress in two rice genotypes that differed in salt tolerance: Pokkali (tolerant) and IR29 (sensitive). A detailed time course analysis pointed out differences in the sense and timing of gene regulation. The tolerant genotype reacted earlier and with more up-regulated genes than the sensitive genotype. At late time points, the tolerant genotype started recovering, whereas the sensitive genotype showed a large number of down-regulated genes and died. The temporal dynamics of transcriptome regulation corresponded to the course of physiological response. The authors underscore that the very early phases of response are crucial for successful adaptation to salt stress. However, since the strong shock stress killed the sensitive genotype before the end of the experiment and prevented its adaptive response, the two transcriptome profiles are only partially comparable because each time point corresponded to very different physiological conditions in the genotypes studied.

In contrast, another comparative study addressed the progressive acclimation to salt stress in rice. Walia et al. [17] analyzed the shoot transcriptome regulation in two rice genotypes with different salt tolerance: FL478 (tolerant) and IR29 (sensitive). These genotypes were compared in a long-term experiment with progressive moderate stress. After 30 days, a time at which plants has become acclimatized to stress, the sensitive genotype showed more stress-induced genes than the tolerant one (448 *versus* 156). The number of down-regulated genes was similar in the two genotypes, but less than 10 genes were commonly regulated (up or down) in both genotypes. The integration of transcriptome data and physiological analyses allowed elucidation of the adaptive response. The tolerant genotype accumulated less sodium in leaves and this could explain the lower number of regulated genes as the result of a locally lower level of ionic stress. Globally, many genes induced only in the sensitive genotype are known as to be salt responsive, consistently with the higher accumulation of sodium in its leaves. This study is interesting in that it addresses the long-term adaptive response to salinity under experimental conditions that are similar to natural salinity. Therefore, the results obtained can complement the information derived from short-term shock experiments.

As in *A. thaliana*, a comparison between different species was applied to study salinity tolerance in rice [18]. Rice genotype IR64 was compared to *Hordeum vulgare* (barley) genotype Haruna-nijyo. Barley is relatively salt tolerant when compared to the other cereals. The transcriptional response was analyzed in seedling at the early phases (1–24 h) of shock salt stress. Leaves and roots were analyzed separately. Barley and rice regulated different sets of genes during early phases of salt stress and only few genes (two up-regulated and 26 down-regulated) were commonly regulated in both species. The commonly suppressed genes were related to fundamental processes: photorespiration, glycolysis, and ATP synthesis. Physiological measurements, together with expression profiles, helped in elucidating the possible role of differentially expressed genes. Barley accumulated more sodium in roots but less sodium in leaves than rice. Concomitant with this, the barley-specific up-regulated genes were involved in osmolyte biosynthesis and ion homeostasis, which are key processes for sodium exclusion at the root level.

A different approach for comparative genomics between rice and *A. thaliana* was applied by Rabbani et al. [19], who performed a transcriptome analysis of rice seedling at early phases of multiple abiotic stresses including cold, drought, salinity, and abscisic acid (ABA) treatment. The comparison of genes regulated by different stresses discriminated general stress-responsive genes and stress-specific genes. An impressive validation work by RNA blot allowed selection of 73 stress-specific genes. The authors carried out a bioinformatics comparison to transcriptome profile of *A. thaliana* in response to the same abiotic stresses [20–22]. The results were consistent with Seki et al. [20, 21] experiments. Fifty percent of the rice stress-inducible genes were common to *A. thaliana* genes regulated by abiotic stresses [20, 21]. The authors proposed these genes as candidates for breeding of plant tolerance to abiotic stresses. However, inconsistencies were found when the comparison was carried out with other *A. thaliana* transcriptome analyses reported by Kreps et al. [23]. The authors attributed these differences to general factors, such as plant species, stress treatment, growth conditions, and array technology. The inconsistencies could be better explained by considering the different tissues analyzed. Rabbani et al. [19] analyzed whole plants, whereas Kreps et al. [23] performed separate analyses of roots and leaves. As reported in many transcriptome experiments (see Tab. 1 for references), the transcriptome regulation in response to abiotic stresses is different in roots and leaves because of different levels of ionic stress, differential timing of stress sensing in early phases, and different cellular processes affected by stress. As a consequence, analyses of whole plants may mask organ-specific information and opposite gene regulation in different organs may result in a lack of detection of gene regulation.

## Genomics of salt stress in perennial species

### *Populus*

*Populus trichocarpa* is the third plant species for which the genome has been sequenced [24] and can serve as model for genomics of woody species and trees [25]. Variability of salt tolerance is present among and within *Populus* species and some of them are interesting for plantation in saline soils. *Populus euphratica* is the most tolerant species and its physiological response to salt stress is well characterized [26–28]. As mentioned above, few experiments have investigated the long-term response to salt stress, which is crucial for the adaptation of long-lived trees. To our knowledge only one published study describes the analysis of the transcriptome response to salinity of plants adapted in natural conditions. Brosché et al. [29] studied the leaf transcriptome of aged *P. euphratica* trees grown in natural conditions in sites differing in soil salinity. Only few genes were differentially regulated among the different sites. The majority of these genes belong to protein metabolism, response to abiotic and biotic stress, and metabolism. These genes could be responsible for the long-term adaptation to different habitats. Contrary to short-term experiments, relatively few genes were involved in signal transduction. Another study on *P. euphratica* [30] investigated the response of young plants to shock salt stress but the treatment duration was extended and included a recovery phase with removal of salt stress. The detailed time course analysis discriminated clusters of genes based on the timing and sense of transcriptional regulation and identified separate groups of genes specifically induced during stress and recovery. A comparative genomics approach was applied to study the transcriptional response of *Populus alba* to salt stress (Beritognolo, 2006 unpublished observations). *P. alba* is moderately salt tolerant and is naturally spread in semi-arid Mediterranean habitats [31]. Two *P. alba* genotypes from contrasting habitats and differing in salt tolerance were compared under shock salt stress and the transcriptome regulation was investigated in leaves. A time course analysis revealed that the number of stress-regulated genes increased from 3 h to 3 days of stress treatment. After 3 days, many genes were significantly down-regulated in both genotypes and up-regulated genes were detected only in the tolerant genotype. A small set of genes was differentially regulated by stress between the two genotypes. Interestingly, two gene clusters were regulated in an opposite sense in the tolerant and sensitive genotype. The genes up-regulated in the tolerant genotype and down-regulated in the sensitive one are candidates to explain the intra-specific variability in salt tolerance.

## Conclusions and perspectives

In the early era of genomics of plant stress tolerance most transcriptome analyses were carried out by shock stress experiments and sampling of tissues at very early stages of treatment, often within the first 24 h. Munns [3] claimed that these shock stress experiments induce an abrupt change and this triggers off a global response that is similar for all shock stresses. As a result, many transcriptome studies identify inducible genes that are not salt specific [21]. Even with these limitations, the explorative studies of transcriptome based on shock stress experiments provided crucial information and uncovered gene networks involved in stress response. Comparative genomics studies do not suffer of these limitations because the general non-stress-specific gene regulation can be subtracted as a common background and the attention is focused on genes that are differentially regulated between the genotypes studied. These genes are good candidates to explain the intra-specific variation of stress tolerance. Comparative analysis of transcriptome is thus a valuable tool to investigate the molecular basis of phenotypic differences between closely related annual species.

There are difficulties in comparing results coming from independent experiments, and caution is required in analyzing this kind of comparative genomics. Difficulties in interpretation and comparison of results also come from the analyses of whole plants, instead of separate organs. The mechanisms for salt exclusion and tolerance in roots and leaves act separately. Analyses of the whole plant can confound the tissue-specific information. Comparison of transcriptome results from independent experiments should be made on a tissue-specific basis. Care should be taken to standardize the experimental conditions and any difference in the experiment lay-out should be considered when conclusions are drawn from comparison of different experiments.

Sahi et al. [9] proposed a two-step flow in the analysis of stress response; first step: shock stress experiment with non killing severity and time course wide enough to encompass sensing, signaling, transcriptional response, and acclimation; and the second step: long term gradual stress to study acclimation and recovery, once the time range of response has been established. Many studies on genomics of stress response published so far belong to the first type but few genomics studies address the long-term adaptation to stress conditions. It is expected that valuable information for identifying salt tolerance genes will come from experiments with more natural stress and comparison between genotypes bringing genetic variability of stress tolerance [3]. The careful choice of contrasting genotypes or species to be compared is fundamental. In the case of salt tolerance, the genotypes studied should be phenotypically characterized, possibly in long-term common gardens trials, to evaluate their degree of tolerance and the different evolutionary strategies for adaptation to salinity. Genetic resources and genotypes selected from natural contrasting habitats not only bring additional genetic variability and different gene complements to extract valuable knowledge but also new genes for adaptation to stresses.

The functional genomics of stress tolerance has made great progress in herbaceous model plants, for which many molecular tools and physiological knowledge have been gathered. *A. thaliana* and rice have been investigated in depth as model plants for dicots and monocots, respectively, but the *Populus* spp. is emerging as model plant for trees. The comparison of genome sequence between *A. thaliana* and *Populus* revealed that the two species share large homology but the *Populus* genome contains more protein-coding genes and some of them are not represented in *A. thaliana*. The functional domains over-represented in *Populus* include genes related to plant defense, meristem development, and transport [24]. Large and long-lived trees have evolved under selective pressure that differs from that of annual plants. Therefore, the *Populus* genome could be a resource to discover new genes that control the adaptation to environmental stresses. We expect that comparative genomics for stress tolerance will prove to be a powerful strategy in *Populus* and that the results achieved in herbaceous species will extend to trees in a medium term.

*Acknowledgements*

The authors thank Muriel Gaudet for her assistance in text editing and Carlo Calfapietra for his critical reading of the manuscript.

## References

1 Munns R, Husain S, Rivelli AR, James RA, Condon AGT, Lindsay MP, Lagudah ES, Schachtman DP, Hare RA (2002) Avenues for increasing salt tolerance of crops, and the role of physiologically based selection traits. *Plant Soil* 247: 93–105
2 Williams WD (1999) Salinisation: A major threat to water resources in the arid and semi-arid regions of the world. *Lakes Reserv Res Manag* 4: 85–91
3 Munns R (2005) Genes and salt tolerance: Bringing them together. *New Phytologist* 167: 645–663
4 Klaus A, Heribert H (2004) Reactive oxygen species: Metabolism, oxidative stress, and signal transduction. *Annu Rev Plant Biol* 55: 373–399
5 Blumwald E (2000) Sodium transport and salt tolerance in plants. *Curr Opin Cell Biol* 12: 432–434
6 Blumwald E, Aharon GS, Apse MP (2000) Sodium transport in plant cells. *Biochim Biophys Acta* 1465: 140–151
7 Hasegawa PM, Bressan RA, Zhu JK, Bohnert HJ (2000) Plant cellular and molecular responses to high salinity. *Annu Rev Plant Physiol Plant Mol Biol* 51: 463–499
8 Zhu JK (2001) Plant salt tolerance. *Trends Plant Sci* 6: 66–71
9 Sahi C, Singh A, Kumar K, Blumwald E, Grover A (2006) Salt stress response in rice: Genetics, molecular biology, and comparative genomics. *Funct Integr Genomics* 6: 263–284
10 Chen W, Chang S, Hudson M, Kwan WK, Li J, Estes B, Knoll D, Shi L, Zhu T (2005) Contribution of transcriptional regulation to natural variations in *Arabidopsis. Genome Biol* 6: R32
11 Inan G, Zhang Q, Li P, Wang Z, Cao Z, Zhang H, Zhang C, Quist TM, Goodwin SM, Zhu J et al (2004) Salt cress. A halophyte and cryophyte *Arabidopsis* relative model system and its applicability to molecular genetic analyses of growth and development of extremophiles. *Plant Physiol* 135: 1718–1737
12 Wang ZL, Li PH, Fredricksen M, Gong ZZ, Kim CS, Zhang C, Bohnert HJ, Zhu JK, Bressan RA, Hasegawa PM et al (2004) Expressed sequence tags from *Thellungiella halophila*, a new model to study plant salt-tolerance. *Plant Sci* 166: 609–616
13 Taji T, Seki M, Satou M, Sakurai T, Kobayashi M, Ishiyama K, Narusaka Y, Narusaka M, Zhu JK, Shinozaki K (2004) Comparative genomics in salt tolerance between *Arabidopsis* and *Arabidopsis*-related halophyte salt cress using *Arabidopsis* microarray. *Plant Physiol* 135: 1697–1709

14 Gong Q, Li P, Ma S, Indu Rupassara S, Bohnert HJ (2005) Salinity stress adaptation competence in the extremophile *Thellungiella halophila* in comparison with its relative *Arabidopsis thaliana*. *Plant J* 44: 826–839
15 Weber M, Harada E, Vess C, Roepenack-Lahaye E, Clemens S (2004) Comparative microarray analysis of *Arabidopsis thaliana* and *Arabidopsis halleri* roots identifies nicotianamine synthase, a ZIP transporter and other genes as potential metal hyperaccumulation factors. *Plant J* 37: 269–281
16 Kawasaki S, Borchert C, Deyholos M, Wang H, Brazille S, Kawai K, Galbraith D, Bohnert HJ (2001) Gene expression profiles during the initial phase of salt stress in rice. *Plant Cell* 13: 889–905
17 Walia H, Wilson C, Condamine P, Liu X, Ismail AM, Zeng L, Wanamaker SI, Mandal J, Xu J, Cui X et al (2005) Comparative transcriptional profiling of two contrasting rice genotypes under salinity stress during the vegetative growth stage. *Plant Physiol* 139: 822–835
18 Ueda A, Kathiresan A, Bennett J, Takabe T (2006) Comparative transcriptome analyses of barley and rice under salt stress. *Theor Appl Genet* 112: 1286–1294
19 Rabbani MA, Maruyama K, Abe H, Khan MA, Katsura K, Ito Y, Yoshiwara K, Seki M, Shinozaki K, Yamaguchi-Shinozaki K (2003) Monitoring expression profiles of rice genes under cold, drought, and high-salinity stresses and abscisic acid application using cDNA microarray and RNA gel-blot analysis. *Plant Physiol* 133: 1755–1767
20 Seki M, Narusaka M, Yamaguchi-Shinozaki K, Carninci P, Kawai J, Hayashizaki Y, Shinozaki K (2001) *Arabidopsis* encyclopedia using full-length cDNAs and its application. *Plant Physiol Biochem* 39: 211–220
21 Seki M, Narusaka M, Ishida J, Nanjo T, Fujita MF, Oono Y, Kamiya A, Nakajima M, Enju A, Sakurai T et al (2002) Monitoring the expression profiles of 7000 *Arabidopsis* genes under drought, cold and high-salinity stresses using a full-length cDNA microarray. *Plant J* 31: 279–292
22 Chen W, Provart NJ, Glazebrook J, Katagiri F, Chang HS, Eulgem T, Mauch F, Luan S, Zou G, Whitham SA et al (2002) Expression profile matrix of *Arabidopsis* transcription factor genes suggests their putative functions in response to environmental stresses. *Plant Cell* 14: 559–574
23 Kreps JA, Wu Y, Chang HS, Zhu T, Wang X, Harper JF (2002) Transcriptome changes for *Arabidopsis* in response to salt, osmotic, and cold stress. *Plant Physiol* 130: 2129–2141
24 Tuskan GA, DiFazio S, Jansson S, Bohlmann J, Grigoriev I, Hellsten U, Putnam M, Ralph S, Rombauts S, Salamov A et al (2006) The genome of black cottonwood, *Populus trichocarpa* (Torr. & Gray). *Science* 313: 1596–1604
25 Taylor G (2002) *Populus*: *Arabidopsis* for forestry. Do we need a model tree? *Ann Bot* 90: 681–689
26 Fung LE, Wang SS, Altman A, Hüttermann A (1998) Effect of NaCl on growth, photosynthesis, ion and water relations of four poplar genotypes. *Forest Ecol Manag* 107: 135–146
27 Chen S, Li J, Wang S, Hüttermann A, Altman A (2001) Salt, nutrient uptake and transport, and ABA of *Populus euphratica*; a hybrid in response to increasing soil NaCl. *Trees* 15: 186–194
28 Chang Y, Chen SL, Yin WL, Wang RG, Liu YF, Shi Y, Shen YY, Li Y, Jiang J, Liu Y (2006) Growth, gas exchange, abscisic acid, and calmodulin response to salt stress in three poplars. *J Integr Plant Biol* 48: 286–293
29 Brosché M, Vinocur B, Alatalo ER, Lamminmaki A, Teichmann T, Ottow EA, Djilianov D, Afif D, Bogeat-Triboulot MB, Altman A et al (2005) Gene expression and metabolite profiling of *Populus euphratica* growing in the Negev desert. *Genome Biol* 6: R101

30 Gu R, Fonseca S, Puskás LG, Hackler LJ, Zvara Á, Dudits D, Pais MS (2004) Transcript identification and profiling during salt stress and recovery of *Populus euphratica*. *Tree Physiol* 24: 265–276

31 Sixto H, Grau JM, Alba N, Alia R (2005) Response to sodium chloride in different species and clones of genus *Populus* L. *Forestry* 78: 93–104

32 Gong Z, Koiwa H, Cushman MA, Ray A, Bufford D, Kore-eda S, Matsumoto TK, Zhu J, Cushman JC, Bressan RA et al (2001) Genes that are uniquely stress regulated in salt overly sensitive (*sos*) mutants. *Plant Physiol* 126: 363–375

33 Maathuis FJM, Filatov V, Herzyk P, Krijger C, Axelsen B, Chen S, Green BJ, Li Y, Madagan KL, Sanchez-Fernandez R et al (2003) Transcriptome analysis of root transporters reveals participation of multiple gene families in the response to cation stress. *Plant J* 35: 675–692

34 Wong CE, Li Y, Whitty BR, Díaz-Camino C, Akhter SR, Brandle JE, Golding GB, Weretilnyk EA, Moffatt BA, Griffith M (2005) Expressed sequence tags from the Yukon ecotype of *Thellungiella* reveal that gene expression in response to cold, drought and salinity shows little overlap. *Plant Mol Biol* 58: 561–574

35 Shiozaki N, Yamada M, Yoshiba Y (2005) Analysis of salt-stress-inducible ESTs isolated by PCR-subtraction in salt-tolerant rice. *Theor Appl Genet* 110: 1177–1186

36 Wang H, Miyazaki S, Kawai K, Deyholos M, Galbraith DW, Bohnert HJ (2003) Temporal progression of gene expression responses to salt shock in maize roots. *Plant Mol Biol* 52: 873–891

37 Buchanan CD, Lim S, Salzman RA, Kagiampakis I, Morishige DT, Weers BD, Klein RR, Pratt LH, Cordonnier-Pratt MM, Klein PE et al (2005) *Sorghum bicolor*'s transcriptome response to dehydration, high salinity and ABA. *Plant Mol Biol* 58: 699–720

38 Kawaura K, Mochida K, Yamazaki Y, Ogihara Y (2006) Transcriptome analysis of salinity stress responses in common wheat using a 22k oligo-DNA microarray. *Funct Integr Genomics* 6: 132–142

39 Ozturk ZN, Talamé V, Deyholos M, Michalowski CB, Galbraith DW, Gozukirmizi N, Tuberosa R, Bohnert HJ (2002) Monitoring large-scale changes in transcript abundance in drought- and salt-stressed barley. *Plant Mol Biol* 48: 551–573

40 Atienza SG, Faccioli P, Perrotta G, Dalfino G, Zschiesche W, Humbeck K, Stanca AM, Cattivelli L (2004) Large scale analysis of transcripts abundance in barley subjected to several single and combined abiotic stress conditions. *Plant Sci* 167: 1359–1365

41 Walia H, Wilson C, Wahid A, Condamine P, Cui X, Close T (2006) Expression analysis of barley (*Hordeum vulgare* L.) during salinity stress. *Funct Integr Genomics* 6: 143–156

42 Kore-eda S, Cushman MA, Akselrod I, Bufford D, Fredrickson M, Clark E, Cushman JC (2004) Transcript profiling of salinity stress responses by large-scale expressed sequence tag analysis in *Mesembryanthemum crystallinum*. *Gene* 341: 83–92

43 Rensink WA, Iobst S, Hart A, Stegalkina S, Liu J, Buell CR (2005) Gene expression profiling of potato responses to cold, heat, and salt stress. *Funct Integr Genomics* 5: 201–207

44 Sreenivasulu N, Miranda M, Prakash HS, Wobus U, Weschke W (2004) Transcriptome changes in foxtail millet genotypes at high salinity: Identification and characterization of a PHGPX gene specifically up-regulated by NaCl in a salt-tolerant line. *J Plant Physiol* 161: 467–477

Biosaline Agriculture and High Salinity Tolerance
Edited by Chedly Abdelly, Münir Öztürk, Muhammad Ashraf and Claude Grignon


# Molecular analyses of a dehydration-related gene from the *DREB* family in durum, wheat and triticale

Arianna Latini[1], Maria Sperandei[1], Sandeep Sharma[1,2], Cristina Cantale[1], Massimo Iannetta[3], Marco Dettori[4], Karim Ammar[5] and Patrizia Galeffi[1]

[1] *ENEA CR Casaccia BIOTEC-GEN, Rome, Italy*
[2] *Banaras Hindu University, Dept. of Genetics and Plant Breeding, Banaras, India*
[3] *ENEA CR Casaccia, BIOTEC-DES, Rome, Italy*
[4] *CRAS, (Centro Regionale Agrario Sperimentale) Cagliari, Sardinia, Italy*
[5] *CIMMYT Centro Internacional de Mejoramiento de Maíz y Trigo, El Batán, México*

## Abstract

Abiotic stresses are the primary cause of crop loss worldwide. They result in average yield losses of more than 50% in major crops. The negative effects of abiotic stresses are thought to be increasing due to global climate change and the resulting erratic weather patterns. Improving crops' ability to tolerate abiotic stresses through conventional breeding has been successful, especially in the case of wheat, as new cultivars better adapting to increasingly difficult growing conditions are being released regularly. However, as many stress-inducible genes have been identified, sequenced, characterized and insights into their functional roles in stress tolerance are being obtained, breeding programs have much to gain by exploring ways to target those stress-related genes that may be useful in their selection. If, or when, the relationship between different alleles or expression patterns of some stress-related genes is demonstrated, perfect markers for assisting breeder in selection for stress-tolerant lines can be readily obtained. Previously, we isolated and characterized the gene designated as *TdDRF1* encoding for a dehydration responsive factor in durum wheat. Results obtained using plant samples of different cultivars in time-course experiments conducted in the greenhouse suggested that the expression profile of *TdDRF1* upon water stress was genotype dependent. In the present paper we report results from field experiments carried at CIMMYT's experimental fields near Obregon in Mexico, in which quantitative RT-PCR was used to monitor the expression profile of the three transcripts produced by the *TdDRF1* gene under stressed (minimally irrigated) and non-stressed (fully irrigated) conditions. Tolerant and susceptible cultivars were analyzed and the results from these field experiments are compared with those from greenhouse testing.

## Introduction

Crop productivity is strongly influenced by various environmental stresses such as drought, high salinity and low temperature. Plants respond and adapt to these stresses at a physiological and cellular level as well as at a biochemical and molecular one [1, 2]. Various genes alter their expression at the transcriptional level in response to these stresses, and their products are thought to play a role in stress tolerance [3–5]. The gene expression regulation is based upon the interaction of transcription factors (*trans*-acting elements) with *cis*-acting elements present in the promoters of stress-inducible genes. Different sets of *cis*-acting elements and *trans*-acting factors have been identified [6, 7]. Yamaguchi-Shinozaki and Shinozaki [6] identified a dehydration-responsive element (DRE) in *Arabidopsis*, as a part of a signal transduction pathway responding to drought, salinity and low temperature, and the DRE-binding (*DREB*) genes, whose products specifically bind to the DRE and induce the expression of stress responsive genes [8]. The isolation and characterization of homologous *DREB* genes is currently underway in many cereals and other plant species [9–13]. We have previously identified and characterized a new *DREB*-related gene in *Triticum durum*, namely *TdDRF1*, which is expressed during the response to dehydration in time-course experiments and produces three forms of transcripts through alternative splicing, designated as *TdDRF1.1*, *TdDRF1.2* and *TdDRF1.3* [14].

In this work we used the real-time RT-PCR (qPCR) [15, 16] to analyze the expression profile of the *TdDRF1* gene in time-course experiments under water-stress conditions in the greenhouse and in experimental fields. Quantitative RT-PCR results suggest a genotype-dependent relationship between expression patterns of the three transcripts and the dehydration stress.

## Materials and methods

### *Plant growth and sample collection*

#### Field experiment

Two *Triticum durum* varieties, CICCIO and GIANNI, and two Triticale varieties, POLLMER and LIRON, were grown at the CIMMYT experimental station in Obregon (Sonora, northern Mexico) under both full-irrigated and minimally irrigated (stressed) conditions. One flag leaf from three different representative plants were harvested at different stages after heading and pooled together, frozen in liquid nitrogen, and stored at –80°C prior to RNA extraction. The sampling schedule was as follows:

$TF_0$: last watering and beginning of time-course experiment
$TF_1$: 5 days after last watering
$TF_2$: 10 days after last watering
$TF_3$: 15 days after last watering
$TF_4$: 20 days after last watering.

## Greenhouse experiment

Only the durum wheat CICCIO and the Triticale POLLMER varieties were used. Plants were grown after germination for approximately 5–6 weeks at 25°C and with a 16-h day length and regularly watered. The time-course water-stress experiment was carried out over 7 days, following the schedule:

$TG_0$: last watering and beginning of time-course experiment
$TG_4$: 4 days after last watering
$TG_7$: 7 days after last watering.

## RNA extraction, cDNA synthesis, real-time PCR

Total RNA was extracted, qualitatively assessed, quantified and reverse transcribed using random hexamers, following standard procedures. Quantitative RT-PCR was carried out on an ABI PRISM 7000 SDS (Applied Biosystems) using a Taq Man detection protocol. Primers and probes were selected to obtain different fragments, each one specific for a single splicing product. In the target PCR reactions, 250 ng reverse-transcribed RNA was used. The 18S rRNA was selected as an endogenous control to normalize the data and the housekeeping gene reactions were performed with 10 ng total RNA. A first preliminary experiment was performed to estimate the PCR efficiencies for both the transcripts and the 18S rRNA.

## Data analysis

Data were collected using the ABI PRISM 7000 SDS software (Applied Biosystem). The expression level of the three target splicing products was measured using the cycle threshold (Ct) value of each experiment and each sample was tested in triplicate. The targets and the 18S reactions were performed as singleplex runs. To determine the PCR efficiency, we tested different template amounts corresponding to 500, 250, 100, 50 and 10 ng of total initial RNA. In the case of the 18S rRNA 5 ng were also examined. The software package *Q-Gene* was used to process the data [17, 18]. The mean normalized gene expression (MNE) was calculated according to this equation: $MNE = (E\ reference)^{CTreference,mean} / (E\ target)^{CTtarget,\ mean}$

## **Results and discussion**

To evaluate the expression of the three different transcripts in response to water stress, qualitative and quantitative PCRs were carried out on several wheat varieties (http://www.economia.uniroma2.it/conferenze/icabr2004/papers/Galeffi%20.P.zip).

However, greenhouse data from one durum wheat (Ciccio) and one triticale (Pollmer) cultivars and field data from two durum wheat (Ciccio, Gianni) and two triticales (Pollmer, Liron) cultivars are presently available in final form. These genotypes were chosen to represent drought-tolerant (Ciccio, Pollmer) and drought-susceptible (Gianni, Liron) varieties for each species. Real-time PCR experiments were set up to detect quantitative differences among the three transcripts and possibly explore genotypic differences in water-stress response.

The results are shown in Figures 1–3. In both greenhouse and field experiments, *TdDRF1.2* generally exhibited the highest expression level (low Ct value), for all varieties at most sampling times, compared to *TdDRF1.3* and *TdDRF1.1*, with the latter being the rarest transcript in all varieties and sampling times.

In the case of durum wheat cultivar Ciccio (Fig. 3) (the only one used in these greenhouse experiments), we observed no real difference in gene expression upon dehydration in the greenhouse, with all three transcripts expressed practically at the same level over time. However, there was a clear change in gene expression for all three transcripts in the field experiments and those changes were somewhat different under control and water-stressed conditions (Fig. 1). Whereas under full-irrigation (control) conditions the abundance of all three transcripts generally increased from $TF_0$ to $TF_2$ and then decreased from $TF_2$ to $TF_4$, the stressed conditions resulted in a sharp increase in gene expression from $TF_0$ to $TF_2$, which then remained practically constant until $TF_4$. For the cultivar Gianni (not included in the greenhouse experiment), a different pattern was observed in the field (Fig. 1). Gene expression appeared to evolve in an opposite fashion in non-stressed and stressed conditions. For all three transcripts, under non-stressed conditions, expression decreased from $TF_0$ to $TF_2$, then increased from $TF_2$ to $TF_4$ to a level close that of the TF0 value. Under stressed conditions, it increased from $TF_0$ to $TF_2$, then decreased from $TF_2$ to $TF_4$. In summary, the expression of all three transcripts was sustained under stress in the drought-tolerant cultivar Ciccio, but not in the drought-susceptible cultivar Gianni.

In the case of triticale, as observed for durum wheat, there was no apparent difference in gene expression of *TdDRF1.2* and *TdDRF1.3* upon dehydration under greenhouse conditions (Fig. 3). However, in this case no detectable level of *TdDRF1.1* transcript could be observed at any stage. Under field conditions, all three transcripts were expressed in very similar ways for a given variety, but the overall pattern differed between the two varieties (Fig. 2). In cultivar Pollmer (drought tolerant), all three transcripts remained constant in their expression from $TF_0$ to $TF_1$ under non-stressed conditions but increased slightly under stress. Subsequently, under non-stressed conditions, the expression of the three transcripts peaked at $TF_2$, decreased moderately at $TF_3$ at a level below that of $TF_0$ to remain relatively constant at $TF_4$. On

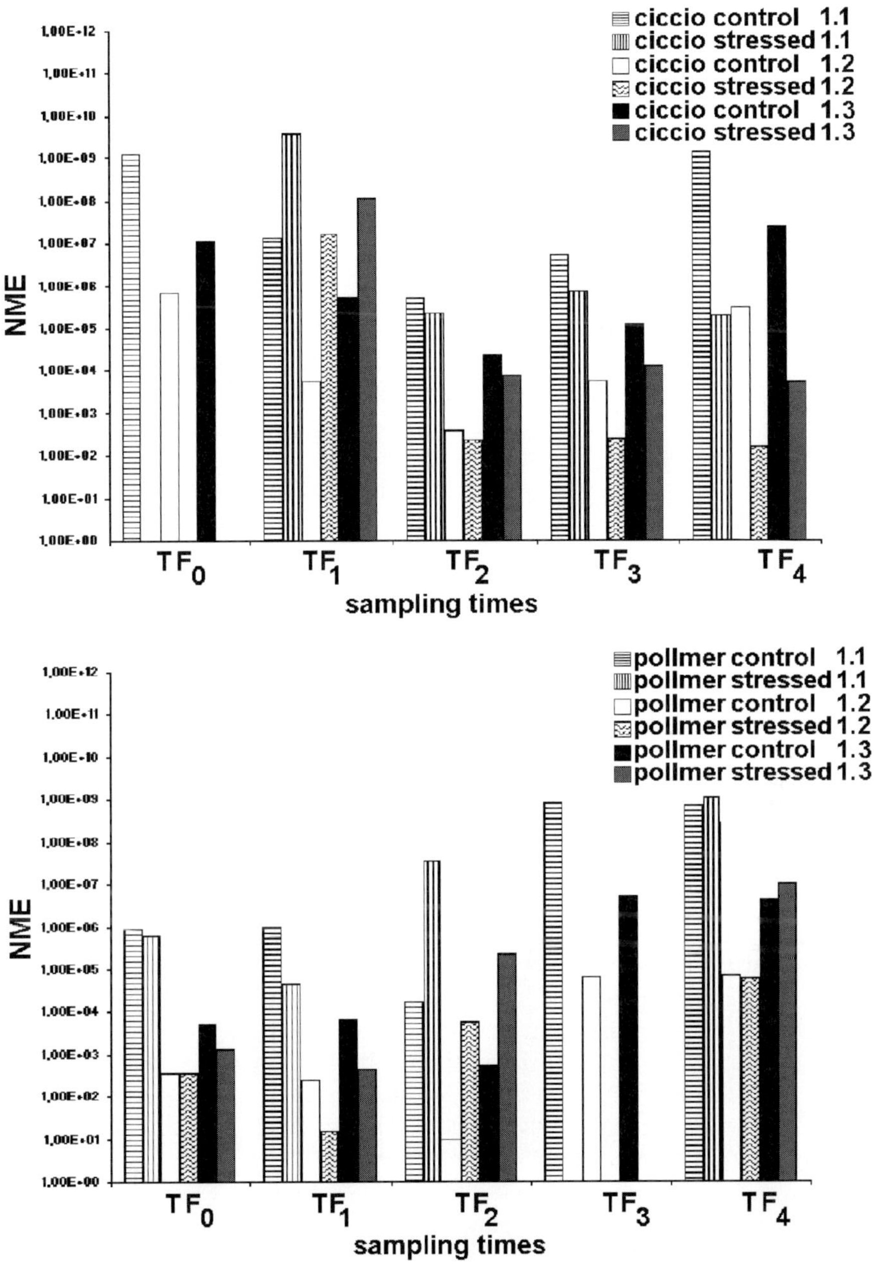


**Figure 1.** Time-dependent expression profile of *TdDRF1* gene in the tolerant genotypes, Ciccio (a) and Pollmer (b), in the field experiments. MNE: mean normalized gene expression. Semi-logarithmic (y-axis *log*-scaled) plot. All replicated samples showed a good reproducibility in Ct values, with standard deviations (SD) ranging from 0.01 to 0.2. Ciccio $TF_0$ and Pollmer $TF_3$ stressed samples were lost during the experimental procedure.

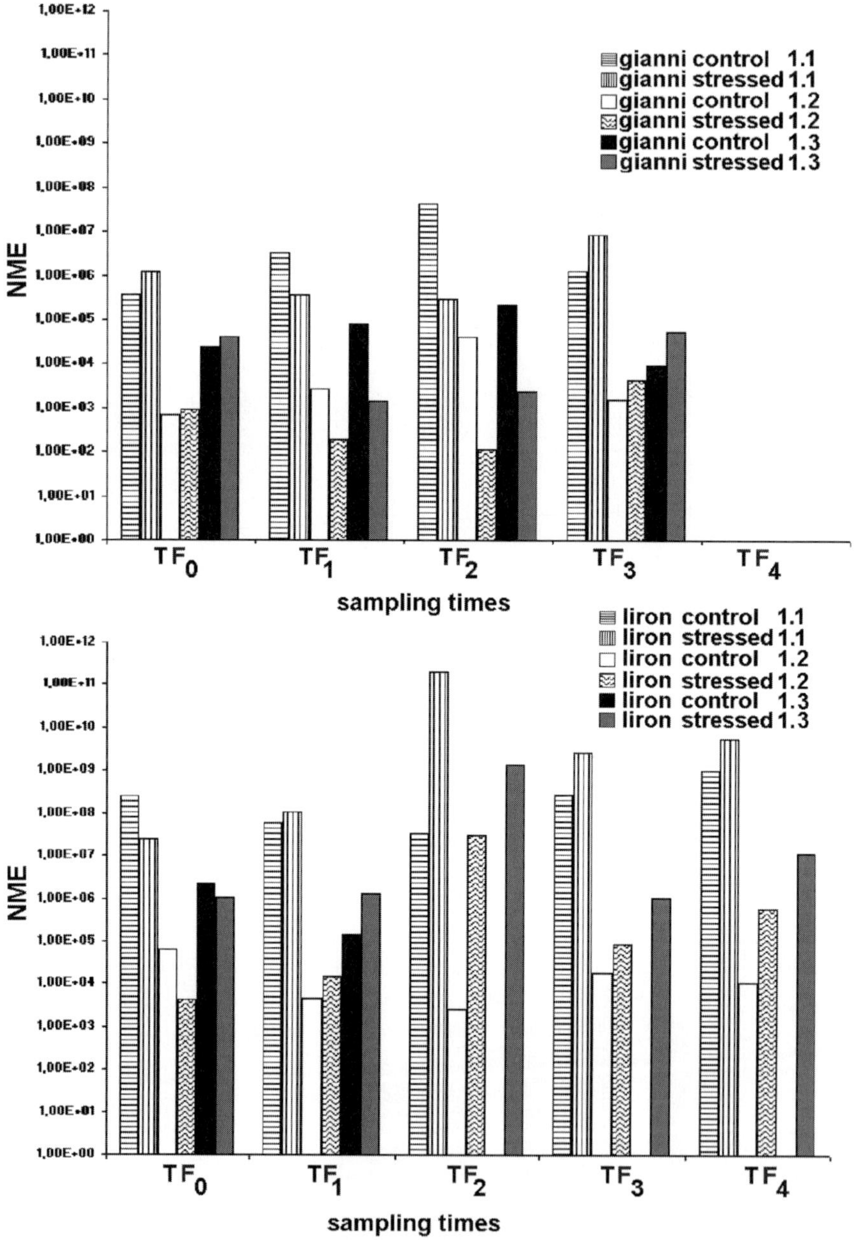


**Figure 2.** Time-dependent expression profile of *TdDRF1* gene in the susceptible genotypes, Gianni (a) and Liron (b), in the field experiments. Semi-logarithmic (y-axis *log*-scaled) plot. All replicated samples showed a good reproducibility in Ct values, with SD ranging from 0.01 to 0.2. Gianni $TF_4$ is not represented in the histogram because the plants were early died before the end of the experiment. Liron 1.3 was not detectable in the $TF_2$, $TF_3$ and $TF_4$ control samples.

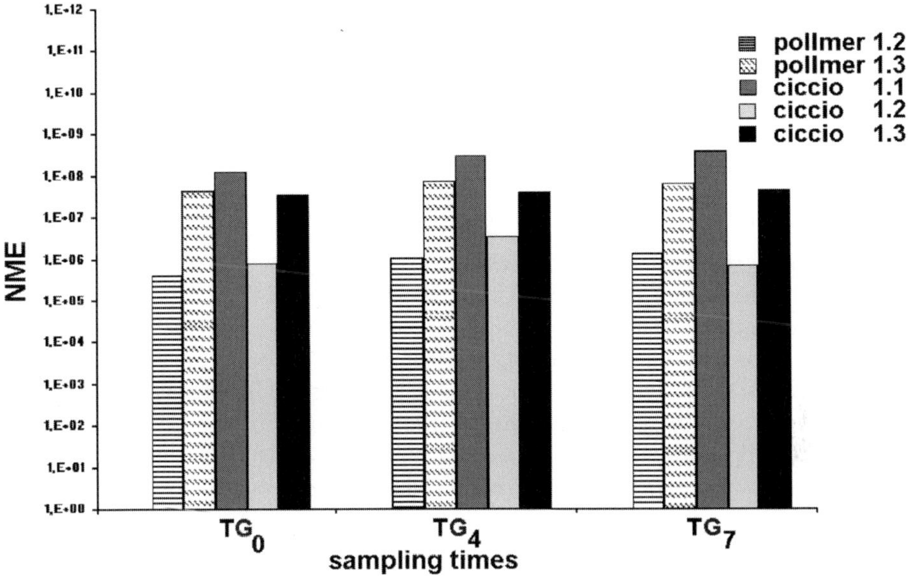


**Figure 3.** Time-dependent expression profile of *TdDRF1* gene in the tolerant genotypes, Ciccio and Pollmer, in the greenhouse experiments. Semi-logarithmic (y-axis *log*-scaled) plot. All replicated samples showed a good reproducibility in Ct values, with SD ranging from 0.01 to 0.2. Pollmer 1.1 was never detectable.

the other hand, under stressed conditions it decreased dramatically from $TF_1$ to $TF_4$, reaching levels that were lower than the lowest levels observed under non-stressed conditions. In summary, water stress affected the expression of *TdDRF1* in the triticale Pollmer by accelerating the peak expression of all its transcripts (observed at $TF_1$ instead of $TF_2$) and by making them substantially more short-lived by preventing a sustained level of expression. In the drought-susceptible variety Liron, a roughly similar scenario was observed with the following quantitative differences: under non-stressed conditions a very slight increase in the expression was apparent (probably not significant) from $TF_0$ to $TF_1$, followed by a constant expression from $TF_1$ to $TF_2$, before a relatively sharp decrease until $TF_4$. Under stressed conditions, there was no expression peak from $TF_0$ to $TF_1$ as in the case of Pollmer, but the expression was either constant or decreased slightly between the two sampling times. Subsequently, as in the case of Pollmer, a sharp decrease was observed until $TF_4$. In durum wheat, genotypic differences in drought tolerance between the two cultivars tested could be hypothetically associated with genotypic differences in sustaining the overall expression of *TdDRF1* transcripts during water stress. On the contrary, it seems more difficult to associate the genotypic difference in drought tolerance between the two triticale varieties and any difference in pattern in *TdDRF1* gene expression. The only difference was the existence of a small early expression peak in the drought-tolerant variety Pollmer, which was absent in the drought-susceptible Liron. However, since that difference was quantitatively small, more in-depth analysis of the data is needed to determine its significance.

## Conclusions

The present results, obtained from both greenhouse and field collected samples, confirm that the *TdDRF1.2* transcript is the most abundant, while *TdDRF1.1* and *TdDRF1.3* transcripts present the lower abundance at most sampling times and in all genotypes. However, it became apparent that samples taken from greenhouse pot experiments have limited value in monitoring the changes in genes expression upon dehydration, and could result in the production of *TdDRF1.1* transcript levels below the detection limit of the present assay. Samples collected from field experiments indicated substantial differences in expression patterns over time associated with genotypes and hydric conditions. In general, for a given variety and a given stress treatment, the three transcripts had very similar patterns of expression over time, albeit at rather different levels. In durum wheat, it was observed that the drought-tolerant variety Ciccio presented a more sustained expression of *TdDRF1* during water stress than the drought-susceptible variety Gianni. Whereas neither triticale cultivars exhibited sustained *TdDRF1* expression under drought, the drought-tolerant Pollmer was characterized by a small (may be not significant) expression peak early in the establishment of water stress; this peak was absent in the drought-susceptible line Liron. These putative associations are noteworthy; however, they are based on preliminary data, which, at present, do not allow the conclusive establishment of a relationship between *TdDRF1* gene expression patterns and field drought tolerance. An additional ten durum wheat genotypes have been sampled in the same field experiments and data are currently being analyzed. Such information should add to our insight into potential role of *TdDRF1* in field drought tolerance.

### *Acknowledgements*

We thank Mr. D. Chiaretti and Mrs. G. Cavicchioni for greenhouse work, Mrs. E. Palmieri for technical laboratory support and Mr. Ian Pace for revising the English text. Dr. S. Sharma was supported by an ENEA-RES Annual Research Fellowships Program 2005. Dr. A. Latini was a fellow supported by Ministero Affari Esteri for a training stage at Cimmyt (Mexico). We wish to thank Dr. M. Cristofaro for his suggestions and criticism and Dr. V. Rosato for the bioinformatics support during the evolving of this work. This work was partially supported by the RIADE project (MIUR), DesertNET (FESR), Frumisis (Mipaf) and MIUR-GENEFUN and MIUR E-GEN Projects.

## References


1 Ingram J, Bartels D (1996) The molecular basis of dehydration tolerance in plants. *Ann Rev Plant Physiol Plant Mol Biol* 47: 377–403

2 Thomashow MF (1999) Plant cold acclimation: Freezing tolerance genes and regulatory mechanism. *Annu Rev Plant Physiol Plant Mol Biol* 50: 571–599

3 Shinozaki K, Yamaguchi-Shinozaki K (2000) Molecular responses to dehydration and low temperature: Differences and cross-talk between two stress signalling pathways. *Curr Opin Plant Biol* 3: 217–223

4 Bray EA (2004) Genes commonly regulated by water-deficit stress in *Arabidopsis thaliana*. *J Exp Bot* 55: 2331–2341

5 Seki M, Narusaka M, Ishida J, Nanjo T, Fujita M, Oono Y, Kamiya A, Nakajima M, Enju A, Sakurai T et al (2002) Monitoring the expression profiles of 7000 *Arabidopsis* genes under drought, cold and high salinity stresses using a full-length cDNA microarray. *Plant J* 31: 279–292

6 Yamaguchi-Shinozaki K, Shinozaki K (1994) A novel *cis*-acting element in an *Arabidopsis* gene is involved in responsiveness to drought, low-temperature, or high salt stress. *Plant Cell* 6: 251–264

7 Liu Q, Kasuga M, Sakuma Y, Abe H, Miura S, Yamaguchi-Shinozaki K, Shinozaki K (1998) Two transcription factors, DREB1 and DREB2, with an EREBP/AP2 DNA binding domain, separate two cellular signal transduction pathways in drought- and low temperature-responsive gene expression, respectively, in *Arabidopsis*. *Plant Cell* 10: 1391–1406

8 Kasuga M, Miura S, Shinozaki K, Yamaguchi-Shinozaki K (2004) A combination of the *Arabidopsis DREB1A* gene and stress-inducible *rd29A* promoter improved drought- and low-temperature stress tolerance in tobacco by gene transfer. *Plant Cell Physiol* 45: 346–350

9 Shen YG, Zhang WK, He SJ, Zhang JS, Liu Q, Chen SY (2003) An EREBP/AP2-type protein in *Triticum aestivum* was a DRE-binding transcription factor induced by cold, dehydration and ABA stress. *Theor Appl Genet* 106: 923–930

10 Dubouzet J, Sakuma Y, Kasuga M, Dubouzet E, Miura S, Seki M, Shinozaki K, Yamaguchi-Shinozaki K (2003) *OsDREB* genes in rice, *Oryza sativa* L., encode transcription factors that function in drought-, high salt- and cold-responsive gene expression. *Plant J* 33: 751–763

11 Jiang Y, Qin F, Li Y, Fang X, Bai C (2004) Measuring specific interaction of transcription factor ZmDREB1A with its DNA responsive element at the molecular level. *Nucleic Acid Res* 32: e101

12 Xue GP, Loveridge CW (2004) *HvDRF1* is involved in abscisic acid-mediated gene regulation in barley and produces two forms of AP2 transcriptional activators, interacting preferably with a CT-rich element. *Plant J* 37: 326–339

13 Li XP, Tian AG, Luo GZ, Gong ZZ, Zhang JS, Chen SY (2005) Soybean DRE-binding transcription factors that are responsive to abiotic stresses. *Theor Appl Genet* 110: 1355–1362

14 Latini A, Rasi C, Sperandei M, Cantale C, Iannetta M, Dettori M, Ammar K, Galeffi P (2007) Identification of a *DREB*-related gene in *Triticum durum* and its expression under water stress conditions. *Ann Appl Biol* 150: 187–195

15 Bustin SA (2002): Quantification of mRNA using real time reverse transcription PCR (RT-PCR): Trends and problems. *J Mol Endocrinol* 29: 23–39

16 Bustin SA, Nolan T (2004) Pitfalls of quantitative real-time reverse transcription polymerase chain reaction. *J Biomol Tech* 15: 155–166

17 Muller PY, Janovjak H, Miserez AR, Dobbie Z (2002) Processing of gene expression data generated by quantitative Real-Time PCR [Erratum in *Biotechniques* (2002) 33: 514]. *Biotechniques* 32: 1372–1374, 1376, 1378–1379

18 Simon P (2003) Q-Gene: Processing quantitative real-time RT-PCR data. *Bioinformatics* 19: 1439–1440

Biosaline Agriculture and High Salinity Tolerance
Edited by Chedly Abdelly, Münir Öztürk, Muhammad Ashraf and Claude Grignon


# Expression analysis of salt stress responsive genes in grapevines

Samia Daldoul[1,2], Michael U. Höfer[2], Claudia Linhard[2], Neila Jallouli[1], Ahmed Mliki[1], Götz M.Reustle[2] and Abdelwahed Ghorbel[1]

[1] *Centre de Biotechnologie de Borj cédria, Laboratoire de Physiologie Moléculaire de la Vigne; B.P.95, 2050 Hammam-Lif, Tunisia*
[2] *RLP-Agroscience GmbH/Alplanta-Institute for Plant research; Breitenweg 71, 67435 Neustadt an der Weinstraße, Germany*

## Abstract

In Tunisia, highly tolerant autochthonous varieties represent a valuable resource for elucidating mechanisms of plant adaptation to salinity. Our aim is to determine which genes in these Tunisian grapevine varieties significantly contribute to the adaptation to increasing salinity. To characterize the complement of salt-responsive genes in Tunisian grapevines, we have constructed subtractive cDNA library from leaves of *Vitis vinifera* var. *Razegui*, a highly salt-tolerant variety. The library was screened for differentially expressed cDNAs, and positive clones were verified. The expression pattern of selected candidate cDNAs was analyzed using a range of *Vitis vinifera* cultivars with different degrees of phenotypic salt-stress tolerance. Expression of these cDNAs was investigated in plants after 6 and 24 h of treatment with 100 mM NaCl in hydroponic culture, and compared to plants grown under control conditions. Here, we describe the analysis of these transcripts and the putative correlation between phenotypic adaptation to salt stress and salt-induced gene expression.

## Introduction

One of the main limiting factors for plant agriculture in semi-arid and arid areas is the presence of salts in the soils and in water used for irrigation. Depending on the plant species and the degree of salinity, plants will respond in different ways [1]. When a certain level of tolerance is exceeded, the plant will eventually die. Thus, creating salt-tolerant plants would be valuable.

About 2 millions hectares of soils are affected by salinity in Tunisia, representing near of 15% of the total surface of the country and 30% of the arable lands [2]. Here we focus on grapevine plants that are of special interest for Tunisia's agriculture. Although commonly used grapevine varieties are slightly salt tolerant, it is well known that salinity causes several forms of damage, affecting yield and quality of vine products. Due to worldwide increasing soil salinity, the identification of genes conferring tolerance to abiotic stresses has been a subject of intensive studies. The large majority of studies have been performed on herbacious species such as tobacco, *Arabidopsis*, barley, and rice, whereas only a few reports on the molecular response to salt stress exist for perennial woody plants. Our working hypothesis is that the expression of specific tolerance-conferring genes is correlated with the phenotypic tolerance, while failure to cope with the salt stress (sensitive variety) is correlated with poor or even no gene activation. Thus, a comparison of early gene expression in salt-tolerant and -sensitive varieties might identify candidate genes that are involved in adaptation to stress [3–5].

Physiological assessment of several Tunisian autochthonous grapevine varieties identified a specific cultivar, *Razegui*, which can tolerate significantly higher salt concentration than elite varieties like *Syrah* [6]. Therefore, the salt-tolerant variety *Razegui* has been chosen for this study to characterize and exploit the major genes involved in the tolerance against salinity. Molecular approaches of the salt response in grapevine are largely confined to studies on individual salt-responsive genes. Genomic approaches to analyze salt-responsive genes in grapevine have not been reported. To gain a better understanding of the genes involved in responses to salt stress in grapevine, we constructed a cDNA subtractive library from control and 24-h NaCl-stressed leaves of *Vitis vinifera* var. *Razegui* using a suppressive subtractive hybridization (SSH) approach [7]. We differentially screened the library to identify genes up-regulated following salt-stress and sequenced some of the induced transcripts.

## Materials and methods

### *Plant material and stress treatment*

Four contrasting grapevine varieties (*Vitis vinifera* L, Var. *Razegui*, *Muscat d'Italie*, *Asli*, *Syrah* and *Galb sardouk*) were grown hydroponically on sand. Six-month-old plants were transferred in aerated nutrient solution (pH 6.0–6.1, electrical conductivity 1.7 ms). The culture system was installed under controlled conditions in a greenhouse (temperature: 28°C; relative humidity: 70%; light: 16 h) to overcome the environmental interactions. Salt-tolerant and salt-sensitive varieties were grown with or without 100 mM NaCl to generate control and stressed plants as starting material for RNA purification: 25 mM NaCl was added to the culture medium each day for 3 days to reach 75 mM, and then to 100 mM NaCl for 24 h. Leaves and roots were harvested separately after 6 h and 24 h at the 100 mM. Some plants were transferred on control medium for 48 h (recovery treatment), and then harvested. These samples were immediately frozen in liquid nitrogen and stored at –80°C until use for SSH library construction. In parallel to this study, a drought-stress experiment was conducted, in which contrasting variety *Razegui* (moderately drought tolerant) and *Galb sardouk* (drought sensitive) were cultivated in pots placed in greenhouse. They were submitted to a severe drought stress by stopping irrigation for 3 weeks before harvest. The greenhouse conditions were as for the salt-stress experiment.

### *RNA isolation and library construction*

Total RNA was isolated from fully developed leaves collected from stressed and control plants as described by Renault et al. [8]. mRNA was purified from total RNA using a Streptavidin Magnetic Beads purification kit (New England Biolabs, Frankfurt A. M., Germany). cDNA library was constructed using the PCR-select cDNA subtraction kit (BD Biosciences, Clontech, Palo Alto, USA) essentially according to the manufacturers protocol (PT1117-1,02/14/2002). Homology search of sequences were performed using the Blast program at the GenBank of the National Center for Biotechnology Information (NCBI; http://www.ncbi.nlm.nih.gov, [9]). The RNAs were analyzed by Northern blotting on nylon membranes (Amersham HybondN$^+$) essentially as described [10]. Temperature was kept at 65°C throughout the hybridization procedure and the washing steps.

## Results and discussion

### *RNA content*

RNAs were isolated from leaves of *Razegui*, *Muscat d'Italie*, *Asli* and *Syrah* under control and stress conditions. For the salt-stress experiment, RNA yields ranged from 107 μg $g^{-1}$ fresh weight (FW) to 480 μg $g^{-1}$ FW (Fig. 1A), depending on variety, organ and stress conditions. For the severe drought-stress experiment (Fig. 1B) we observed a significant decline in RNA content during the stress treatment.

The differential regulation of selected and sequenced genes was validated by performing Northern blot analysis and using a set of contrasting varieties differing in their salt-stress tolerance. A 26S ribosomal RNA gene hybridized in parallel showed no significant differences when control and stress leaves were compared, indicating equal loading of RNA samples.

### *Stress-responsive genes*

To study global gene expression under salt stress, we constructed an SSH cDNA library from leaf tissue of salt-stressed grapevines seedlings [11].

The leaf 24-h cDNA library prepared from poly(A+) RNA derived from the salt-stress-tolerant variety *Razegui* contained, respectively, about 3840 recombinant clones with insert sizes ranging from 150 to 900 bp, and an aver-

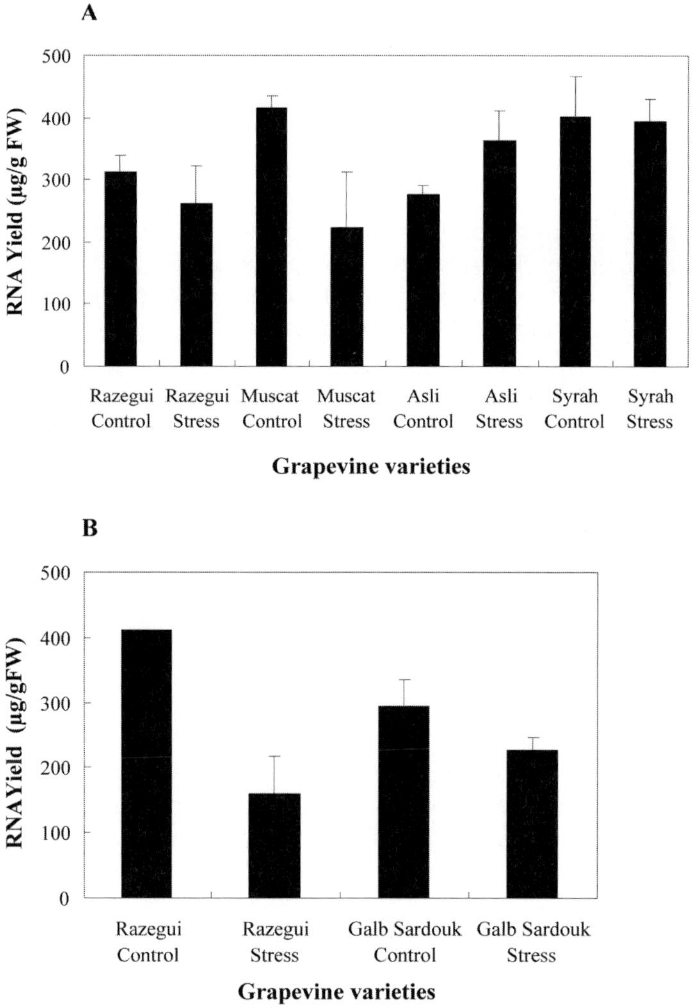


**Figure 1.** Amount of total RNA obtained from grapevines leaves under salt-stressed (A) and drought-stressed (B) conditions. Values are means of three different preparations.

age insert size of about 500 bp. PCR products derived from 1536 SSH cDNA clones were spotted in duplicate onto nylon membrane and screened with subtracted and non-subtracted probes [11].The macroarray analysis revealed that 42% of the clones were differentially expressed. Sequencing of selected SSH cDNAs showed high homology to the abiotic stress-related genes already reported in the literature (Tab. 1).

We analyzed the expression pattern of a seed imbibition gene, isolated from the library. This transcript was up-regulated early (6 h) in the tolerant

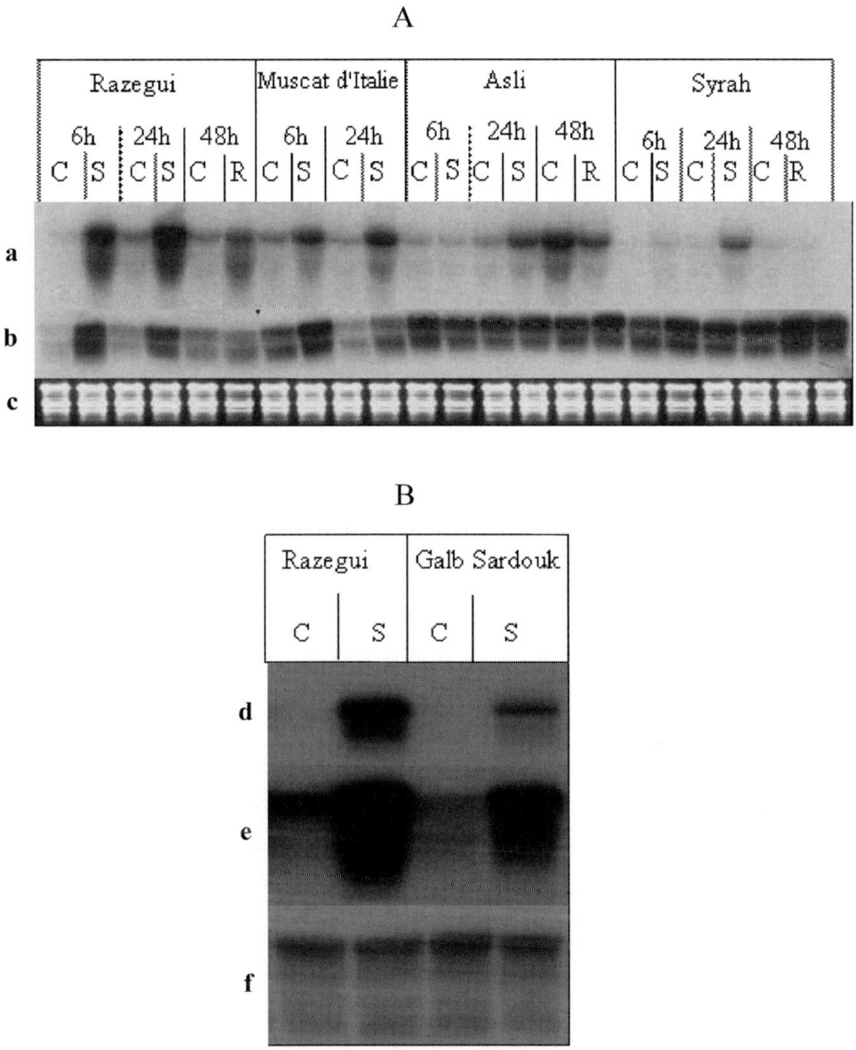


**Figure 2.** Northern analysis of two cDNAs encoding proteins overexpressed in response to salt stress and drought. (A) Salt stress: (a) cDNA homologous to *C. arietinum* seed imbibition protein; (b) cDNA homologous to *A. thaliana* leucine-rich repeat transmembrane protein kinase; (c) gel stained with ethidium bromide. (B) Drought: (d) seed imbibition protein; (e) leucine-rich repeat transmembrane protein kinase; (f) 26S ribosomal protein; C, Control; S, stressed (6 or 24 h in 100 mM NaCl or 3 weeks of drought); R, 48-h recovery from salt stress.

varieties *Razegui* and *Muscat*, but its differential expression was delayed (24 h) in the sensitive variety *Syrah*. After an additional 48 h of recovery, the expression of this gene decreased for *Razegui* (*Syrah*; Fig. 2A, a). This gene was also induced by drought stress (Fig. 2B, d).

**Table 1.** cDNAs selected from the 24-h cDNA library for sequence analysis.

| Clones | Gene or protein | % identity | Species |
|---|---|---|---|
| 1A2A1 | Putative leucine-rich repeat transmembrane protein kinase | 76% | *Arabidopsis thaliana* |
| 1A1H6 | Putative ethylene response factor 3a | 98% | *Vitis aestivalis* |
| 1B2B7 | Protein kinase MK | 84% | *Mesembryanthemum crystallinum* |
| 1B2D5 | Putative seed imbibition protein | 81% | *Cicer arietinum* |
| 3B1B6 | Nam-like protein 3 | 78% | *Petunia × hybrida* |
| 3A1G6 | RD22-like protein | 100% | *Vitis vinifera* |
| 1A1D7 | Histidine-containing phosphotransfer protein | 82% | *Catharanthus roseus* |
| 3B2C3 | Syntaxin-related protein Nt-Syr1 | 65% | *Nicotiana tabacum* |

Another selected cDNA, homologous to a leucine-rich repeat transmembrane protein kinase, was overexpressed in the tolerant variety *Razegui* and *Muscat* after 6 and 24 h of salt stress (Fig. 2A, b) as compared to control. This gene was also induced by drought stress (Fig. 2B, e).

## Perspectives

We are currently screening both the 6-h NaCl and the 24-h NaCl cDNA libraries for differentially expressed cDNAs to evaluate the differential expression patterns. This study should identify the full complement of cDNAs involved in salt-stress tolerance of *Razegui*. The rationale of this approach is the assumption that genes overexpressed in response to salt encode proteins that participate in the adaptation to salt stress.

## References

1 Verslues PE, Agarwal M, Katiyar-Agarwal S, Zhu J, Zhu JK (2006) Methods and concepts in quantifying resistance to drought, salt and freezing, abiotic stresses that affect plant water status. *Plant J* 45: 523–539

2 Hachicha M, Mtimet A (1994) Les sols salés et la salinisation en Tunisie. *Sols de Tunisie* 15: 271–341

3 Munns R (2005) Genes and salt tolerance: Bringing them together. *New Phytologist* 167: 645–663

4 Kawasaki S, Borchert C, Deyholos M, Wang H, Brazile S, Kawai K, Galbraith D, Bohnert H (2001) Gene expression profiles during the initial phase of salt stress in Rice. *Plant Cell* 13: 889–906

5 Amtmann A, Bohnert HJ, Bressan RA (2005) Abiotic stress and plant Genome evolution. Search for new models. *Plant Physiol* 138: 127–130

6 Hamrouni L, Ben Abdallah F, Abdelly C, Ghorbel W (2002) Evaluation de la tolérance des vignes tunisiennes cultivées *in vitro. Proceeding du XXVIIeme Congrès Mondial de la Vigne et du vin*, Bratislava-Solvénie, 24–28 June 2002, 102–112

7 Diatchenko L, Lau YFC, Campbell AP, Chenchik A, Moqadam F, Huang B, Lukyanov S, Lukyanov K, Gurskaya N, Sverdlov ED, Siebert PD (1996) Suppression subtractive hybridization : a method for generating differentially regulated or tissue-specific cDNA probes and libraries. *Proc Natl Acad Sci USA* 93: 6025–6030

8 Renault AS, Deloire A, Letinois I, Kraeva E, Tesniere C, Ageorges A, Redon C, Bierne J (2000) Beta-1,3-glucanase gene expression in grapevine leaves as a response to infection with *Botrytis cinerea. Am J Enol Viticult* 51: 81–87

9 Altschul SF, Gish W, Miller W, Myers EW, Lipman DJ (1990) Basic local alignment search tool. *J Mol Biol* 215: 403–410

10 Höfer MU, Santore UJ, Westhoff P (1992) Differential accumulation of the 10-, 16- and 23-kDa peripheral components of the water-splitting complex of photosystem II in mesophyll and bundle-sheath chloroplasts of the dicotyledonous $C_4$ plant *Flaveria trinervia* (Spreng.) C. Mohr. *Planta* 186: 304–312

11 Hoefer MU, Daldoul S, Linhard C, Ghorbel W, Krczal G, Reustle GM (2005) Isolation and characterization of salt stress-associated cDNAs from a stress-tolerant Mediterranean *V. vinifera* variety. In: W Qiu W, LG Kovacs (eds): *Proceedings of the International Grape Genomics Symposium*. The International Grape Genome Program and Missouri State University, USA, 131–136

Biosaline Agriculture and High Salinity Tolerance
Edited by Chedly Abdelly, Münir Öztürk, Muhammad Ashraf and Claude Grignon


# Molecular biology and transport properties of grapevine $Na^+/H^+$ antiporter

Mohsen Hanana[1,2], Olivier Cagnac[2], Toshio Yamaguchi[2], Saïd Hamdi[3], Abdelwahed Ghorbel[1] and Eduardo Blumwald[2]

[1] *Laboratoire de Physiologie Moléculaire de la Vigne, CBBC, BP 901, Hammam-lif 2050, Tunisia*
[2] *Plant Reproductive Biology, Extension Drive Center, Room 1201, University of California Davis, CA 95616, USA*
[3] *Laboratoire de Génomique Fonctionnelle et de la Qualité de la Baie de Raisin, Université de Bordeaux1, 71 Av. Edouard Bourleaux, BP 81, Bordeaux 33883, France*

## Abstract

$Na^+/H^+$ antiporters are involved in the transport of sodium and hydrogen ions across membranes and contribute in pH regulation of actively metabolizing cells. They play a primary role in homeostasis and are found in every biological kingdom, from bacteria to humans to higher plants. In plants, vacuolar $Na^+/H^+$ antiporters use the proton electrochemical gradient generated by the vacuolar $H^+$-translocating enzymes, $H^+$-ATPase, and $H^+$-$PP_i$ase to couple the downhill movement of $H^+$ with the uphill movement of $Na^+$. Moreover, it has been shown that they compartmentalize $Na^+$ into the vacuoles for detoxification and improve consequently the salt tolerance in yeasts and plants. Recently, genes encoding these $Na^+/H^+$ antiporters have been identified and studied using a molecular genetic approach in the model systems *Arabidopsis* or *Saccharomyces cerevisiae*. We describe here the identification, cloning, molecular characterization and functional properties in yeast heterologous system of a vacuolar $Na^+/H^+$ antiporter from grapevine. To identify a $Na^+/H^+$ antiporter from grapevine we applied a candidate gene approach. A 1.83-kb genomic sequence adjacent to the *VvNHX1* gene was isolated using the thermal asymmetric interlaced-PCR. Histochemical localization of β-glucuronidase gene (GUS) activity was directed by *VvNHX1* promoter-GUS fusion in transgenic *Arabidopsis*. To determine the subcellular localization of the VvNHX1 protein by heterologous expression in yeast and transient expression in onion epidermal cells, chimera constructions were prepared using a modified green fluorescent protein mGFP6. An RT-PCR approach was used to examine the VvNHX1 mRNA levels in different organs and tissues of grapevine plants. To assess VvNHX1 transport properties, *VvNHX1* was expressed in the *nhx1* mutant TY001 (that lacks the endogenous Nhx1 $Na^+/H^+$ antiporter) and the rates of $H^+$-coupled transport was measured by fluorescence quenching. Rates of cation-dependent proton movements in vacuoles isolated from yeast expressing *VvNHX1* were measured.

## Introduction

Salt tolerance is a complex trait, and the long list of salt stress-responsive genes seems to support this [1]. In addition, salt stress results in perturbation of ionic steady state, and the re-establishment of homeostasis in cell is dependent on transmembrane transport proteins [2]. In particular, $Na^+/H^+$ antiporters are involved in the transport of sodium and hydrogen ions across membranes and contribute in pH regulation. They play a primary role in homeostasis and are found in every biological kingdom, from bacteria to humans to higher plants [3–6]. Moreover, vacuolar $Na^+/H^+$ antiporters compartmentalize $Na^+$ into the vacuoles for detoxification and improve consequently the salt tolerance in yeasts and plants [2, 7–10].

Recently, genes encoding these $Na^+/H^+$ antiporters have been identified and studied using a molecular genetic approach in the model systems *Arabidopsis* or *Saccharomyces cerevisiae* [8, 11, 12]. Comparative studies have resulted in the suggestion that most basic biological functions of the vacuole of *S. cerevisiae* are equivalent to those of the vacuole of plant cells. Thus, yeast cells proved to be an excellent model for studying the transport properties and physiological function of cation transporters.

The present study describes the identification, cloning and functional characterization in a yeast heterologous system of a vacuolar $Na^+/H^+$ antiporter from grapevine and its cation/proton exchange properties. The subcellular localization, the expression of the encoded protein and the features of its isolated promoter were also analyzed to gain a better understand the physiological role of this vacuolar membrane protein, which is mainly expressed in the tissue of the berry.

## Materials and methods

### *Identification and cloning of the VvNHX1 gene*

To identify a $Na^+/H^+$ antiporter from grapevine we applied a candidate gene approach. The first step was to design degenerate primers from conserved regions of different plant $Na^+/H^+$ antiporter sequences, and then, to amplify the gene, PCR reactions were performed on a berry cDNA library at the véraison stage. The conditions for amplification were 94°C for 3 min followed by 30 cycles at 94°C for 1 min, 54°C for 30 s, 72°C for 2 min, and 72°C for 10 min. The amplified fragments were purified from agarose gels, ligated into the pGEMT-easy vector (Promega, Madison, WI) and sequenced. On the basis of the sequences of these fragments and to get both ends of the cDNA, additional nested primers and universal SP6 and T7 primers were used for asymmetric PCR. The assembly of contiguous sequences provided us with a full-length cDNA (2.053 kb) named VvNHX1 with an open reading frame of 1626 nucleotides (GenBank accession no. AY634283). The VvNHX1 ORF was inserted into the *EcoRI/Sal*I site of the pDR196+URA3 yeast expression vector. This construction was used for complementation studies and transport assays in yeast.

## *Promoter region identification and construction of VvNHX1 promoter-GUS*

A 1.83-kb genomic sequence adjacent to the *VvNHX1* gene was isolated using thermal asymmetric interlaced (TAIL)-PCR as described by Liu et al. [13].

A 1.56-kb promoter region just upstream of the ATG start codon of the *VvNHX1* gene was amplified from genomic DNA by PCR with the following oligonucleotides incorporating *Eco*RI and *Nco*I sites: 5´-CGAATTCTTTCAAGTCTCAAAGGAAAC-3´ and 5´-CATGCCATGG-TCAGTTCCCCTAACAAAGC-3´. The PCR fragment was verified by sequencing and then cloned into the *Eco*RI/*Nco*I site of the binary promoter-less vector pCAMBIA 1381Z (http://www.cambia.org/) to obtain a transcriptional fusion of the *VvNHX1* promoter and the β-glucuronidase gene (GUS) coding sequence, with the hygromycin selection gene. As a positive control, the CaMV 35S promoter was sub-cloned into the same vector and used for comparative purposes.

## *GUS assays*

*Arabidopsis thaliana* (isolate WS) plants were transformed with *Agrobacterium* strain GV3101 harboring the above vector constructs using the floral dipping method [14]. Young seedlings from transgenic plants selected on 0.5× MS medium [15] supplemented with hygromycin (30 mg/L) were used for histochemical detection of GUS expression. Seedlings and dissected tissues were fixed in acetic acid/alcohol (3:1 v/v) for 1 h at room temperature and then incubated in staining solution (100 mM sodium phosphate buffer, 5 mM potassium ferricyanide, 5 mM potassium ferrocyanide, 10 mM EDTA, 1 mg/mL X-gluc) at 37°C overnight as described by Jefferson et al. [16]. After staining, tissues were rinsed in 70% ethanol to remove chlorophyll and excess stain, and mounted for microscopy.

## *VvNHX1-GFP localization*

To analyze the subcellular localization of the VvNHX1 protein transient expression in grape berry cells, chimera constructions were prepared using a green fluorescent protein mGFP6 modified by the introduction of two-point mutations (conversions of Phe64 to Leu, F64L, and Ser65 to Thr, S65T) to change the emission and the excitation characteristics of the protein [17]. The cDNA fragments carrying the *VvNHX1* and *mGFP6* coding regions were amplified and fused by PCR and then sub-cloned into the pGEMT-easy vector using the *Spe*I and *Eco*RI restriction sites. Sequence fidelity of VvNHX-mGFP6 was confirmed by sequencing. This construction was cloned in the pRT105 plant expression vector [18]. As a positive control, the mGFP6 was cloned into the same vector for comparative purposes.

The grape berry cell suspension derived from Cabernet Sauvignon berries was maintained at 25°C on an orbital shaker (100 rpm) by weekly subculture in a medium supplemented with 58 mM sucrose [19].

Protoplasts were isolated from berry cells cultured as described previously [20] on the 4th day after subculture in fresh medium. Protoplast preparation and transformation using a polyethylene glycol-based technique were performed essentially according to the method of Neuhaus and Boevink [21]. Observation was made 48 h after transformation, and protoplasts were maintained at 23°C in the dark without shaking.

A drop of cell suspension was spread on poly-L-lysine-treated coverslip and placed on glass slide. Images were acquired using a Leica DMR fluorescence microscope.

### *Yeast strains, media, and growth conditions*

The *S. cerevisiae* yeast strains used in the complementation studies, wild type and mutant (Δ*ena1-2::HIS3*, Δ*nhx1::HIS3*) were isogenic to W303-1B (*MAT*α, *ade2-1*; *ura3-1*, *can1-100*, *leu2-3*, *trp1-1*, *his3-11,15*). The wild-type strain was transformed with lithium acetate method [22] with an empty pDR196 vector. The mutant was either transformed with the empty pDR196 plasmid or the pDR196 plasmid containing VvNHX1 cDNA. For complementation studies, yeasts were grown at 30°C in YPD (1% yeast extract, 2% peptone, 2% D-glucose) or APG (synthetic minimal medium containing 10 mM arginine, 8 mM phosphoric acid, 2% glucose, 2 mM $MgSO_4$, 1 mM KCl, 0.2 mM $CaCl_2$, and trace minerals and vitamins). The pH was adjusted to 5.5 and 6.5 for APG and YPD, respectively.

Exponentially grown cells were harvested and adjusted to $A_{600}$ of 2.0 with double distilled water. Serial dilutions of 0.1, 0.01 and 0.001 were made, and 10 μl of each culture was spotted onto APG and YPD plates [5]. Hygromycin, NaCl and LiCl were added as indicated. Growth of colonies was assessed after 2 days at 30°C.

For vacuolar transport assay, the TY001 (*MAT*α *pep4-3 leu2 trp1 ura3-52 prb1-1122 nhx1*Δ*::TRP1*) *nhx1* yeast mutant (vacuolar protease deficient) was transformed with either the empty pDR196 plasmid, or the pDR196 plasmid containing *VvNHX1*.

Intact vacuoles were isolated as described by Ohsumi and Anraku [23] with the modifications according to Yamaguchi et al. [5].

### *Computer analysis*

Sequence alignment was analyzed by the CLUSTAL X multiple sequence alignment algorithm [24]. Structural and topological predictions were done by TopPred2 (http://bioweb.pasteur.fr/seqanal/interfaces/toppred.html) and TMHMM (http://www.cbs.dtu.dk/services/TMHMM-2.0/), DAS (http://www.sbc.su.se/~miklos/DAS/), PRED-TMR (http://o2.biol.uoa.gr/PRED-TMR) and TMpred (http://www.isrec.isb-sib.ch/software/TMPRED_form.html) ProtParam (Expasy) programs. Sequence analysis was performed using DNA-STAR; and BLAST search was performed on the NCBI platform [25]. The promoter motives and *cis*-regulatory elements of the promoter sequence were analyzed with the programs PLACE Signal Scan (http://www.dna.affrc.go.jp/htdocs/PLACE/signalup.html) and MatInspector (http://www.genomatix.de/matinspector/).

## Results

### *Cloning and molecular characterization of* VvNHX1

The cloned cDNA of *VvNHX1* is 2053 nucleotides long, with a 5'-untranslated region of 60 nucleotides, a predicted ORF of 1626 nucleotides, and a 3'-untranslated region of 427 nucleotides. The predicted ORF encodes a protein of 541 amino acids with a calculated molecular mass of 60 kDa, which shared similarities with other vacuolar-type $Na^+/H^+$ antiporters. BLAST analysis allowed the identification of conserved domains related to cations exhange activity.

### *Structural characterization of VvNHX1*

As in other higher plant NHX1 proteins, VvNHX1 contained the highly conserved ten putative transmembrane segments (TM) revealed by the transmembrane prediction programs (TMHMM, DAS, PRED-TMR, TMpred, SMART and toppred) and a putative amiloride-binding domain ($^{84}$LFFIYLLPPI$^{93}$) in TM3. These transmembrane segments with around 23 residues each were likely mostly organized in an $\alpha$-helix structure, which represented more than 50% of the whole protein. In addition, the presence of several domain interaction sites mainly localized at the C terminus, 102 residues in length, is a great of importance for the regulation of the activity of the protein, its structural stability and its interaction with other proteins. Indeed, several putative glycosylation sites and more than 30 putative phosphorylation sites were predicted in the VvNHX1 sequence especially in the C terminus, which suggests a very high level of regulation. In TM1, there was a predicted membrane lipoprotein lipid attachment site that could be involved in the structural stability and attachment of the protein inside the membrane. Moreover, TM5 and TM6, which seemed not span the tonoplast for AtNHX1 according to Yamagushi et al. [5], are considered critical for antiporter transport activity.

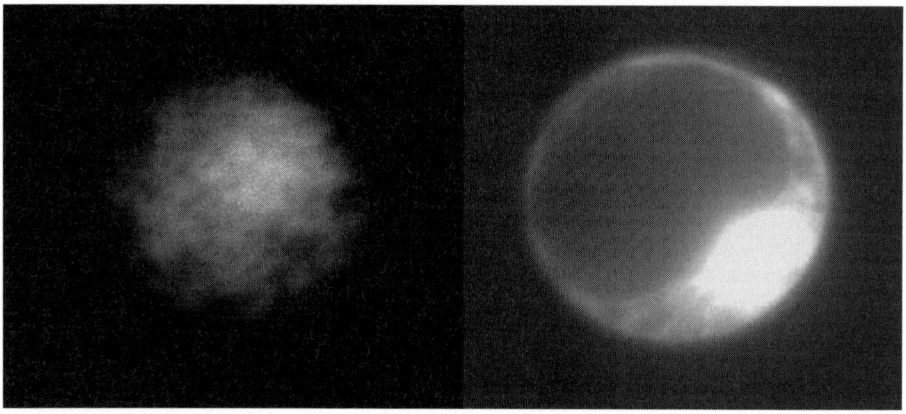

**Figure 1.** Subcellular localization of VvNHX1. Left: VvNHX1-GFP fusion protein. Right: GFP protein (control).

*Subcellular localization of VvNHX1*

In transformed berry cells the VvNHX1-GFP protein was mainly present in the vacuole compartment (Fig. 1) but the control GFP protein seems to be diffused in the cytosol and highly concentrated inside the nucleolus. Thus, VvNHX1 localizes to the vacuole of berry cells.

*Promoter expression and histochemical localization of GUS activity*

In *VvNHX1* promoter-*GUS Arabidopsis* seedlings, GUS staining was observed in the stem and the leaves, particularly in vascular tissues (not shown). Relatively weak GUS expression was observed in roots (not shown). The high level of promoter activity in the sense orientation is indicative of the promoter competence of the cloned *VvNHX1* regulatory region. Thus, GUS activity analysis confirmed that the 1.56-kb cloned fragment was a functional promoter for *VvNHX1*.

*DNA sequence and characterization of the 5-flanking region of* VvNHX1

The 1.83-kb genomic region adjacent to the *VvNHX1* gene was isolated by the TAIL-PCR and its nucleotide sequence was determined. This latter was screened for potentially important *cis*-acting and regulatory elements in both strands. The potential *cis*-acting and regulatory elements found within this region are indicated in Table 1. Nucleotide sequence analysis of the 1.83-kb promoter region revealed multiple putative response elements upstream from the transcription initiation site (Tab. 1).

**Table 1.** Potential *cis*-acting and regulatory elements in *VvNHX1* promoter region (1.8 kb).

| Regulatory elements | Positive strand | Negative strand |
|---|---|---|
| Circadian/light inducibility elements | 43 | 30 |
| Myb/Myc box | 24 | 28 |
| Poly-adenylation signal | 19 | 16 |
| Floral development activation element | 17 | 14 |
| Mesophyll expression module | 11 | 18 |
| TATA box | 10 | 17 |
| ABA-inducible transcriptional activator | 10 | 8 |
| CAAT box | 7 | 9 |
| Zinc-finger motif regulating salt tolerance | 5 | 4 |

The first TATA and CCAAT boxes were located around 300 nucleotides upstream the transcription initiation site. A cluster of responsive elements for growth regulation factors is present immediately upstream of the transcription start site. These include abscisic acid (ABA), ethylene, and gibberellic acid-responsive elements and cytokinin-dependent binding protein. Further upstream, several other potentially *cis*-acting elements were identified such as Myb and Myc boxes, light-responsive elements and motives required for dehydration and salt tolerance. Moreover, putative *cis* elements for floral development, pollen-specific expression and mesophyll expression module were also identified in this region. The presence of all these regulatory and *cis*-acting elements in this region suggests a role for growth and drought and salt tolerance with a tissue-specific expression for the *VvNHX1* gene.

### *Functional characterization of VvNHX1*

Simple eukaryotic yeast *S. cerevisiae* cells proved to be an excellent model for studying the transport properties and physiological function of alkali-metal-cation transporters, and the existence of mutant strains lacking their own transport systems provided an efficient tool for a molecular study of alkali-metal-cation transporters from higher eukaryotes upon their expression in yeast cells [26, 27]. To gain information about the function of *VvNHX1*, full-length cDNA was introduced into yeast double mutant *ena1-2nhx1* that lacks the plasma membrane sodium efflux pump and the vacuolar $Na^+/H^+$ antiporter. The *VvNHX1* ORF was placed under the control of PMA1 (plasma membrane $H^+$-ATPase) constitutive promoter for the expression of heterologous proteins in yeast. Studies were carried out to determine whether the *VvNHX1* gene product could functionally complement the yeast double mutant by suppressing its observed phenotypes. The VvNHX1 product should function more or less in the same manner as vacuolar-type yeast $Na^+/H^+$ antiporter, NHX1.

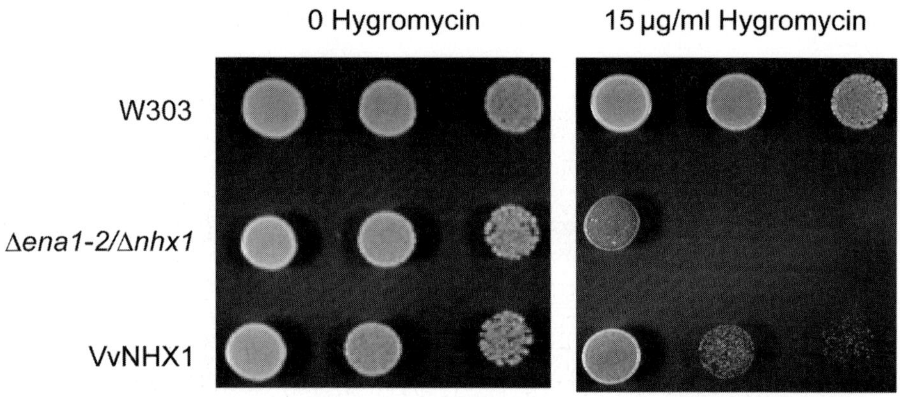


**Figure 2.** Complementation of hygromycin-sensitive phenotype by VvNHX1 of yeast mutant *ena1-2nhx1*. Three serial dilution (1:10, 1:100, 1:1000) of saturated cultures were spotted onto YPD solid media at pH 6.5 supplemented with hygromycin as indicated.

Growth of the Δ*ena1-2*/Δ*nhx1* yeast mutant is sensitive to 15 µg/mL hygromycin, a concentration that did not inhibit the growth of the wild-type W303. The double mutant containing *VVNHX1* could grow partially under this concentration (Fig. 2). Hygromycin is an organic cation that is toxic to cells upon accumulation in the cytosol in response to electrochemical proton gradient [27]. Since $Na^+/H^+$ exchange is electroneutral and assuming that VvNHX1 does not directly catalyze the transport of hygromycin, VvNHX1 could detoxify hygromycin only *via* proton exchange across the tonoplast and subsequently pH regulation.

For $Na^+$ complementation, different pH values were tested and the best result was obtained for pH 6.5. VvNHX1 suppressed somewhat $Na^+$ sensitivity of the double mutant (Fig. 3), which means that VvNHX1 could protect the cytosol from $Na^+$ by carrying it inside the vacuole. Other pH conditions were not suitable to show a recovery of growth with the expression of VvNHX1. The same result was obtained for $Li^+$, indicating that VvNHX1 can catalyze not only $Na^+$ but also $Li^+$ transport into vacuoles. Although the growth of the transformed yeast was not restored to the level of the control strain for all the treatments, the partial complementation realized by VvNHX1 indicates that this latter localized in yeast tonoplasts and functioned as a yeast NHX1.

### *Transport assays*

The assays used here to functionally characterize VvNHX1 are based on work done by Yamagushi et al. [6] with the same yeast mutants to functionally characterize AtNHX1, the *Arabidopsis* homolog of VvNHX1. The yeast

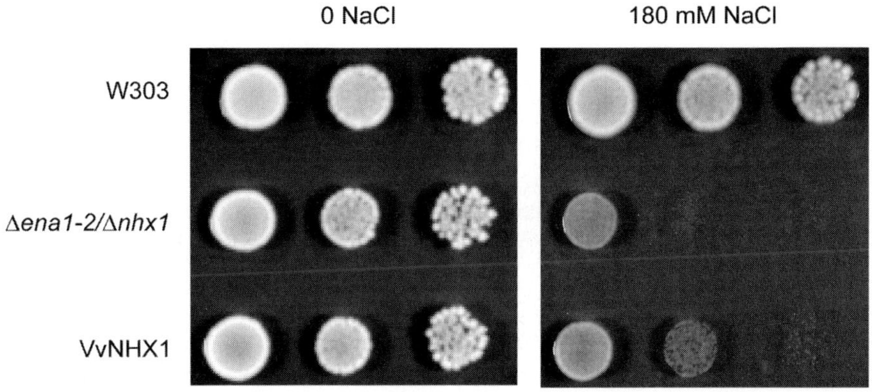


**Figure 3.** Complementation of the salt-sensitive phenotype by *VvNHX1* of yeast mutant *ena1-2nhx1*. Three serial dilution (1:10, 1:100, 1:1000) of saturated cultures were spotted onto YPD solid media at pH 6.5 supplemented with NaCl as indicated.

mutant strain makes it possible to assess the contribution of this vacuolar sodium proton antiporter to ion transport by comparing transport in vacuoles isolated from TY001 *nhx1* mutant and TY001 harboring *VvNHX1*. This was assayed by the measurement of rates of cation-dependent proton efflux using fluorescence quenching of acridine orange. We postulated that TY001 *nhx1* mutant would display a lower cation-dependent proton exchange, which would be observed as a decrease in the rate of quench recovery [6]. For both $Na^+$ and $K^+$, VvNHX1 exhibited simple Michaelis-Menten transport kinetics (not shown).

These results support the conclusion that VvNHX1 is a functional cation ($Na^+$, $K^+$) -dependent proton exchanger across the yeast tonoplast.

## Conclusion

We have cloned and characterized *VvNHX1*, a gene encoding a vacuolar cation/$H^+$ antiporter from *Vitis vinifera* cv. Cabernet Sauvignon. VvNHX1 belongs to the vacuolar NHX protein family and showed high similarity with other known vacuolar antiporters.

Topological analysis suggests that VvNHX1 protein contains ten transmembrane segments and a hydrophilic C terminus. VvNHX1 contains several putative interaction domains at the C terminus that could mediate post-translational modification processes, influence protein stability, or interact with other proteins or signaling molecules.

The expression of *VvNHX1-GFP* chimera in berry cells demonstrated that VvNHX1 localized to the vacuoles.

The isolation and analysis of the 5-flanking region of *VvNHX1* showed that it is a functional promoter region of *VvNHX1* and that it contains several putative *cis*-acting and regulatory elements of gene expression.

The functional characterization of VvNHX1 was realized by complementation studies in yeast double mutant lacking the plasma membrane sodium efflux pump and the vacuolar $Na^+/H^+$ antiporter. The expression of *VvNHX1* partially complements the salt- and hygromycin-sensitive phenotypes of an ena1-4 nhx1 yeast strain.

## *Acknowledgments*

We would like to thank Dr. Barrieu François (Université de Bordeaux1, France) for kindly providing the Grapevine cDNA library. This work was supported by grant from the Ministry of Scientific Research, Technology and Competency Development in Tunisia.

## References

1 Zhu JK (2000) Genetic analysis of plant salt tolerance using arabidopsis. *Plant Physiol* 124: 941–948

2 Hasegawa PM, Bressan RA, Zhu JK, Bohnert HJ (2000) Plant cellular and molecular responses to high salinity. Ann Rev *Plant Physiol Plant Mol Biol* 51: 463–499

3 Padan E, Venturi M, Gerchman Y, Dover N (2001) $Na^+/H^+$ antiporters. *Biochim Biophys Acta* 1505: 144–157

4 Orlowski J, Grinstein S (1997) $Na^+/H^+$ exchangers of mammalian cells. *J Biol Chem* 272: 22373–22376

5 Yamaguchi T, Apse MP, Shi H, Blumwald E (2003) Topological analysis of a plant vacuolar $Na^+/H^+$ antiporter reveals a luminal C terminus that regulates antiporter cation selectivity. *Proc Natl Acad Sci USA* 100: 12510–12515

6 Yamaguchi T, Aharon GS, Sottosanto JB, Blumwald E (2005) Vacuolar $Na^+/H^+$ antiporter cation selectivity is regulated by calmodulin from within the vacuole in a $Ca^{2+}$ and pH-dependent manner. *Proc Nal Acad Sci USA* 102: 16107–16112

7 Nass R, Cunningham KW, Rao R (1997) Intracellular sequestration of sodium by a novel $Na^+/H^+$ exchanger in yeast is enhanced by mutations in the plasma membrane $H^+$-ATPase – Insights into mechanisms of sodium tolerance. *J Biol Chem* 272: 26145–26152

8 Gaxiola RA, Rao R, Sherman A, Grisafi P, Alper SL, Fink GR (1999) The *Arabidopsis thaliana* proton transporters, AtNhx1 and Avp1, can function in cation detoxification in yeast. *Proc Natl Acad Sci USA* 96: 1480–1485

9 Blumwald E (2000) Sodium transport and salt tolerance in plants. *Curr Opin Cell Biol* 12: 431–434

10 Apse MP, Blumwald E (2002) Engineering salt tolerance in plants. *Curr Opin Biotech*nol 13: 146–150

11 Apse MP, Sottosanto JB, Blumwald E (2003) Vacuolar cation/$H^+$ exchange, ion homeostasis, and leaf development are altered in a T-DNA insertional mutant of AtNHX1, the *Arabidopsis* vacuolar $Na^+/H^+$ antiporter. *Plant J* 36: 229–239

12 Zorb C, Noll A, Karl S, Leib K, Yan F, Schubert S (2005) Molecular characterization of $Na^+/H^+$ antiporters (ZmNHX) of maize (*Zea mays* L.) and their expression under salt stress. *J Plant Physiol* 162: 55–66
13 Liu YG, Chen Y, Zhang Q (2005) Amplification of genomic sequences flanking T-DNA insertions by thermal asymmetric interlaced polymerase chain reaction. *Methods Mol Biol* 286: 341–348
14 Clough SJ, Bent AF (1998) Technical advance: Floral dip: A simplified method for *Agrobacterium*-mediated transformation of *Arabidopsis thaliana*. *Plant J* 16: 735–743
15 Murashige T, Skoog F (1962) A revised medium for rapid growth and bio-assays with tobacco tissue cultures. *Physiol Plant* 15: 473–497
16 Jefferson RA, Kavanagh TA, Bevan MW (1987) GUS fusion: β-Glucuronidase as a sensitive and versatile gene fusion marker in higher plants. *EMBO J* 6: 3901–3907.
17 Haseloff J (1999) GFP variants for multispectral imaging of living cells. *Methods Cell Biol* 58: 139–151
18 Töpfer R, Matzeit V, Gronenborn B, Schell J, Steinbiss HH (1987) A set of plant expression vectors for transcriptional and translational fusion. *Nucleic Acids Res* 15: 5890
19 Decendit A, Ramawat KG, Waffo P, Deffieux G, Badoc A, Mérillon JM (1996) Anthocyanins, catechins, condensed tannins and piceid production in *Vitis vinifera* cell bioreactor cultures. *Biotechnol Lett* 18: 659–662
20 Atanassova R, Leterrier M, Gaillard C, Agasse A, Sagot E, Coutos-Thévenot P, Delrot S (2003) Sugar-regulated expression of a putative hexose transport gene in grape. *Plant Physiol* 131: 326–334
21 Neuhaus JM, Boevink P (2001) The green fluorescent protein (GFP) as reporter gene in plant cells. In: CR Haves, B Satiat-Jeunemaitre (eds): *Plant Cell Biology*. Oxford University Press, Oxford, 127–142
22 Gietz RD, Schiestl RH, Willems AR, Woods RA (1995) Studies on the transformation of intact yeast cells by the LiAc/SS-DNA/PEG procedure. *Yeast* 11: 355–360
23 Ohsumi Y, Anraku Y (1981) Active transport of basic amino acids driven by a proton motive force in vacuolar membrane vesicles of *Saccharomyces cerevisiae*. *J Biol Chem* 256: 2079–2082
24 Thompson JD, Gibson TJ, Plewniak F, Jeanmougin F, Higgins DG (1997) The CLUSTAL_X windows interface: flexible strategies for multiple sequence alignment aided by quality analysis tools. *Nucleic Acids Res* 25: 4876–4882
25 Altschul SF, Gish W, Miller W, Myers EW, Lipman DJ (1990) Basic local alignment search tool. *J Mol Biol* 215: 403–410
26 Quintero FJ, Blatt MR, Pardo JM (2000) Functional conservation between yeast and plant endosomal $Na^+/H^+$ antiporters. *FEBS Lett* 471: 224–228
27 Aharon GS, Apse MP, Duan SL, Hua XJ, Blumwald E (2003) Characterization of a family of vacuolar $Na^+/H^+$ antiporters in *Arabidopsis thaliana*. *Plant Soil* 253: 245–256
28 Darley CP, Van Wuytswinkel OC, Van der Woude K, Mager WH, De Boer AH (2000) *Arabidopsis thaliana* and *Saccharomyces cerevisiae* NHX1 genes encode amiloride sensitive electroneutral $Na^+/H^+$ exchangers. *Biochem J* 351: 241–249

Biosaline Agriculture and High Salinity Tolerance
Edited by Chedly Abdelly, Münir Öztürk, Muhammad Ashraf and Claude Grignon


# Opposite lipid signaling pathways tightly control proline accumulation in *Arabidopsis thaliana* and *Thellungiella halophila*

Mohamed Ali Ghars[1,2], Elodie Parre[1], Anne-Sophie Leprince[1], Marianne Bordenave[1], Delphine Lefebvre-De Vos[1], Luc Richard[1], Chedly Abdelly[2] and Arnould Savouré[1]

[1] *Physiologie Cellulaire et Moléculaire des Plantes, UMR 7180 CNRS, Université Pierre et Marie Curie-Paris 6, case 156, 4 place Jussieu, 75252 Paris cedex 05, France*

[2] *Adaptation des Plantes aux Stress Abiotiques. Centre de Biotechnologie à la technopôle de Borj Cedria, BP 901, 2050 Hammam-Lif, Tunisia*

## Abstract

Throughout evolution, plants have developed various strategies to tolerate water stress. Among them, the accumulation of proline has been reported in a wide range of species. The metabolic pathway of this compatible solute is relatively well characterized in *Arabidopsis thaliana*. However, the signaling cascades involved in its regulation remain largely unknown. *Thellungiella halophila*, which is a close relative of *Arabidopsis*, tolerates extreme salinity up to 500 mM NaCl. In this work, the involvement of lipid signaling pathways in the regulation of proline accumulation was investigated in these two species upon water stress. A pharmacological approach has been performed using specific inhibitors of key signaling elements. The effects of these inhibitors have been investigated on proline accumulation. The present data show that phospholipases D (PLDs) are negative regulators of proline anabolism under normal conditions in *A. thaliana*. When such PLD-mediated regulation is abolished by 1-butanol, plants show a higher proline responsiveness to osmotic stress. In contrast to *Arabidopsis*, 1-butanol does not have any effect on proline accumulation in *T. halophila* under non-stress conditions. However, upon water constraints, 1-butanol reduces rather than increase proline accumulation. Our data suggest the involvement of a PLD-mediated signaling pathway in the tight regulation of proline metabolism that acts in a opposite way in *A. thaliana* and *T. halophila*. On the other hand, phospholipases C exert a positive control on proline accumulation in *A. thaliana* upon salt stress and a negative control in *T. halophila* upon water stress and non-stress conditions. In conclusion, we provide experimental evidence that positive and negative regulators are involved in the fine regulation of proline metabolism upon moderate water stress. Our study has defined a critical role of lipid signaling pathways in proline accumulation in *A. thaliana* and in *T. halophila*. Thus, in *Arabidopsis*, our data indicate that PLC-based signaling is a committed step in proline biosynthesis upon salinity but not upon hyperosmotic stress.

## Introduction

Among environmental factors, drought and salinity represent major constraints for plants [1]. To achieve water-stress tolerance, plants have developed various adaptive responses including physiological and biochemical changes. The most notable ones include the reduction of cell growth and a decrease in the intracellular water potential, which results from the accumulation of compatible osmolytes such as proline; however, the exact role of proline in stress tolerance remains puzzling. Several studies proposed that proline may act as a hydroxyl radical scavenger, as a factor maintaining balance between photosynthetic apparatus and NADPH utilization in carbon fixation, and as sink of reducing power resulting from metabolic disturbance [2]. Accumulation of proline may also be a consequence of stress injury. Recently, proline has been proposed to be a stress-related signal molecule because L-proline is able to trigger the induction of *ProDH* gene transcription by bZIP transcription factors [3, 4].

In plants, proline is synthesized from either glutamate or ornithine [5]. In response to water stress, proline accumulation is dependent on phosphorylation and reduction of glutamate by the pyrroline-5-carboxylate (P5C) synthetase to produce P5C. This compound is then reduced to proline by the P5C reductase. Concomitantly, proline degradation, which is catalyzed by the sequential action of the mitochondrial proline dehydrogenase (ProDH) and the P5C dehydrogenase, is abolished. Understanding the mechanisms by which a plant transduces stress signals to trigger proline accumulation is important to elucidate the role of this amino acid in the acquisition of stress tolerance.

A central role for lipid signaling pathways in plant responses to abiotic stresses is clearly emerging from current research in various plant model systems [6]. Several studies have demonstrated that hyperosmotic stress increases phosphatidylinositol 4,5-bisphosphate [PI(4,5)$P_2$], probably by activating the corresponding kinases [7, 8]. Osmotic stress is also known to activate phosphoinositide-specific phospholipases C (PLCs), which in turn hydrolyze PI(4,5)$P_2$ to generate two potential secondary messengers: the inositol 1,4,5-trisphosphate (Ins$P_3$) and the 1,2-diacylglycerol (DAG) [8–10]. Ins$P_3$ is known to diffuse into the cytosol and to activate receptors coupled to calcium channels that release local calcium concentration from intracellular stores. In animals, DAG stays within the membrane and allosterically activates the protein kinase C. Lastly, phospholipase D (PLD) can hydrolyze structural glycerophospholipids, resulting in the synthesis of phosphatidic acid (PA). PA plays a central role in the biosynthesis of major phospholipids and is also considered as a secondary messenger [11]. Increase in PA levels is observed in response to osmotic stress in plants [12–14].

To date, only parts of the signaling pathways involved in the regulation of proline homeostasis have been identified in *Arabidopsis*, but they remain poorly understood. We have shown that PLDs are negative regulators of proline biosynthesis and that plants present higher proline responsiveness to hyperosmotic stress when such PLD-mediated signaling pathway is abolished [14]. Moreover, the overexpression of a calmodulin isoform markedly triggered the expression of *P5CS1* and increased the proline content threefold in transgenic plants [15]. Plants may discern between ionic and non-ionic osmotic stresses through lipid signaling pathways, and PLCs are essential for proline accumulation upon salt stress [16]. It seems likely that calcium acts as a molecular switch to trigger downstream signaling components [16]. These results together with other works [6] indicate that PLDs and PLCs play a major role in water-stress signaling.

The salt cress *Thellungiella halophila*, which is a close relative of *A. thaliana* and possesses many of its genetic and experimental advantages, has recently been described to tolerate high salt concentrations and to naturally accumulate high level of proline [17, 18].

Therefore, in this study, we used both *Arabidopsis* and *Thellungiella* as models to discern possible lipid signaling pathways to regulate proline metabolism upon stress and non-stress conditions.

## Materials and methods

### *Plant material*

*A. thaliana* (L.) Heynh. ecotype Columbia and *T. halophila* seeds were surface-sterilized and grown on 0.5× Murashige-Skoog (MS) agar medium [19] in 14-cm diameter petri dishes. After a period at 4°C to raise dormancy, seedlings were placed under continuous illumination with 60 μmol photons $m^{-2}$ $s^{-1}$ at 22°C.

### *Salt and hyperosmotic stress and inhibitory treatments*

Twelve-day-old seedlings were removed from 0.5× MS agar plates and put onto a liquid 0.5× MS medium supplemented with 1-butanol (0.5% v/v) or U73122 (100 μM dissolved in DMSO) or with either water or DMSO as control for 1 h. Then, an equal volume of either 0.5× MS, 200 mM NaCl or 400 mM mannitol was added to the medium.

After 24-h treatment, seedlings were collected and immediately frozen in liquid nitrogen and stored at –80°C prior to analysis.

### In vivo *PLD activity*

Lipids were labeled, extracted and then separated as described [14]. Briefly, phospholipids were metabolically labeled by incubating 12-day-old seedlings for 24 h with [$^{33}$P]orthophosphate (53 MBq $L^{-1}$). Primary butyl alcohols were then added to the medium supplemented with NaCl or mannitol to measure PLD activity by transphosphatidylation through phosphatidylbutanol (PBut) formation. After extraction, lipids were separated by TLC.

### *Proline determination*

Free proline content was measured using L-proline as standard according to [20].

## Results

### *Water stress triggers a higher proline accumulation in* T. halophila *than in* A. thaliana

Upon 24-h treatment with either NaCl or mannitol, proline accumulation was observed in *A. thaliana* (Fig. 1A) and to a higher extent in *T. halophila* (Fig. 1B). While high salt concentration (400 mM) led to the death of *A. thaliana* seedlings, *T. halophila* accumulated tenfold higher proline level compared to control.

### *PLD activity presents contrasted responses in* T. halophila *and* A. thaliana

We investigated the role of PLDs in the regulation of proline metabolism in *T. halophila* and *A. thaliana*. PLD activity can lead to the synthesis of phosphatidylalcohol in the presence of a primary alcohol. This reaction, referred to as transphosphatidylation, is not only a hallmark of PLD activity but also a way to block the production of PA in functional studies. PLD activity is measured by the accumulation of PBut. We have previously shown that 1-butanol triggered proline accumulation under non-stress conditions in *A. thaliana*, while it did not have any effect on proline level upon water stress [14].

In response to 24-h treatment with either 400 mM NaCl or 400 mM mannitol, proline was accumulated up to tenfold compared with control in *T. halophila* seedlings (Fig. 1B). Interestingly, proline level decreased in 1-butanol-treated plants upon stress but did not change in normal growth condition. The fact that 1-butanol inhibited proline accumulation upon salt and hyperosmotic stress conditions may be explained by a stimulation of PLDs by 1-butanol or through a diminution of PA signaling from PLDs.

To analyze the biological function of PLDs in *T. halophila*, PLD activity was measured by transphosphatidylation reactions in the presence of primary butyl alcohols. The seedlings were incubated with [$^{33}$P]orthophosphate for 24 h

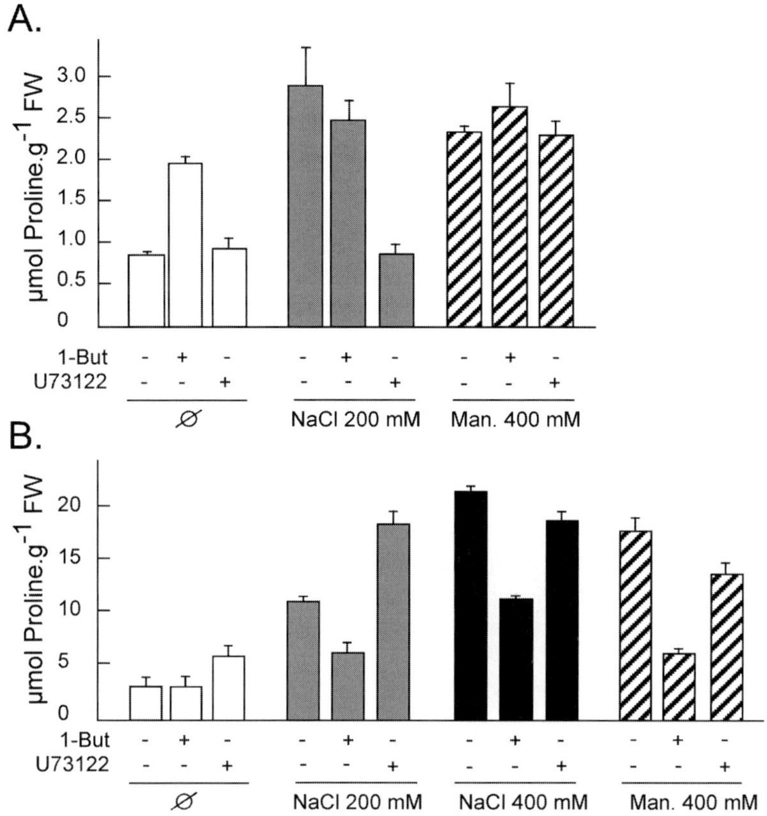


**Figure 1.** Proline accumulation is differently modulated by 1-butanol and U73122 upon NaCl and mannitol stresses in *Arabidopsis thaliana* (A) and *Thellungiella halophila* (B). Two-week-old seedlings grown on 0.5× Murashige and Skoog (MS) medium were treated for 24 h with NaCl or mannitol in the presence or absence of 0.5% of 1-butanol or 100 μM U73122. The results are shown as the means ± SE.

to label structural phospholipids and were subsequently treated with NaCl or mannitol. Lipids were then extracted and separated by TLC. Because PBut was formed after the addition of 1-butanol, the amount of PBut reflected a cumulative PLD activity under each treatment.

PA levels were lower in seedlings upon moderate salt stress but higher upon severe salt stress or mannitol stress compared with controlf plants, suggesting a differential PLD activity under the nature of water stress. Under non-stress conditions, a decrease in PA level was observed after the addition of 1-butanol, indicating that PBut was formed at the expense of PA. However, upon salt and hyperosmotic treatments, the increase in PBut level was accompanied by an increase of PA synthesis, suggesting a stimulation of PLDs by primary alcohol.

*PLCs are positive regulators of proline biosynthesis in* A. thaliana *and a negative regulator in* T. halophila

We assessed the effect of U73122, a specific PLC inhibitor, in *A. thaliana* and *T. halophila* seedlings treated for 24 h with either NaCl or mannitol (Fig. 1). When U73122 was added, a significant decrease in proline biosynthesis was observed in *A. thaliana* seedlings grown in 200 mM NaCl as well as in T. halophila seedlings grown under more severe water-stress conditions (400 mM NaCl or 400 mM mannitol). Interestingly, an increase in proline level was observed under non-stress conditions in *T. halophila* treated with U73122. In *A. thaliana*, this PLC inhibitor did not have any effect on proline accumulation whatever control or hyperosmotic conditions.

## Discussion

In the present work, we compared the water-stress responses of the glycophyte *A. thaliana* to the halophyte *T. halophila* with regard to lipid signaling pathways, which have previously been shown to be involved in the regulation of proline metabolism in *Arabidopsis*.

This pharmacological study allowed us to propose an integrated cellular model on the tight regulation of proline metabolism in response to salt stress in which PLDs and PLCs are involved. PLDs exert a negative control on proline accumulation in *A. thaliana* [14]. This negative control is raised upon water constraints (NaCl or mannitol) or with 1-butanol. In contrast to *Arabidopsis*, 1-butanol had no effect on proline accumulation in non-stress conditions in *T. halophila*. However, upon water stress, the contribution of 1-butanol did not lead to an additional accumulation of proline, but rather a reduction of proline content. Our data suggest the implication of PLD signaling in the regulation of proline metabolism in both *A. thaliana* and *T. halophila* but in opposite fashions.

PLCs exert a positive control on proline accumulation upon salt stress in *A. thaliana*. Interestingly, a negative control of proline accumulation mediated by PLCs operates in *T. halophila* whatever non-stress and moderate salt stress conditions are applied.

This work clearly showed that proline metabolism is tightly regulated by lipid signaling pathways in plants, which act in an opposite way in *A. thaliana*, a glycophyte, and *T. halophila*, a halophyte. Such regulation may partially explain why higher proline response is observed in the latter species. We have also demonstrated that lipid signaling pathways are involved in sensing ionic and non-ion stresses.

## References


1 Boyer JS (1982) Plant productivity and environment. *Science* 218: 443–448

2 Hare P, Cress W, van Staden J (1998) Dissecting the roles of osmolyte accumulation during stress. *Plant Cell Environ* 21: 535–553

3 Nakashima K, Satoh R, Kiyosue T, Yamaguchi-Shinozaki K, Shinozaki K (1998) A gene encoding proline dehydrogenase is not only induced by proline and hypoosmolarity, but is also developmentally regulated in the reproductive organs of *Arabidopsis*. *Plant Physiol* 118: 1233–1241

4 Weltmeier F, Ehlert A, Mayer CS, Dietrich K, Wang X, Schütze K, Alonso R, Harter K, Vicente-Carbajosa J, Dröge-Laser W (2006) Combinatorial control of *Arabidopsis* proline dehydrogenase transcription by specific heterodimerisation of bZIP transcription factors. *EMBO J* 25: 3133–3143

5 Delauney AJ, Verma DPS (1993) Proline biosynthesis and osmoregulation in plants. *Plant J* 4: 215–223

6 Meijer HJ, Munnik T (2003) Phospholipid-based signaling in plants. *Annu Rev Plant Biol* 54: 265–306

7 DeWald DB, Torabinejad J, Jones CA, Shope JC, Cangelosi AR, Thompson JE, Prestwich GD, Hama H (2001) Rapid accumulation of phosphatidylinositol 4,5-bisphosphate and inositol 1,4,5-trisphosphate correlates with calcium mobilization in salt-stressed *Arabidopsis*. *Plant Physiol* 126: 759–769

8 Pical C, Westergren T, Dove SK, Larsson C, Sommarin M (1999) Salinity and hyperosmotic stress induce rapid increases in phosphatidylinositol 4,5-bisphosphate, diacylglycerol pyrophosphate, and phosphatidylcholine in *Arabidopsis thaliana* cells. *J Biol Chem* 274: 38232–38240

9 Drobak BK, Watkins PA (2000) Inositol(1,4,5)trisphosphate production in plant cells: An early response to salinity and hyperosmotic stress. *FEBS Lett* 481: 240–244

10 Takahashi S, Katagiri T, Hirayama T, Yamaguchi-Shinozaki K, Shinozaki K (2001) Hyperosmotic stress induces a rapid and transient increase in inositol 1,4,5-trisphosphate independent of abscisic acid in *Arabidopsis* cell culture. *Plant Cell Physiol* 42: 214–222

11 Testerink C, Munnik T (2005) Phosphatidic acid: A multifunctional stress signaling lipid in plants. *Trends Plant Sci* 10: 368–375

12 Frank W, Ratnadewi D, Reski R (2005) *Physcomitrella patens* is highly tolerant against drought, salt and osmotic stress. *Planta* 220: 384–394

13 Munnik T, Meijer HJ, Ter Riet B, Hirt H, Frank W, Bartels D, Musgrave A (2000) Hyperosmotic stress stimulates phospholipase D activity and elevates the levels of phosphatidic acid and diacylglycerol pyrophosphate. *Plant J* 22: 147–154

14 Thiery L, Leprince AS, Lefebvre D, Ghars MA, Debarbieux E, Savouré A (2004) Phospholipase D is a negative regulator of proline biosynthesis in *Arabidopsis thaliana*. *J Biol Chem* 279: 14812–14818

15 Yoo JH, Park CY, Kim JC, Heo WD, Cheong MS, Park HC, Kim MC, Moon BC, Choi MS, Kang YH et al (2005) Direct interaction of a divergent CaM isoform and the transcription factor, MYB2, enhances salt tolerance in *Arabidopsis*. *J Biol Chem* 280: 3697–3706

16 Parre E, Ghars MA, Leprince AS, Thiery L, Lefebvre D, Bordenave M, Richard L, Mazars C, Abdelly C, Savoure A (2007) Calcium signaling *via* phospholipase C is essential for proline accumulation upon ionic but not nonionic hyperosmotic stresses in *Arabidopsis*. *Plant Physiol* 144: 503–512

17 Inan G, Zhang Q, Li P, Wang Z, Cao Z, Zhang H, Zhang C, Quist TM, Goodwin SM, Zhu J et al (2004) Salt cress. A halophyte and cryophyte *Arabidopsis* relative model system and

its applicability to molecular genetic analyses of growth and development of extremophiles. *Plant Physiol* 135: 1718–1737

18 Ghars MA, Parre E, Debez A, Bordenave M, Richard L, Leport L, Bouchereau A, Savouré A, Abdelly C (2008) Comparative salt tolerance analysis between *Arabidopsis thaliana* and *Thellungiella halophila*, with special emphasis on $K^+/Na^+$ selectivity and proline accumulation. *J Plant Physiol* (in press)

19 Murashige T, Skoog F (1962) A revised medium for rapid growth and bio assays with tobacco tissue culture. *Physiol Plant* 15: 473–497

20 Bates LS, Waldren RP, Teare ID (1973) Rapid determination of free proline for water-stress studies. *Plant and Soil* 39: 205–207

Biosaline Agriculture and High Salinity Tolerance
Edited by Chedly Abdelly, Münir Öztürk, Muhammad Ashraf and Claude Grignon


# Cadmium and copper genotoxicity in plants

Chamseddine Mediouni[1,2], Guy Houlné[2], Marie-Edith Chaboute[2], Mohamed Habib Ghorbel[1] and Fatma Jemal[1]

[1] *Unité de Recherche „Nutrition et Métabolisme Azotés et Protéines de Stress“ UR 99/09-20, Faculté des Sciences de Tunis, Université Tunis EL-MANAR, 1060 Tunisie*

[2] *Equipe Signalisation de la réparation de l'ADN, Institut de Biologie Moléculaire des Plantes, 12 rue du Général Zimmer, 67084 Strasbourg Cedex, France*

## Abstract

Heavy metal contamination in soils is easily transmitted to human through plants *via* the food chain. A major concern is to understand the plant response to heavy metal soil contamination to develop phytoremediation. Two plant models have been investigated in our study, the tomato, which is of agronomical importance, Arabidopsis, which is used as a model for molecular genetics. Heavy metal toxicity is described to induce oxidative stress linked to oxidation of proteins and membrane lipids but also to alterations of DNA damage response. We have investigated the metabolic response of cadmium and copper in parallel in both plant models and analyzed the transcriptional response of Arabidopsis RNR genes coding for isoforms of ribonucleotide reductase, an essential enzyme involved in DNA synthesis. Both Cd and Cu had a dose-dependent effect on plant growth. We also observed a rapid increase of catalase activity upon Cd or Cu treatments in tomato and Arabidopsis. At the transcriptional level, treatment with Cd resulted in a biphasic induction of two RNR genes in Arabidopsis; the first induction peak could be paralleled to the increase of the catalase activity.

## Introduction

Heavy metal contamination in soil is mainly due to industrial emission, application of sewage sludge and phosphate fertilizers [1], and represents a serious problem in crop and vegetable production. Heavy metals, such as cadmium and copper can be readily taken up by plant roots and transported to different plant organs [2]. Heavy metal uptake and accumulation in plants poses a serious health issue to humans through the food chain [3].

The presence of excessive amounts of heavy metals in soil commonly elicits stress symptoms in plants, such as growth reduction, disturbances in mineral nutrition and carbohydrate metabolism [4]. Up to now relatively little information has been available on the plant response to the same rate of cadmium or copper exposure. The few available reports comparing Cu and Cd effects suggest that plants are more sensitive to the former metal [5].

The toxicity of heavy metals has been related to their ability to induce oxidative stress [6]. Reactive oxygen species (ROS) may lead to unspecific oxidation of proteins and membrane lipids, resulting in an increased concentration of thiobarbituric acid-reacting substances (TBARS), which are products of lipid peroxidation [7]. A large accumulation of TBARS has been reported after Cu [8] or Cd [9] exposure. Plants have developed several systems to protect themselves against oxidative stress. Antioxidative enzymes play an important role in this type of protection. Catalase (CAT) removes hydrogen peroxide [10]. The effects of Cd and Cu on the activity of this enzyme and their involvement in the defense of plant tissues against metal-induced damage remain controversial. Several studies have shown a Cu- or Cd-dependent increase in the activity of CAT [11, 12]. Others report that CAT activity can be inhibited under heavy metals exposure [13, 14].

ROS are generated continuously by a diverse range of intracellular oxidative events and can alter the biological activity of enzymes, modulating intracellular signaling, and exerting damaging effects on biological macromolecules, including DNA [15]. Cells are equipped with antioxidants, ROS-eliminating enzymes, and efficient DNA repair mechanism. However, when levels of ROS exceed the capacity for defense or antioxidant systems, oxidative damage occurs to DNA, which often leads to DNA mutations, double-strand breaks and chromosomal deletions [16].

In all cellular organisms, ribonucleotide reductase (RNR) is an enzyme essential to the cell providing in a limiting step the deoxyribonucleoside triphosphates (dNTPs) required for DNA replication and DNA repair [17]. RNR is an $\alpha2\beta2$-type enzyme coomprised of two large subunits (R1) and two small subunits (R2). In *Arabidopsis thaliana*, RNR1 gene is unique, whereas RNR2 genes belong to a small multigenic family (ATRNR2,3, ATRNR2,3 bis and AtRNE2,5). In response to DNA damage, different RNR genes are specifically induced, according to the genetics applied.

The objectives of the present study were to examine in two plant models the effects of Cd and Cu. We characterized the response of antioxidative enzyme CAT. At the transcriptional level, we analyzed the response of two *RNR* genes (ATRNR2,3 bis and AtRNR2,5) encoding ribonucleotide reductases under Cd exposure.

## Materials and methods

### *Plant materials and growth conditions*

The seeds of tomato (*Lycopersicon esculentum* Mill. cv. Ibiza F1) were sterilized in 10% hydrogen peroxide for 20 min. They were then thoroughly washed with distilled water and germinated on moistened filter paper at 25°C in the dark. Selected, uniform seedlings were transferred to continuously aerated nutrient solutions containing 0.5 mM $KH_2PO_4$; 1.25 mM $Ca(NO_3)_2$; 1

mM $KNO_3$; 0.5 mM $MgSO_4$; 50 μM Fe-K-EDTA; 5 μM $MnSO_4$,; 1 μM $ZnSO_4$; 1 μM $CuSO_4$; 30 μM $H_3BO_3$; and 1 μM $(NH_4)_6$ $MO_7O_{24}$. After an initial growth period of 10 days, either $CdCl_2$ or $CuSO_4$ was added to the medium. Plants were grown and treated in a growth chamber (26°C/70% relative humidity during the day, 20°C/90% during the night). A 16-h (daily) photoperiod was used with a light irradiance of 150 μmol $m^2$ $s^{-1}$ at the canopy level. After 7 days of heavy metal treatment, plants were assorted into shoots and roots. Roots were rapidly washed three times in 1 L distilled water. Samples of shoots and roots were desiccated at 60°C and dry weight was determined. Leaves samples used for TBARS content and CAT activity determination were weighted and stored in liquid nitrogen.

*Arabidopsis thaliana* (Colombia 0) seeds were grown in liquid Murashige and Skoog (MS) medium supplemented with sucrose (10 g $L^{-1}$) in petri dishes at 21°C with a 16-h (daily) photoperiod. Eighteen-day-old seedlings were then transferred to petri dishes supplemented with 25 μM $CdCl_2$ or $CuSO_4$. Samples collected for CAT activity determination and RNA extraction are weighted and stored in liquid nitrogen until use.

*Arabidopsis* hydroponic culture was performed according to Tocquin et al. [18]. The same treatment with Cd or Cu (25 μM) was applied to 18-day-old seedlings. Samples collected were used to estimate TBARS content.

### *Lipid peroxide determination*

Lipid peroxidation was determined by measuring the concentration of thiobarbituric acid-reacting substances (TBARS), as described by Buege and Aust [19]. TBARS concentration was calculated using an extinction coefficient of 155 $mM^{-1}$ $cm^{-1}$.

### *Protein determination and enzyme assay*

Protein content in plant material preparations was determined using the method of Bradford [20], with bovine serum albumin (BSA) as a standard. CAT activity was determined according to Aebi [21].

### *RNA extraction and RT-PCR analysis*

For semiquantitative RT-PCR, total RNA was isolated with Tri-reagent (Sigma Aldrich). For cDNA synthesis 1 μg of total RNA was used. PCR was conducted with 1 μL of RT reaction at 94°C for 4 min for 1 cycle followed by 94°C for 1 min, 56°C for 40 s and 72°C for 50 s for 34 cycles. Primers used for PCR amplification were 5'-TCGCTTGTCTACTCTACACG-3' and 5'-CCGCGTCGCAGACGATTGA-3' for AtRNR 2,3 bis, and 5'-CTGGA-

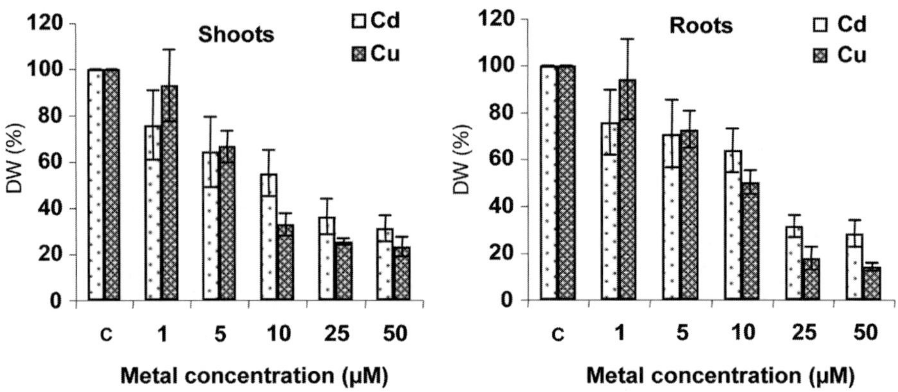


**Figure 1.** Effect of Cd and Cu on tomato growth. Biomass (dry weight, DW) of shoots or roots from tomato seedlings not treated (C) or treated for 7 days with various concentrations of $CdCl_2$ or $CuSO_4$. Data (in % of the control C) are means ± SE ($n = 7$). The DW (mg) of the untreated plants (C) is: 0.232 ± 0.037 for shoots and 0.035 ± 0.005 for roots.

CAGCCGAAGAAGTC-3' and 5'-AGTGACGTTCGTCGTTGGTT-3' for AtRNR 2,5. PCR products were analyzed by gel electrophoresis and quantified using Quantity One Bio-Rad software.

## Results

### *Cd and Cu effects on plant growth*

Growing tomato plants for 7 days on nutrient solution supplemented with different concentrations of $CdCl_2$ or $CuSO_4$ resulted in the inhibition of shoot and root growth (Fig. 1). Decrease in biomass production occurred despite the low metal concentrations and depended on the nature and the concentration of the metal. With 1 µM Cd treatment, growth of shoots and roots was reduced by 24% compared to the untreated plants. The same concentration of Cu did not lead to a significant decline in the biomass production of either plant organ. Treatment of tomato plants with 25 µM Cu reduced the growth of shoots and roots to 74% and 82% of the control, respectively. With Cd at the same concentration, the growth was 64% (shoots) and 68% (roots) of the control.

Chlorosis and necrosis of leaves, and reduced length and browning of roots were the main visual toxicity symptoms in tomato and *Arabidopsis* plants under Cd and Cu exposure. These effects were observed, especially at metal concentrations exceeding 5 µM, and were more marked with Cu (data not shown).

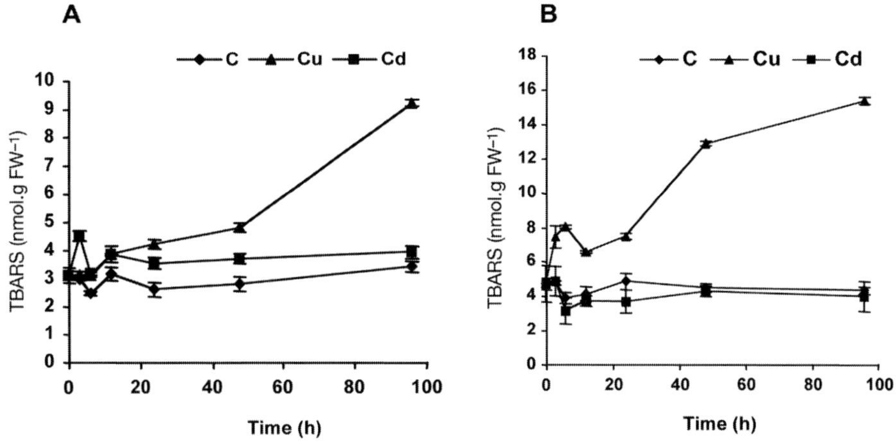


**Figure 2.** Effect of Cd and Cu on lipid peroxidation. Thiobarbituric acid reactive substances (TBARS) in leaves of tomato (A ) and *Arabidopsis* plants (B) not treated (control, C) or treated with 25 μM $CdCl_2$ or $CuSO_4$.

## *Evaluation of oxidative stress and associated defense responses upon Cd and Cu treatment*

### Effects on TBARS content

Effects of Cd and Cu on TBARS contents in leaves of tomato plants and in *Arabidopsis thaliana* plants are reported in Figure 2. In tomato leaves, Cd treatment induced an immediate increase of the lipid peroxidation products. A maximal accumulation was detected after 3 h, the TBARS amount then diminished but remained slightly higher than the control level (Fig. 2A). On Cu exposure, TBARS content increased from 12 h and it continued to augment until day 4. In *Arabidopsis thaliana* plants, lipoperoxide product accumulation was detected 3 h after Cu treatment and reached a maximal level at day 4. In contrast, Cd did not have a significant effect on TBARS accumulation during this period (Fig. 2B).

### Effects on CAT activity

In control plants, the CAT activity only showed a limited fluctuation during seedling growth. Cd and Cu treatment increased CAT activity 3 h after application (Fig. 3A). Augmented CAT activity was maintained through exposure time, especially with Cu. A similar early induction of CAT activity by Cd and Cu treatment was observed in *Arabidopsis* plants (Fig. 3B). While CAT activity remained high with Cu treatment until day 4, Cd-induced CAT

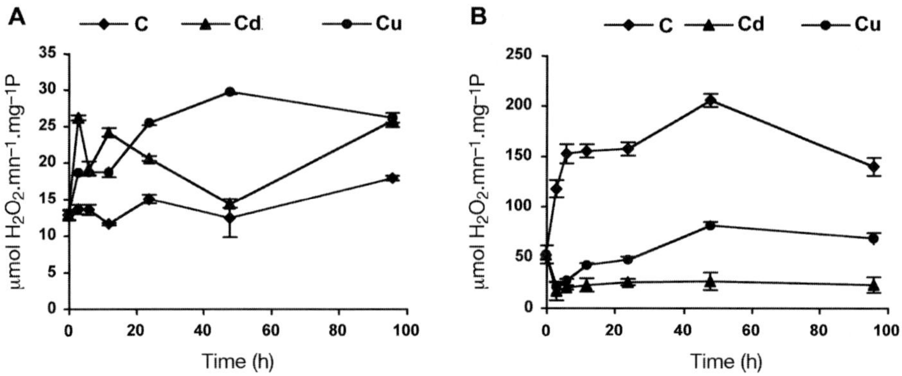


**Figure 3.** Effect of Cd and Cu on catalase activity. Catalase activity in leaves of tomato (A ) and *Arabidopsis* plants (B) not treated (control, C) or treated with 25 µM $CdCl_2$ or $CuSO_4$.

activity decreased through exposure time, and returned to the control level at 48 h.

### *Transcriptional response of RNR genes in Arabidopsis upon Cd treatment*

The effects of Cd on *AtRNR2,3 bis* and *AtRNR2,5* gene expression are shown in Figure 4. Treatment with Cd led to a biphasic activation of the expression of these genes. The first peak of transcript abundance appeared, at 3 h after Cd addition for *AtRNR2,5* and at 6 h for *AtRNR2,3 bis*. A second peak was observed after, respectively, 24 and 48 h.

## Discussion

Both Cd and Cu reduced the growth of all tomato organs. At low concentrations, Cd was more efficient than Cu. This could be due to the fact that Cu is an essential oligonutrient for plant growth. It is involved in numerous physiological functions as a component of several plant enzymes [22]. On the contrary, Cd is known to be a non-essential element for growth. However, at high concentrations Cu was more toxic than Cd.

The reduction of growth could be a consequence of heavy metals interference with a number of metabolic processes associated with normal development, like photosynthesis, nitrogen metabolism, and lipid metabolism. Therefore, difference in heavy metals toxicity could be due to a difference in their effect in such metabolisms. Indeed, $NO_3^-$ uptake inhibition, and reduction of lipid content, of photosynthetic $O_2$ evolution, and of PSII photochemical yield have been described following exposure to Cu [23, 24].

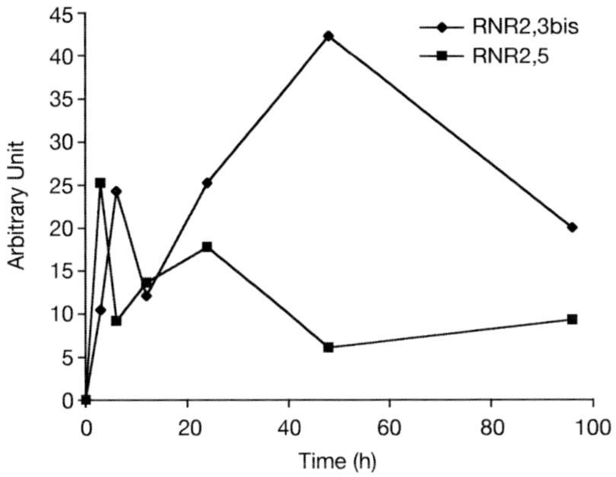


**Figure 4.** Effects of Cd on RNR gene expression in *Arabidopsis*. Evolution of *RNR* 2,3 bis and *RNR* 2,5 transcript abundance after addition of 25 μM Cd to the medium.

Toxicity of heavy metals may also be related to their ability to disorganize cell membrane through lipid peroxidation. Our results suggest that Cd and especially Cu induce oxidative stress in tomato and *Arabidopsis* plants, resulting in accumulation of lipid peroxides. The early Cu-induced accumulation of peroxidation products could be due to an increase in the activity of lipoxygenase (LOX), an enzyme known to initiate the process of lipid peroxidation. The increase in the capacity of this enzyme after heavy metal treatment has been shown to reach a maximal level after 24 h [25]. Moreover, the increase in LOX activity, in *A. thaliana*, appeared to be more enhanced with Cu than with Cd [26]. Accumulation of ROS has been described in several reports. $H_2O_2$ was observed to be greatly accumulated 24 h after heavy metal application [25]. Cu induced an increase in TBARS level beyond 24 h, which could be a consequence of Cu-induced ROS accumulation rather than Cu-induced LOX activation. In addition, Cu is a most powerful catalyzer of free radical formation and Cu ions themselves can initiate directly oxidative breakdown of polyunsaturated lipids [27]. Conversely, Cd is not a transition metal and hence does not directly generate damaging oxygen species.

Concomitantly with the Cu- and Cd-induced TBARS accumulation, an early increase in CAT activity in tomato leaves and in *Arabidopsis* plants was observed (Fig. 3). Since CAT is implicated in scavenging of ROS [28], the heavy metal-induced rise of its activity is a circumstantial evidence for an enhanced ROS production. Such an increase in the activity of CAT enzyme has been reported in several plant species [11, 12]. During all the experimental period, except at 3 h, Cu appeared to be the highest stimulator of CAT activity in both plant species tested. This suggests that Cu was more efficient than Cd in triggering the formation of ROS, which could explain the higher TBARS accumulation observed in the presence of Cu.

Like other environmental mutagens, such as UV light, heavy metals-induced ROS can induce DNA damage [29]. To protect the genetic integrity of organisms, DNA damage is countered in cells by DNA repair, including a mismatch repair pathway, homologous or illegitimate recombination processes [30]. Either DNA repair pathway requires dNTPs, which are provided by RNR. In fact, $CdCl_2$ treatment resulted in a biphasic activation of *AtRNR2,3 bis* and *AtRNR2,5* genes. The first peak of *RNR* gene induction coincided with increased CAT activity induced by Cd, and so could be a response for ROS-induced DNA damage in *A. thaliana* plants. The second peak of induction might be related to stress signaling.

## Conclusion

Cadmium and copper induce a severe growth inhibition and lead to a lipid peroxidation, more pronounced at high Cu levels. Cd- and Cu-induced lipid peroxidation was correlated with the ability of these metals to stimulate CAT activity in both tomato and *Arabidopsis*. The induction of *RNR* genes under Cd exposure was concomitant to the increase of CAT activity, mainly at early time points of treatment. It is possible that *RNR* expression was triggered by Cd-induced ROS production.

## References

1 Wagner GJ (1993) Accumulation of cadmium in crop plants and its consequence to human health. *Adv Agron* 51: 173–212

2 Mediouni C, Benzarti O, Tray B, Ghorbel MH, Jemal F (2006) Cadmium and copper toxicity in tomato seedlings (*Lycopersicon esculentum*). *Agron Sustain Dev* 26: 227–232

3 Shah K, Dubey RS (1998) A 18 kDa cadmium inducible protein complex from rice: Its purification and characterization from rice (*Oryza sativa* L.) roots tissues. *J Plant Physiol* 152: 448–454

4 Dong J, Wu F, Zhang G (2006) Influence of cadmium on antioxidant capacity and four microelement concentrations in tomato seedlings (*Lycopersicon esculentum*). *Chemosphere* 64, 1659–1666

5 Xia JR, Li YJ, Lu J, Chen B (2004) Effects of copper and cadmium on growth, photosynthesis and pigment content in *Gracilaria lemaneiformis*. *Bull Environ Contamin Toxicol* 73: 979–986

6 Mishra S, Srivastava S, Tripathi RD, Govindarajan R, Kuriakose SV, Prasad MNV (2006) Phytochelatin synthesis and response of antioxidants during cadmium stress in *Bacopa monnieri* L. *Plant Physiol Biochem* 44: 24–37

7 Cho UH, Seo NH (2005) Oxidative stress in *Arabidopsis thaliana* exposed to cadmium is due to hydrogen peroxide accumulation. *Plant Sci* 168: 113–120

8 Groppa MD, Tomaro ML, Benavides MP (2001) Polyamines as protectors against cadmium or copper-induced oxidative damage in sunflower leaf discs. *Plant Sci* 161: 481–488

9 Smeets K, Cuypers A, Lambrechts A, Semane B, Hoet P, Van Laere A, Vangronsveld J (2005) Induction of oxidative stress and antioxidative mechanisms in *Phaseolus vulgaris* after Cd application. *Plant Physiol Biochem* 43: 437–444

10 Foyer CH, Noctor G (2000) Oxygen processing in photosynthesis: Regulation and signaling. *New Phytol* 146: 359–388
11 Lombardi L, Sebastiani L (2005) Copper toxicity in *Prunus cerasifera*: Growth and antioxidant enzymes responses of *in vitro* grown plants. *Plant Sci* 168: 797–802
12 Zhang H, Jiang Y, He Z, Ma M (2005) Cadmium accumulation and oxidative burst in garlic (*Allium sativum*). *J Plant Physiol* 162: 977–984
13 Drazkiewicz M, Skorzynska-Polit E, Krupa Z (2004) Copper-induced oxidative stress and antioxidant defence in *Arabidopsis thaliana*. *Biometals* 17: 379–387
14 Chaoui A, El Ferjani E (2005) Effects of cadmium and copper on antioxidant capacities, lignification and auxin degradation in leaves of pea (*Pisum sativum* L.) seedlings. *C R Biol* 328: 23–31
15 Sagun KC, Carcamo JM, Golde DW (2006) Antioxidants prevent oxidative DNA damage and cellular transformation elicited by the over-expression of c-MYC. *Fund Mol Mechan Mutagen* 593: 64–79
16 Halliwell B, Gutteridge JMC (1989) *Free radicals in biology and medicine.* Clarendom, Oxford
17 Nordlund P, Reichard P (2006) Ribonucleotide reductase. *Annu Rev Biochem* 75: 681–706
18 Tocquin P, Corbesier L, Havelange A, Pieltain A, Kurtem E, Bernier G, Périlleux C (2003) A novel high efficiency, low maintenance, hydroponic system for synchronous growth and flowering of *Arabidopsis thaliana*. *BMC Plant Biol* 3: 2
19 Buege AJ, Aust SD (1972) Microsomal lipid peroxidation. *Methods Enzymol* 52: 302–310
20 Bradford MM (1976) A rapid and sensitive method for the quantitative determination of microgram quantities of protein utilizing the principle of protein-dye binding. *Anal Biochem* 72: 248–254
21 Aebi H (1984) Catalase *in vitro*. *Methods Enzymol* 105: 121–126
22 Marschner H (1995) *Mineral nutrition of higher plants*, 2nd edn. Academic Press, London, 337–347
23 Mosulén S, Domínguez MJ, Vigara J, Vílchez C, Guiraum A, Vega JM (2003) Metal toxicity in *Chlamydomonas reinhardtii*. Effect on sulfate and nitrate assimilation. *Biomol Eng* 20: 199–203
24 Ouariti O, Boussama N, Zarrouk M, Cherif A, Ghorbal MH (1997) Cadmium- and copper-induced changes in tomato membrane lipids. *Phytochemistry* 45: 1343–1350
25 Weckx JEJ, Clijsters HMM (1997) Zn phytotoxicity induces oxidative stress in primary leaves of *Phaseolus vulgaris*. *Plant Physiol Biochem* 35: 405–410
26 Skorzynska-Polit E, Krupa Z (2003) The activity of lipoxygenase in *Arabidopsis thaliana* (L.) Heynh – A preliminary study. *Cell Mol Biol Lett* 8: 279–284
27 De Vos CHR, Bookum VMT, Vooijs R, Schat H, Dekok LJ (1993) Effect of copper on fatty acid composition and peroxidation of lipids in roots of copper tolerant and sensitive *Silene cucubalus*. *Plant Physiol Biochem* 31: 151–158
28 Mishra S, Srivastava S, Tripathi RD, Govindarajan R, Kuriakose SV, Prasad MNV (2006) Phytochelatin synthesis and response of antioxidants during cadmium stress in *Bacopa monnieri* L. *Plant Physiol Biochem* 44: 24–37
29 Lindahl T (1993) Instability and decay of the primary structure of DNA. *Nature* 362: 709–715
30 Hartwig A, Asmuss M, Ehleben I, Herzer U, Kostelac D, Pelzer A, Schwerrdtle T, Bürkle A (2002) Interference by toxic metal ions with DNA repair process and cell cycle control: molecular mechanisms. *Environ Health Perspect* 110: 797–799

Biosaline Agriculture and High Salinity Tolerance
Edited by Chedly Abdelly, Münir Öztürk, Muhammad Ashraf and Claude Grignon


# Effect of salinity on growth, leaf-phenolic content and antioxidant scavenging activity in *Cynara cardunculus* L.

Falleh Hanen, Riadh Ksouri, Wided Megdiche, Nejla Trabelsi, Mondher Boulaaba and Chedly Abdelly

*Laboratoire d'Adaptation des Plantes aux Stress Abiotiques, Centre de Biotechnologie à la Technopole de Borj Cédria (CBBC), BP 901, 2050 Hammam-lif, Tunisia*

## Abstract

*Cynara cardunculus* L. (cardoon) is a medicinal plant widespread in arid and semi-arid regions where high salinity frequently occurs. Cardoon leaves are known for their high content of natural bioactive molecules, notably polyphenols, that exhibit pharmacological activities such as antioxidant, antibacterial, and metal chelating activity. We studied the effect of different salt concentrations on plant growth, phenolic content and superoxide scavenging activity in locally grown *C. cardunculus* L. leaves (El Jem locality). No significant effect on leaf growth (leaf biomass, length and number) was found at moderate salinity (25–50 mM NaCl). However, these parameters were severely reduced (–30 to –90% as compared to the control) at 150 mM NaCl. Leaf phenolic content was significantly increased at 25–50 mM NaCl, and decreased at 150 mM NaCl. The superoxide anion scavenging capacity of leaf extracts was stimulated by salt treatment, with a maximum at the highest NaCl level. Our findings indicate that the two studied characteristics of *C. cardunculus* leaves (polyphenol content and $O_2^-$ quenching capacity) were not modified in parallel with increasing salinity, and that only the polyphenol content was correlated with leaf growth.

## Introduction

Soil salinity is usually high in arid regions where evaporation is greater than the amount of precipitation infiltrating the soil. Excess salt in soil causes growth inhibition of most plants species, and thus limits crops production [1]. However, some moderately or highly salt-tolerant plants can survive in salty environments. These species are able to avoid ion toxicity and maintain water uptake in the presence of high salt concentration [2]. They are also able to avoid other disorders caused by salinity, e.g., the formation of reactive oxygen species (ROS). ROS are compound able to react with biological molecules, such as DNA, proteins, or lipids, generating mutations and damaging membranes, resulting in cell and tissue injuries [3]. Plant species localized in these hostile conditions are able to withstand ROS, owing to their powerful antioxidant systems operating *via* two main mechanisms, primary and secondary antioxidants [4]. Primary antioxidants can inhibit or delay oxidation by scavenging ROS, and by donation of hydrogen atoms or electrons. Secondary antioxidants function by binding metal ions, scavenging oxygen, converting hydroperoxides to non-radical species, absorbing UV radiation or deactivating singlet oxygen [5, 6]. Among the various natural secondary antioxidants, polyphenols have received much attention [7, 8]. In fact, phenolics are stress metabolites exhibiting a wide range of physiological properties, such as anti-oxidant, anti-allergic, anti-artherogenic, anti-inflammatory, anti-microbial, anti-thrombotic, cardioprotective and vasodilatory effects [9].

*Cynara cardunculus* L. (Asteraceae), commonly named "cardoon", is a Mediterranean species that endures extreme pedoclimatic conditions in arid and semi-arid regions [2, 10]. It is a highly productive, multipurpose, robust plant. Cardoon flowers are traditionally used for cheese preparation [11, 12], while leaves are known for their therapeutic potential in folklore as having diuretic, choleretic, cholagogic, anti-diabetic and hepatoprotective properties [13]. However, data relating to the polyphenol compounds and their anti-oxidant and antibacterial activities of this species are scarce [11, 14]. The aim of the present study was to explore the effects of salinity on *C. cardunculus* L. leaf characteristics (growth, total leaf polyphenol content and capacity to quench superoxide anion).

## Materials and methods

### *Plant materials and growth conditions*

Seeds of *C. cardunculus* L. were sampled from the El Jem locality (semi-arid bioclimatic stage), sterilized and scarified in sulfuric acid ($H_2SO_4$) for 15 min. They were then placed to germinate in the dark at 25°C. Four-day old seedlings were placed into pots filled with inert and sterilized sand under greenhouse conditions (25±5°C temperature, 60±10% relative humidity). They were then irrigated with a standard nutrient solution containing: 2.5 mM $Ca(NO_3)_2$; 1 mM $MgSO_4$; 2 mM KCl; 0.25 mM $CaH_2(PO_4)_2$, 12.5 μM $H_3BO_3$; 1 μM $MnSO_4$; 1 μM $ZnSO_4$; 0.25 μM $CuSO_4$; 0.2 μM $(NH_4)_6Mo_7O_{24}$; 10 μM Fe-ethylene-diamine-di-*o*-hydroxyphenylacetic acid [2]. After 42 days, the plants were challenged with salt (25, 50, 75, 100, 150 mM NaCl) for 109 days. At the end of this period, leaves (seven replicates per treatment) were collected to study main growth parameters (leaf biomass, length and number), and extracted with pure methanol for assays of phenolic content and superoxide anion scavenging activity.

### *Analysis of total phenolic content*

Total phenolic content of leaf methanolic extract was assayed spectrophotometrically according to Dewanto et al. [15], using the Folin-Ciocalteu reagent.

Total phenolic content was expressed as mg gallic acid equivalents per gram of dry weight (mg GAE $g^{-1}$ DW). All samples were analyzed in three replications.

### *Superoxide anion scavenging activity*

Superoxide scavenging capacity of leaf methanolic extract was assessed using the method described by Duh and Yen [16]. A phenazine methosulfate-NADH mixture was used to generate superoxide radical anions, which oxidized NBT to formazan. After incubation in the presence of different aliquots of the leaf extract, formazan concentration was spectrophotometrically measured. The $IC_{50}$ value was defined as the amount of antioxidant necessary to reduce the generation of superoxide radical anions by 50%. The $IC_{50}$ values were expressed as mg $ml^{-1}$. A lower $IC_{50}$ value corresponds to a higher antioxidant activity of leaf methanolic extract. The inhibition percentage of superoxide anion generation was calculated as $100 \times [(A_0 - A_1)/A_0]$, where $A_0$ is the absorbance of the control, and $A_1$ is the absorbance of the sample. All samples were analyzed in three replications.

## Results

### *Leaf growth parameters*

Moderate salt treatments (25–50 mM NaCl) did not modify total leaf biomass, as compared to controls in salt-free medium (about 2 g DW per plant). At higher salt concentrations in the culture medium, leaf biomass was reduced,

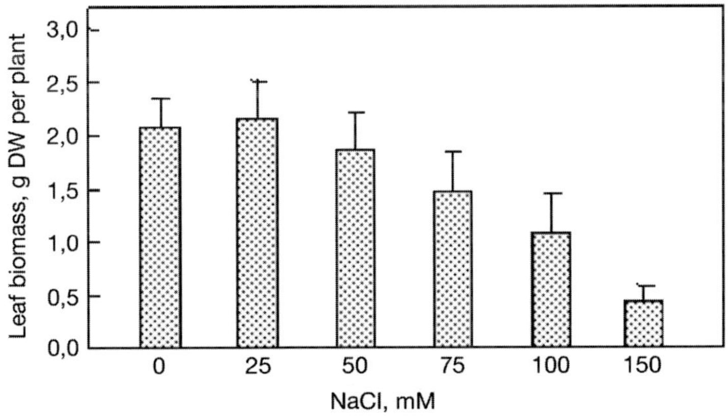


**Figure 1.** Effect of salt treatment on total leaf biomass. Five-month-old plants were irrigated with salt solution for 109 days. Mean of seven replicates and SE at $p = 0.05$.

down to 20% of control at 150 mM NaCl (Fig. 1). The leaf number per plant was much less sensitive to salt concentration than leaf biomass (Fig. 2). It ranged from six in control plants to five at 100 mM NaCl and three at 150 mM. The mean leaf length was limited by salinity (Fig. 3). It was reduced to 26% of control at 150 mM NaCl, which compared to leaf biomass reduction (20%).

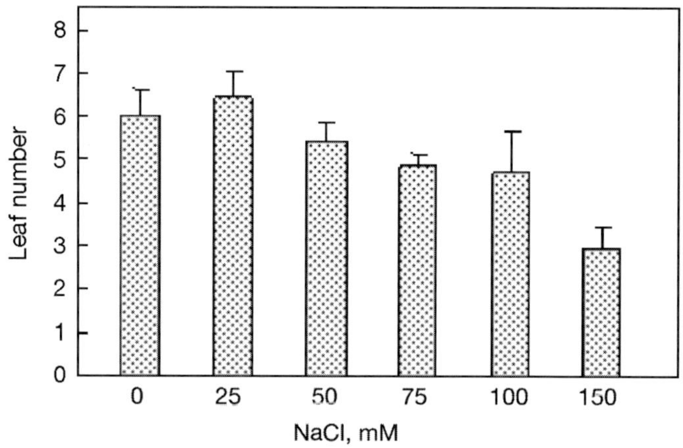


**Figure 2.** Effect of salt on leaf number. Five-month-old plants were irrigated with salt solution for 109 days. Mean of seven replicates and SE at $p = 0.05$.

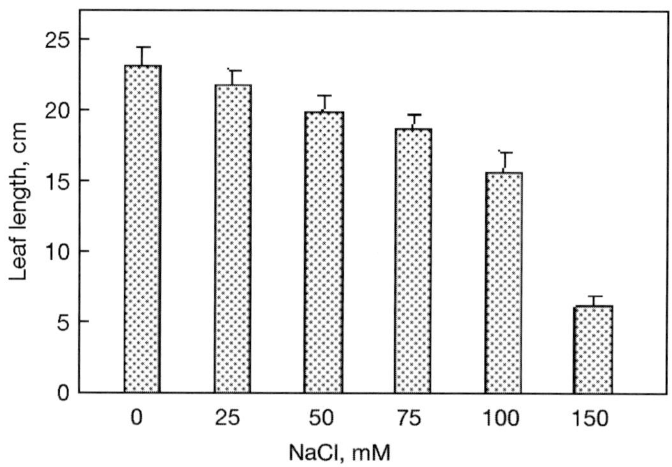


**Figure 3.** Effect of salt treatment on individual leaf length. Five-month-old plants were treated wth salt solution for 109 days. Mean of seven replicates and standard error at $p = 0.05$.

*Total phenolic content*

The total phenolic content in leaves of control plants was about 1.5 mg GAE $g^{-1}$ DW. It was significantly increased by moderate salt treatment (Fig. 4) with a maximum at 50 mM NaCl, and decreased at 150 mM NaCl.

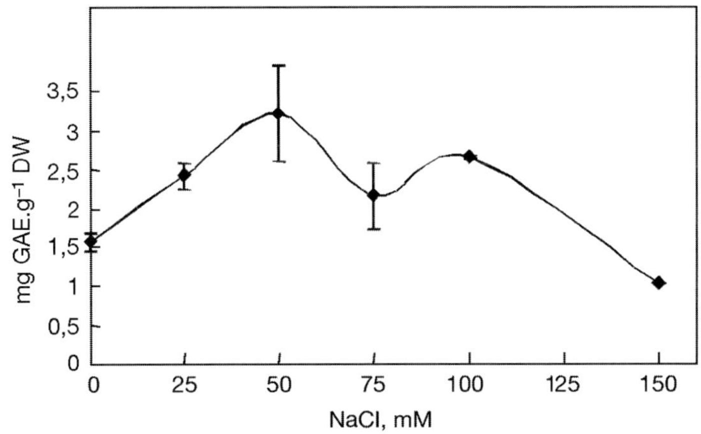


**Figure 4.** Effect of salt treatment on leaf total phenolic content. Five-month-old plants were treated with salt solution for 109 days. Phenolic content (PC) is expressed as mg gallic acid equivalent (GAE). Mean of seven replicates and standard error at $p = 0.05$.

*Superoxide anion scavenging activity*

The $IC_{50}$ values of the leaf extracts shown in Table 1 are the concentration of antioxidant required for 50% scavenging of $O_2^-$ after 5 min. Thus, smaller $IC_{50}$ values correspond to higher antioxidant activities. $IC_{50}$ value decreased as the salinity was increased in the culture medium, indicating that the antioxidant capacity of the leaf extract was remarkably stimulated by salt treatment. The $IC_{50}$ value of 150 mM NaCl-treated plants (0.03 mg $ml^{-1}$) indicated that the superoxide anion scavenging activity was more than 50-fold higher than in the extract from control plants.

## Discussion

The significant decrease in plant biomass caused by moderate concentrations of salt (100 mM NaCl) is in accordance with the results of Benlloch-González et al. [2] on the same species. The high leaf polyphenol content that we mea-

**Table 1.** Effect of salt treatment on leaf antioxidant content. Five-month-old plants were treated with salt solution for 109 days. The antioxidant concentration is expressed as the amount of methanolic leaf extract required for 50% scavenging $O_2^-$ in the reaction mixture ($IC_{50}$ in mg $ml^{-1}$). Mean of seven replicates and SE at $p = 0.05$.

| Salt concentration (mM NaCl) | Scavenging $O_2^-$ activity ($IC_{50}$ value mg $ml^{-1}$) |
|---|---|
| 0 | 1.56 ± 0.16 |
| 25 | 0.83 ± 0.13 |
| 50 | 0.42 ± 0.09 |
| 75 | 0.46 ± 0.08 |
| 100 | 0.07 ± 0.02 |
| 150 | 0.03 ± 0 |

sured 3.56 mg GAE $g^{-1}$ DW probably reflects a characteristic of Asteraceae. Djeridane et al. [17] reported that the amount of phenolics in leaf extracts from several other species belonging to this family (ca. 3–5 mg GAE $g^{-1}$ DW leaf extracts) was higher than in plants from other families. The high polyphenol accumulation in Asteraceae may be related to hard climate conditions of their habitats (high temperature, high solar exposure, and salinity). The leaf phenolic content was dependent on the salt concentration in the medium. Low salt levels (≤50 mM NaCl) stimulated by twofold the production of polyphenols. At higher salinity levels, the accumulation of these compounds declined. Our results agree with several reports of increasing polyphenols content at moderate salinity [18–20]. They suggest that the leaf polyphenols protect *C. cardunculus* leaves against the oxidative stress generated by salt aggression, supporting the hypothesis that secondary metabolites may play a role in cardoon salinity tolerance.

In plants challenged by salt, the total leaf biomass was systematically diminished, but the polyphenol concentration in leaf tissues presented a complex pattern, augmenting from 0 to 50 mM NaCl, and decreasing at higher salt concentration (Fig. 5). As a result, the amount of polyphenols per plant was maximal at 50 mM NaCl. In contrast, the concentration of antioxidative activity in leaf tissues augmented over the whole range of salt concentration while the biomass decreased (Fig. 6).

These opposite effects resulted in an almost constant antioxidant activity per plant between 0 and 75 mM NaCl, and a large increase at higher salt concentrations. For instance, at 150 mM NaCl, the total leaf biomass per plant was almost 5-fold lower than that of control plants. This large reduction of the plant foliage resulted from a 51% diminution in leaf number and a 58% diminution in individual leaf biomass. In parallel, the specific antioxidative activity (per unit leaf biomass) was 52-fold higher than in control plants. The total antioxidant activity per plant, estimated from these values, was 10-fold higher than in control. Two important conclusions may be drawn from these results. Firstly, the conditions for induction of polyphenol accumulation and antioxidative activity differed. Polyphenols were synthesized only in response

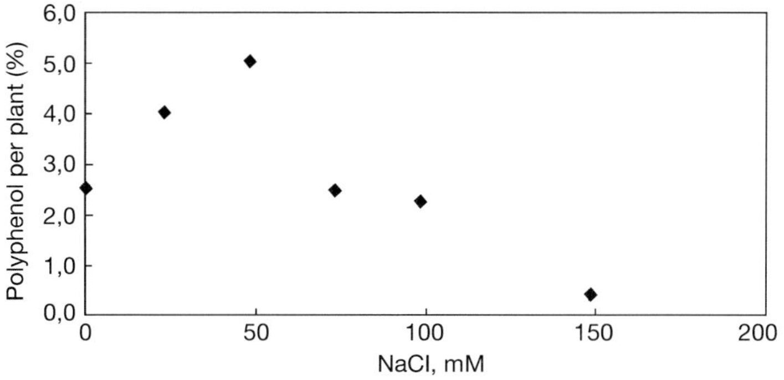


**Figure 5.** Effect of salt treatment on the amount of polyphenols per plant. Five-month-old plants were treated with salt solution for 109 days (mean of seven replicates).

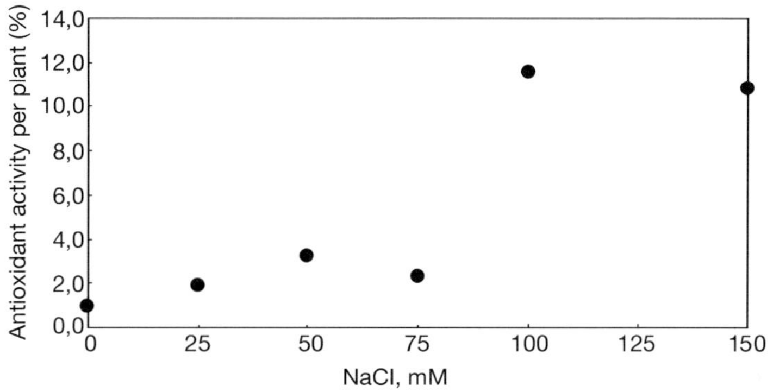


**Figure 6.** Effect of salt treatment on antioxidant activity per plant. Five-month-old plants were treated with salt solution for 109 days (mean of seven replicates).

to moderate salt stress. The induction of antioxidative activity occurred only in severe conditions of salt stress, sufficient to inhibit leaf growth by ca. 50%. Secondly, considering the yield in polyphenols, the highest harvests would be obtained with a trade-off between polyphenol synthesis and biomass production. When considering the yield of antioxidative substances, it may be more interesting to specifically look for environmental conditions favoring the induction of this activity than to optimize biomass production of source plants.

In summary, *C. cardunculus*, which can be easily be propagated under moderately high salinity (>75 mM NaCl) may be an interesting source of polyphenols. Furthermore, treating this plant with higher salt concentration specifically increase the production of the antioxidative substances that are not yet identified, but potentially useful. This species might be an interesting model for tapping natural biochemical mechanisms for controlled production of various antioxidants.

## References


1 Horie T, Schroeder JI (2004) Sodium transporters in plants. Diverse genes and physiological functions. *Plant Physiol* 136: 2457–2462

2 Benlloch-Gonzàlez M, Fournier JM, Ramos J, Benlloch M (2005) Strategies underlying salt tolerance in halophytes are present in *Cynara cardunculus*. *Plant Sci* 168: 635–659

3 Abdi S, Ali A (1999) Role of ROS modified human DNA in the pathogenesis and etiology of cancer. *Cancer Lett* 142: 1–9

4 Maisuthisakul P, Suttajit M, Pongsawatmanit R (2007) Assessment of phenolic content and free radical-scavenging capacity of some Thai indigenous plants. *Food Chem* 100: 1409–1418

5 Gordon MH (2001) The development of oxidative rancidity in foods. In: J Pokorny, N Yanishlieva, MH Gordon (eds): *Antioxidants in food: Practical applications*. Woodhead Publishing Limited, Cambridge, 7–21

6 Gordon MH (2001) Measuring antioxidant activity. In: J Pokorny, N Yanishlieva, MH Gordon (eds): *Antioxidants in food: Practical applications*. Woodhead Publishing Limited, Cambridge, 71–84

7 Espin JC, Soler-Rivas C, Wichers HJ, Garcia-Viguera C (2000) Anthocyanin-based natural colorants: A new source of antiradical activity for foodstuff. *J Agric Food Chem* 48: 1588–1592

8 Luo XM, Basile MJ, Kennelly EJ (2002) Polyphenolic antioxidants from the fruit of *Chrysophyllum cainito* L. (Star Apple). *J Agric Food Chem* 50: 1379–1382

9 Balasundram N, Sundram K, Samman S (2006) Phenolic compounds in plants and agri-industrial by-products: Antioxidant activity, occurrence, and potential uses. *Food Chem* 99: 191–203

10 Raccuia SA, Cavallaro V, Melilli MG (2002) Intraspecific variability in *Cynara cardunculus* var. Sylvestris Lam. Sicilian populations: Seed germination under salt and moisture stresses. *J Arid Environ* 56: 107–116

11 Valentao P, Fernandes E, Carvalho F, Andrade PB, Seabra RM, Bastos ML, (2002) Antioxidatives properties of cardoon (*Cynara cardunculus* L.) infusion against superoxide radical, hydroxyl radical, and hypochlorous acid. *J Agric Food Chem* 50: 4989–4993

12 Roseiro LB, Barbosa M, Ames JA, Wilbey RA (2003) Cheesemaking with vegetable coagulants – The use of *Cynara* L. for the production of ovine milk cheeses. *Int J Dairy Technol* 56: 76–85

13 Križková L, Mučaji P, Nagy M, Krajčovič J (2004) Triterpenoid cynarasaponins from *Cynara cardunculus* L. reduce chemically induced mutagenesis *in vitro*. *Phytomedicine* 11: 673–678

14 Grancai D, Nagy M, Suchy V, Novomesky P (1994) Cynarin from the fresh flower buds of *Cynara cardunculus*. *Fitoterapia* 65: 282

15 Dewanto V, Wu X, Adom KK, Liu RH (2002) Thermal processing enhances the nutritional value of tomatoes by increasing total antioxidant activity. *J Agric Food Chem* 50: 3010–3014

16 Duh PD, Yen GC (1999) Antioxidant activity of water extract of Harng Jyur (*Chrysanthemum morifolium* Ramat.) varieties in soybean oil emulsion. *Food Chem* 66: 471–476
17 Djeridane A, Yousfi M, Nadjemi B, Boutassouna D, Stocker P, Vidal N (2006) Antioxidant activity of some Algerian medicinal plants extracts containing phenolic compounds. *Food Chem* 97: 654–660
18 Parida AK, Das AB, Sanada Y, Mohanty P (2004) Effects of salinity on biochemical components of the mangrove, *Aeceras corniculatum. Aquat Bot* 80: 77–87
19 De Abreu N, Mazzafera P (2005) Effect of water and temperature stress on the content of active constituents of *Hypericum brasilienne* Choisy. *Plant Physiol Biochem* 43: 241–248
20 Navarro JM, Flores P, Garrido C, Martinez V (2006) Changes in the contents of antioxidants compounds in pepper fruits at different ripening stages, as affected by salinity. *Food Chem* 96: 66–73

Biosaline Agriculture and High Salinity Tolerance
Edited by Chedly Abdelly, Münir Öztürk, Muhammad Ashraf and Claude Grignon


# Contributions to the 2006 meeting on sustainable utilization of cash crop halophytes in Tunis

Helmut Lieth[1], Brigitte Herzog[2] and Jens Varnskuehler[3]

[1] *Institute of Environmental Systems Research, University of Osnabrück, Wipperfürther Str. 147, 51515 Kürten-Dürscheid, Germany*
[2] *Daimlerstr. 26, 74189 Weinsberg, Germany*
[3] *Plant Science Services GmbH, Heerstr. 49, 49492 Westerkappeln, Germany*

## Abstract

In recent years the utilization of halophytes has received increasing attention. Climate changes, which the senior author predicted some 30 years ago, have in the meantime reached such dimensions that economists, politicians and the general public are now ready to react to this warning given earlier by a few ecologists. The present meeting in Tunis discussed the major topics that the members of International Society for Halophyte Utilization (ISHU) could recommend for action to politicians and to the scientific community. These topics comprised: salinity tolerance types and levels for species of interest; selection of potential cash crop halophytes; genomic analyses; greenification of deserts and Sabkha landscapes; and future research and development topics. Several new points for the halophyte research group were: the discussion of genomic and proteomic analyses and their value for salinity-tolerance improvement of cash crops; the use of satellite remote sensing for the assessment of net primary production (NPP) from the normalized difference vegetation index (NDVI). The classical method to calculate average annual NPP from meteorological and soil fertility data was compared with the correlation of annual NPP values from satellite remotely assessed by NDVI; and the salinity tolerance tests and the multiplication of suitable plants with tissue cultures were explained in greater details. For the presentation of these topics, we include here information obtained from other authors, which are used with their permission.

## Salinity tolerance types and levels for species of interest

Until now the physiological and ecological work on halophytes has mainly concerned with analyses of the kinds of environmental constraints that they could endure [1, 2]. These are summarized in the present chapter.

The following is an outline of the topics of importance relating to key problems that need to be tested for halophytes: different kinds of tolerance; different levels of tolerance; different strategies of tolerance, i.e., excluders and includers; different chemical composition; different types of culture media, i.e., water, swamp, or dry soil; and different kinds of plants, i.e., algae, ferns, herbs, grass, woody, or trees.

These topics cover the most of the problems that need to be understood if halophytes are to be converted into cash crops. Some of these points will only be described briefly because they are fully discussed in the cited literature. Others would need a whole chapter to fully explain the new aspects. The need for new research is the subject of the last section here.

### *The different kinds of tolerance*

Under salinity tolerance we understand the tolerance of plants mainly to NaCl, which is the prime component of seawater salinity. However, in inland deserts, besides NaCl, we find especially regions with a large portion of gypsum ($CaSO_4{\cdot}2\,H_2O$). Other salt-producing elements, compounds, cations and anions are found in special systems, such as hot springs and direct saltwater contact with hot lava, as in under (sea)water volcanoes. Such conditions require special types of physiological pathways, which we cannot cover in this presentation. Here we deal only with three types: salinity, sodicity and alkalinity. These types are discussed in relation to their utilization possibilities in wetlands and uplands, where they are especially important with respect to the quantity of saline components in the water, as explained in more detail below.

### *The different levels of salinity tolerance*

The different types of salt in the soil water are important for halophytes. There are several thresholds that are accepted as being indicative for a species, plant or animal to be called salinity tolerant. For plants, this can be determined according to the table provided by [3]. This table is reproduced here in Table 1. The comparison is useful because of the notorious uses of μ Siemens by all agronomists, of equivalence and mol $m^{-3}$ or kg by the physiologists and of the percentage of NaCl equivalent in the soil and water salinity. The different kinds of units are measured by different instruments. What we compare is mainly the levels of NaCl and not alkalinity and sodicity in the salinity, which needs to be distinguished in upland ecosystems.

### *The different strategies of salinity tolerance*

Salinity tolerance can be achieved in two major ways. The first is the rejection of NaCl at the root or rhizome surface of the plants. These plants possess low concentrations of salt in their biomass, and are called excluders. The second is the storage of salinity in special cells or glands inside or on the surface of the plant body. These are called includers.

Some plants even excrete the salt at special openings in their leaves, from where it can fall down to the ground when the leaf is wetted by rain or can be

**Table 1.** Conversion table for salinity and halophyte research (extract from [1])

| ‰ g/l | % | ppm | EC NaCl at 20°C | | EC Seawater | meq/l mol/m³ NaCl |
|---|---|---|---|---|---|---|
| | | | dS/m EC/ 1000 mmhos | µS/cm | µS/cm at 20°C | |
| 1 | 0.1 | 1000 | 1.818 | 1818 | 1429 | 17.11 |
| 2 | 0.2 | 2000 | 3.636 | 3636 | 2857 | 34.22 |
| 3 | 0.3 | 3000 | 5.455 | 5455 | 4286 | 51.33 |
| 4 | 0.4 | 4000 | 7.273 | 7273 | 5714 | 68.45 |
| 5 | 0.5 | 5000 | 9.091 | 9091 | 7143 | 85.56 |
| 6 | 0.6 | 6000 | 10.909 | 10909 | 8572 | 102.67 |
| 7 | 0.7 | 7000 | 12.727 | 12727 | 10000 | 119.78 |
| 8 | 0.8 | 8000 | 14.546 | 14546 | 11429 | 136.89 |
| 9 | 0.9 | 9000 | 16.364 | 16364 | 12857 | 154.00 |
| 10 | 1.0 | 10000 | 18.182 | 18182 | 14286 | 171.12 |
| 15 | 1.5 | 15000 | 24.274 | 24274 | 21429 | 256.67 |
| 20 | 2.0 | 20000 | 30.606 | 30606 | 28572 | 342.23 |
| 25 | 2.5 | 25000 | 36.634 | 36634 | 35715 | 427.79 |
| 30 | 3.0 | 30000 | 42.356 | 42356 | 42858 | 513.35 |
| 35 | 3.5 | 35000 | 47.774 | 47774 | 50001 | 598.90 |
| 40 | 4.0 | 40000 | 52.887 | 52887 | 57144 | 684.46 |
| 45 | 4.5 | 45000 | 57.696 | 57696 | 64287 | 770.02 |
| 50 | 5.0 | 50000 | 62.199 | 62199 | 71430 | 855.58 |
| 55 | 5.5 | 55000 | 66.398 | 66398 | 78573 | 941.14 |
| 60 | 6.0 | 60000 | 70.292 | 70292 | 85716 | 1026.69 |
| 65 | 6.5 | 65000 | 73.881 | 73881 | 92859 | 1112.25 |
| 70 | 7.0 | 70000 | 77.165 | 77165 | | 1197.81 |
| 75 | 7.5 | 75000 | 80.145 | 80145 | | 1283.37 |
| 80 | 8.0 | 80000 | | | | 1368.93 |
| 85 | 8.5 | 85000 | | | | 1454.48 |
| 90 | 9.0 | 90000 | | | | 1540.04 |
| 95 | 9.5 | 95000 | | | | 1625.60 |
| 100 | 10.0 | 100000 | | | | 1711.16 |

Low salinity    Transition    High salinity

Three levels of salinity tolerance are indicated by shading in this table: low, transition and high salinity.
Low levels to salinity, i.e., to 0.7–0.8% per liter in the soil water are assumed to be limiting for barley; tolerance to interim/transition levels between 0.8% and 2.5% salinity is found with sugar beet and several halophytes. From 3% to 6.5% we speak about "above seawater salinity tolerance", and tolerance to these levels is found for a number of halophytes that can be expected to convert to cash crops. For *Sesuvium portulacastrum* and *Tamarix aphylla* we have found the tolerance levels experimentally to be much higher than 6.5%.

shed off when the plant is shaken by dry winds. All types of halophytes can be useful in production systems. We cover some important aspects in the final section on research and develpoment (R&D), including systems with sodicity and alkalinity in the soil water.

In most ecosystems different salinity tolerance levels exist between plants and animals. Most land plants and animals tolerate low salinity levels in their environment, but are sensitive to higher levels. This is not the case in seawater. Fish and many seafood species can tolerate 3–5% salinity in the water, as do the plants in this environment. This differs in natural land-based ecosystems where we can find plants that tolerate high salinities in the soil water, but many animals tolerate only low salinities in their drinking water. Therefore, careful selection of animals is required if we want to feed them with halophytic biomass containing high levels of salt in their dry or wet biomass.

### *Different chemical composition*

Most of our investigations have so far been performed with NaCl salinity. This is the salinity that we face mostly frequently in agriculture. Many inland deserts and sabkhas, however, have sodic or alkaline soils, which pose different problems for the plant's physiology and require special treatment. The surface of sodic soils are often covered with gypsum, and large portions of sodium bicarbonate often appear in alkaline soils, which can cover the plant root with calcium carbonate or phosphate which is also present in the soil. Careful experiments are required to determine how to best use such soils, but there are plants that can grow in such soils. For these conditions we need to create special systems to develop sustainability, especially if agrosystems are planned.

### *Different types of cultivations*

In addition to considering the growth of plants with different salinity tolerance levels, thought must also be given to the different kinds of water availability. If we consider the ecosystem as sink or pathway for capturing solar energy, we need to compare its structure on its function as a conductor for the energy. This can be done in several ways and each ecosystem develops the fastest way to conduct the energy [4]. It either binds it in fossil deposits or uses the energy to release water as vapor into the air and from there into outer space. The comparison of energy bound in biomass with the energy transmitted in evaporation of water, developing reflection systems to conduct the energy out of the ecosystem or developing other pathways remains to be investigated.

From the viewpoint of management, we can separate the different types of culture media into groups of cultivation in liquid solution of different salinity levels, i.e., in swampy environments that simulate salt marshes, and systems that can be established on dry land. The specific problems for each cultivation medium can be solved. All media are possible, but the management system required to reach economic feasibility is different for each system. Many of them still need intensive research. These are discussed below.

### *Different kinds of plants*

The most appropriate plants for a system are selected depending on the culture medium. Algae, ferns and herbs are mostly grown in wet systems either submerged or emerging. Herbs, grasses and woody shrubs are mainly used in swampy systems. Grasses, shrubs and trees of different sizes are usually selected for upland systems.

The aim for each cultivation of halophytes must be the long-term sustainability of the system, to develop an economically feasible system for human utilization.

## Selection of potential cash crop halophytes

In our previous experiments we found it necessary to select promising species from natural habitats. Depending on average annual temperature and precipitation patterns we selected these species from salt marshes and mangroves, along coastlines of the oceans. To do this, it was necessary to determine the salinity tolerance levels of each species using certain methods that would provide us with quick results. The possible methods are: tissue culture; germination tests; quick check lab; and field plots.

We have used the last 3 methods in many experiments, but have added the tissue culture as means of a quick check and for mass propagation of suitable species. We discuss this method below in more detailed because we assume that it will become more prominent in the near future. The other methods are detailed in [5] and in other references mentioned.

### *Micropropagation of potential cash crop halophytes from tissue cultures*

The cash crop halophytes concept described in [3] may be an economically viable approach to convert wasteland areas into productive agrosytems, enabling local people to gain income. This concept will only work if halophytes produce marketable crops on a sustainable basis.

Plenty of wild plants are identified as tolerating saline irrigation. A few varieties have already been selected for maximum yield or biomass production. Future selection of promising species of halophytes as cash crop has to screen for mutations with superior agricultural value. Using classical in-bred technologies 10–15 generations have to be back-crossed and selected *in vivo* in a controlled manner to achieve homologous alleles.

**Table 2.** List of plants successfully established so far in our lab; data as 29/07/2007

| Plant genera | Stage I | Stage II | Stage III |
|---|---|---|---|
| *Atriplex canescens* | X | | |
| *Atriplex nummularia* | X | | |
| *Batis maritima* L. | X | X | |
| *Crithmum maritimum* | X | X | |
| *Kosteletzkya virginica* (L.) Presl. | X | X | |
| *Leptochloa fusca* | X | X | |
| *Limoniastrum multiflorum* | X | X | X |
| *Sesuvium portulacastrum* L. | X | X | X |
| *Sporobolus* spec. | X | X | X |
| *Tamarix aphylla* | X | | |
| *Tamarix gallicum* | X | | |

Another well-established technology in agri- and horticulture is the cloning of promising genotypes by cuttings, graftings or divisions. The disadvantage of this method is the time required to multiply beneficial mutation in one plant. It takes a long time to produce sufficient quantities of planting material for economical culture. For *Miscanthus sinensis* the initial stock planted for biofuel purposes requires as many as 10 000 plants per hectare.

Such levels of propagation can easily be obtained with the technique described in this review. We show with a number of salinity-tolerant species that this technique works with halophytes.

## *Materials and methods*

The principal of micropropagation is cloning of plants by microcutting in optimized ambient abiotic environment, e.g., in sterile media and controlled light and temperature regimes. The list of plants included in this study is shown in Table 2. The different steps of this technology are shown in Table 3.

We presented this procedure as a poster together with J. Varnskuehler at the 2006 ISHU meeting, where we described the successful application of micropropagation to 11 plant species representing 8 families.

Fresh growing shoots of greenhouse-grown stock of halophytes from the Botanical Garden of the University of Osnabrück were collected and immediately transferred to the lab. Subsequently they were cut into single nodes and washed in household detergent solution for 15 min. After rinsing in 70% ethanol-water solution for 60 s the explants were disinfected for 15 min in 300 ml 1.6% $Ca(ClO)_2$ solution with few drops Tween 20® as detergent. After chlorine treatment, explants were rinsed three in 300 ml autoclaved tap water.

In a laminar-flow cabinet the single nodes were isolated by cutting off all obviously damaged material using sterile blades. The clean explants were then

**Table 3.** Production of plant material for the plantation.

| Stage | Step |
|---|---|
| I | 1. Selection of healthy mother plants or special pre-treatment to cure infested stock<br>2. Surface disinfection<br>3. Inoculation on media containing all macro- and micronutrients, organics and sugar, addition of plant hormones |
| II | 1. Sprouting of sleeping buds<br>2. Propagation by subculturing aseptically every 3–6 weeks<br>3. Upscaling of stock |
| III | 1. Rooting *in vitro* or *ex vitro*<br>Developing routine multiplication of plants |
| IV | 1. Developing a healthy transportation and shipping routine<br>Delivery of young plants to nurseries as stock or directly to field |

placed into single plastic tubes containing 40 ml sterile media, and put in a growth room at 22±2°C temperature with a 16:8-h light:dark regime under cool fluorescent lighting.

The media used were modified standard media as described in [6] supplemented with 0.5 mg/l 6-benzylaminopurine as cytokinin and increasing concentrations of NaCl. A successful example is shown in Figure 1 for *Leptochloa fusca*, a tall grass species widely used in Pakistan as a fodder plant that grows on waterlogged saline soils and propagates only vegetatively.

*Results*

Using this standard protocol for initiation of sterile cultures for 15 cash crop halophytes we produced actively growing axenic cultures of all sample plants. In some cases the addition of 0.5% NaCl to the nutrient solution resulted in more vigorous growth as compared to controls.

## Genomic analyses

Independently from the activities of the halophyte research groups, the international crop improvement community has also developed ideas to improve the salinity tolerance of the important cash crops like rice, wheat and maize by determining genes for salinity tolerance in organisms for which they have the full genome sequence [7] (see Table 4, taken with permission of the author from the internet presentation by [8]).

The scientists working on halophytes were surprised that the first analyses were done with *Arabidopsis thaliana*, a small glycophytic weed species. While this particular case needs no further discussion, it seems to be necessary for

**Figure 1.** *Leptochloa fusca*, a tall plant without seeds, propagated by tissue culture.

the halophyte and saline ecosystem group to include the concept of genome analyses into their research interests. The next meetings of the ISHU should certainly include experiments dealing with gene changes, like those discussed in the papers of session 2 of the meeting in Tunis, in their conferences sessions.

To prepare for future inclusion of genomics, proteomics and high-throughput-screening, we suggest visiting the homepage of the CDCh workshop (2003) in [8]. We have extracted some relevant information on salinity tolerance from this presentation under the title AtGI/The Tiger *Arabidopsis thaliana* Gene Index for the query "salt" [9].

The "Boehringer charts" are very informative for understanding of the importance of the salinity tolerance genes in plants. These charts and the content of the database in [8] cannot be included here, but should become part of the required reading for all graduate courses for students entering halophyte and saline systems research.

From the present experiences with salinity markers, it appears that many genes can be responsible for salinity tolerance or even demands. This shows that our concept of converting existing halophytes into cash crops is still valid, especially as we know from our modeling of sustainability that we need, in the long run, seawater level-salinity tolerance from new crops to be able feed a

**Table 4.** Sequenced organisms.

| Organism | Genome size | Completion date |
|---|---|---|
| Bacteriophage FX174 | 5 kb | 1980 |
| *Haemohphyllus influenzae* | 1.8 Mb | 1995 |
| *Sacharomyces cerevisiae* | 12.1 Mb | 1996 |
| *Clostridium elegans* | 97 Mb | 1998 |
| *Arabidopsis thaliana* | 100 Mb | 2000 |
| *Drosophila melanogaster* | 180 Mb | 2000 |
| *Homo sapiens* | 3000 Mb | 2001 |

population of 6 billion humans and their animal populations. For these populations we may need all the fresh water we can catch from natural rainfall and still have to produce additionally drinking water from desalinization, using seawater and wastewater for plant production.

Our research on the salinity tolerance may help the scientists undertaking common cash crop research in another way. The tolerance mechanisms of halophytes may provide much relevant information for locating genes for future genetic manipulations to increase the yield and quality of biomass.

We expect this to become very prominent for plants with industrial exploitations.

## Greenification of desert and sabkha landscapes

The original goal for the scientists cooperating in the ISHU was the greenification of deserts and wastelands with halophytes by irrigating such landscapes with wastewater and seawater. Their first international congress in Al Ain (1990) brought together the full range of problems with saline irrigation and several strategies for sustainable use of wastelands and reclaimation or protection of arable land after irrigation. Water had increased soil salinity beyond the tolerance of their crops. This generated the desire to use suitable halophytes as catch crops to reduce the salt level in the soil. A model for this was found in Tunisian Sabkhas where glycophytic *Medicago* species were grown between saltgrasses. Under these conditions, the salt-tolerant grasses absorbed large amounts of salts and so reduced the salinity of the remaining soil water to the level at which the glycophytes could grow.

This observation, together with a large number of physiological observations on different halophytes, widened the research goals, including now the use of plant species including or excluding salt in the soil water, i.e., plants which keep the salt inside their biomass and those that excrete it, plants reducing the evapotranspiration and, with it, biomass yield, plants accumulating large portions of below-ground biomass, and plants which grow extremely fast under wet or swampy conditions.

**Figure 2.** Sabkha in Tunisia. The vegetation cover of sabkhas depends on the soil type and the salinity level of the soil (photo by Brigitte Herzog).

For each type of species, structure and physiology were considered and suitable applications were envisioned and appropriate cropping techniques were planned. For several plants and cropping techniques solutions have been found as described in the first section here.

Geographical studies have shown in the meantime the extend of sabkhas and other saline environments, which allow us to compare species composition and salinity tolerance strategies of species growing in different soil types and water supplies (Figs. 2, 3) . These topics are especially important for inland deserts or semi-desert landscapes with low rainfall. The ISHU intends to devote several future conference sessions to such systems, e.g., at the meeting in Uzbekistan.

The senior author has in several papers explained the primary production in the context of physical laws of energy flow on the earth surface. Biologists have paid little attention to the fact that the total energy flow on the surface of the earth is the main driving force for evolution and not just the biological part alone [4]. More studies are required in this respect because the concepts shed new light upon the possibility to use saline water for crop irrigation.

The reference list includes several volumes dealing with sabkha systems, those describing mangroves and salt marshes of the world and those dealing with inland saline ecosystems in various ways [11–13]. There are plenty

**Figure 3.** The open loamy soil in the sabkha shows typically the cracks in the surface that will close again if the soil becomes soaking wet. This is independent of the salinity level of the soil water (photo by Brigitte Herzog).

of problems with immediate practical relevance. We expect that the 2006 meeting has added new topics in this respect, besides the strong emphasis on papers supporting future studies on genomic analyses.

## Future R&D topics

In the proceeding sections we have shown the kind of problems the members of the ISHU have discussed. To focus upon the most promising research topics, these can be sorted into general levels: species, production level and total system level. From each level we have selected some topics that we assume will need attention first: at the species level: different levels of salinity tolerance, and different strategies of salt tolerance; at the production level: different possibilities for the use of biomass; food, feed, wood, ornamentals, $CO_2$ sink and industrial raw material; extraction of chemicals and production of biofuel; and mineralization of the irrigation water; at the system level: health aspects of using halophytes; economic potential of cash crop halophytes; organization of life and work in saline systems; planning the use of desert landscape with saline irrigation; and using seawater production systems on land.

*Species level*

At present it is difficult to estimate the performance of a potential cash crop species under field conditions and under different water availability levels. Under upland conditions, the performance of species with strategies of salt exclusions and salt intakes can show quite different results with respect to the salinity tolerance. This is in general especially important for inland sabkhas and rain-fed dry lands. It seems very important to get an agreement between the internationally research community working on halophytic plants to include all parameters that are needed to prepare models of the potential for certain halophytes to serve in the future as cash crop or catch crop under large scale production. First systems like this have been undertaken by Hodges, Sato, Lieth and Yensen (see their homepages [14–17]). In all cases the initial value shown were from the performance of these species in their natural environment. More species are needed for the future utilization in agricultural or forestry systems. There is a growing demand for such species. The more we know about the physiological demands and the chemical composition of these species, the more their value will increase.

*Production level*

The practical use of halophytic plants is probably as prominent as the utilization of glycophytes. The only difference is that we are dealing here with plants that either demand or tolerate salinity levels in water or soil. In Table 5 the different utilizations of halophytes are shown from economical and ecological viewpoints. Research groups have developed that each work on 1 of the 12 categories; however, they have little contact with each other. The ISHU has undertaken the task of providing the means of communication between these groups, in workshops and in seminars. From the categories listed in Table 5, food, feed, $CO_2$ sink, ornamentals and industrial raw materials were discussed during the 2006 conference.

*Food*

The ISHU, together with other research groups, has collected over the last few decades a long list of halophytes that can be used for human consumption. Some of these have been available on local markets for a long time, and these species can be cropped in other areas of the world, where the environmental conditions throughout the year or in parts of the year allow the cropping of such species. Research into the transferability of such species should include the possible ecological impacts of these species upon the local indigenous flora and fauna besides the practical problems of cropping.

**Table 5.** Already existing utilization of halophytes and possibilities under investigation (according to [1])

| | |
|---|---|
| 1. Food | Starch, protein, fat, vitamins |
| 2. Feed | Starch, protein, minerals |
| 3. Wood | Firewood, building, furniture, crates |
| 4. Chemicals | Industrial chemicals, pharmaceuticals, plastics |
| 5. Landscaping | Roadsides, housing areas, turf |
| 6. Ornamental | Potting plants, gardening |
| 7. $CO_2$ sequestration | Desert greenification, aforestation |
| 8. Tertiary treatment | Water, soil |
| 9. Industrial raw material | Fiber, biomass, ash |
| 10. Unconventional irrigation water | Seawater, brackish water, wastewater |
| 11. Environmental protection | Coastlines, dune fixation, erosion prevention |
| 12. Wildlife support | Species diversity, maintenance |

### *Feed*

A large number of species and types of saline vegetation are used as feed for livestock and wildlife. Changing large sections of the landscapes by planting halophytes can have an impact upon the environment in many respects. It is necessary, therefore, to discuss such changes and conduct cost/benefit calculations along with introduction of the new species as standard crops.

The use of halophytes as feed for fish deserves completely separate attention. Many fish and seafood species consume saline plants. Many of them live in water with normal seawater salinities or even higher than that. Several of these are candidates for growing in closed or open systems in deserts or semi-deserts if we can provide the necessary amount of water. A group that is presently coordinated by the systems research group in Osnabrück together with colleagues from Uzbekistan analyses this potential. More information is available from the authors of this paper. Further discussions are planned for the next ISHU conference.

### *Wood and $CO_2$ sink*

The current worldwide discussion about the increase of $CO_2$ in the air is fired by the two slogans "deforestation and fossil fuel burning". At present the discussion focuses on the reduction of fossil fuel consumption by the population and industry; while this is certainly correct, it will not solve the problem of $CO_2$ increase, because a large part of the human populations lives on standards far below the accepted living standards of the population in industrial countries.

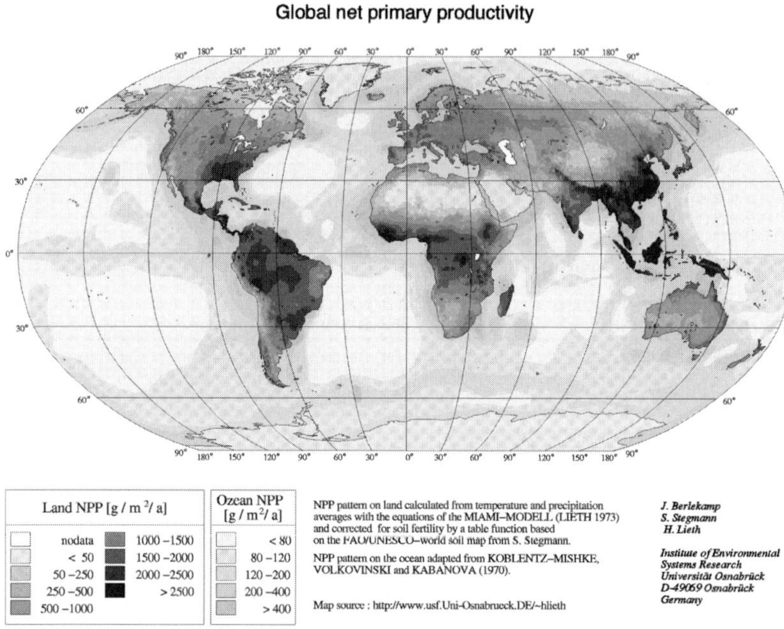


**Figure 4.** The world pattern of net primary production (NPP) developed from regression models with climate variables and soil fertility data [16].

We must therefore work further on planting halophytes in areas where the climatic conditions reduce the net primary production (NPP) to low levels. These regions are shown in the two world maps presented in Figures 4 and 5.

Much emphasis in the long-lasting $CO_2$ sequestration is attached to wood production. Again, while this is certainly correct, one should not forget that many herbaceous halophytes produce large amounts of biomass that can be used as $CO_2$ storage for a long period of time. Systems of this kind have been developed during the tenure time of the European Community Concerted Action project no. CA 39. Similar projects have been successfully implemented by Sato [15], Yensen [17], Loughland [18], the ICBA [19], and Khan [20].

## *Industrial raw material*

The production of raw material (biomass for industrial usage) from halophytes in suitable desert regions with unconventional irrigation (seawater and wastewater) could increase the $CO_2$ sequestration well above the levels of all "reductionists" proposals together. In many ways biomass can be used instead

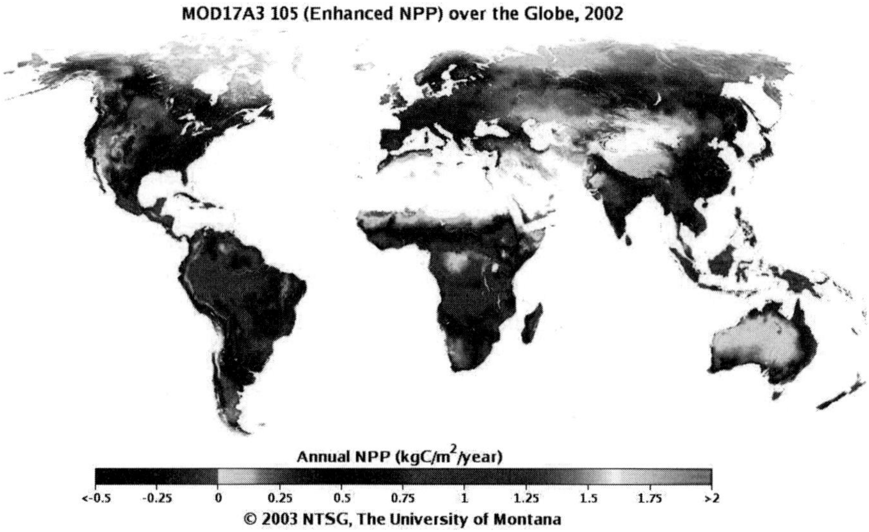


**Figure 5.** Global terrestrial NPP over 110 million km² for 2002, computed from MODIS (Moderate Resolution Imaging Spectroradiometer) data [10]. The colored version of the map is available from the author. On the map the different NPP values in the deserts of the world are clearly visible.

of fossil fuel. The engineering community is asked to develop new processing methods for using halophytes, either novel ways or methods similar to those currently being used with glycophytes.

### *Extraction of chemicals*

Many plants produce chemicals that can be used for industry, medicine, food and feed. Many halophytes have special physiological pathways, which end in the production or accumulation of particular organic or inorganic chemicals. The most important ones discovered in the past were valuable oils in the seeds of halophytes and heavy metal accumulations in plants growing on salinity-polluted mine spoils and natural salinity flats. More research is necessary to discover further cases in respective halophytes and in methods to produce these plants on a large scale, sufficient for industrial exploitation. Under present-day conditions, it is difficult to comprehend why large amounts of paper are still produced from wood, when the total residue from sugar and banana production would satisfy the entire world paper demand. We have plenty of halophytes that could fill any gap if there were any.

*Ornamentals*

A large amount of fresh water is used for the production of ornamental plants. The majority is used for landscaping, gardening and the provision for cut flowers. Among the halophytes are many flowers with the potential to become runners in the flower shops. Halophytic plants can also be used in desert areas for roadside greenification with wastewater irrigation and seawater utilization. As a first large scale example is a project that we helped to initiate in the UAE. This project was initially started together with HH. Sheikh Zayed bin Sultan and is presently continued under HH. Sheikh Sultan bin Zayed al Nahjan on Samaliah Island by R. Loughland ([18] and personal communication).

The major problem for the general use of halophytic species in gardening and landscaping is the lack of sufficient nurseries for the production of species adapted to the local environmental conditions, as well as the taste of the local population. In Karakalpakstan, the salinity tolerant shrubs/trees species of the genus *Tamarix* are reserved for utilization in graveyards, although they would be well suited for the greenification for roadsides and residences where they could be irrigated with saline water. Some of the *Tamarix* species provide good wood and some are splendid ornamental plants.

*Acknowledgements*

We acknowledge the help of Prof. Dr. F. Kreuzaler, RWTH Aachen, and his students in locating and using the entries of the Erlangen Data Base.

## References

1 Lieth H, Al Masoom AA (eds) (1993) Towards the rational use of high salinity tolerant plants. 1: Deliberations about high salinity tolerant plants and ecosystems. *Tasks Veg Sci* 27
2 Lieth H, Al Masoom AA (eds) (1993) Towards the rational use of high salinity tolerant plants. Vol. 2. Agriculture and forestry under marginal soil water conditions. *Tasks Veg Sci* 28
3 Lieth H, Mochtchenko M (eds) (2002) *Halophyte uses in different Climates, Cashcrop Halophytes for future halophyte growers*. Backhuys Publishers, Leiden
4 Lieth H (2005) External and internal controls of ecosystems. Ecosystems as accelerators of the energy flow from the earth surface. In: W Eisenberg, D Fey, KM Meiss (eds): *Naturwissenschaftlich-technische Systeme im Focus der Fremd- und Selbstorganisation*. A. Sommerfeld-Gesellschaft, Leipzig, 55–64
5 Lieth H, Mochtchenko M (eds) (2003) *Cash crop halophytes recent studies.* Kluwer Academic Publisher, Dordrecht
6 Murashige T, Skoog F (1962) A revised medium for rapid growth and bioassays with tobacco tissue cultures. *Physiol Plant* 15: 473–497
7 Lliang GH, Skinner DZ (eds) (2004) *Genetically modified crops*. Food Products Press, New York

8 v. Homeyer A (2003) *Biological databases*. 135 PPTs. Available at: http://www.2.chemie.uni-erlangen.de
9 The TIGR *Arabidopsis thaliana* Gene Index for "salt" query. Available at: http://www.AtGI
10 Running W, Nemani RR, Heinsch FA, Zhao M, Reeves M (2004) A continuous satellite-derived measure of global terrestrial primary production. *Bioscience* 54: 547–560
11 Barth H-J, Böer B (eds) (2001) Sabkha ecosystems. I: The Arabian Peninsula and adjacent countries. *Tasks Veg Sci* 36
12 Khan MA, Böer B, Barth H-J, Kust GS (eds) (2007) Sabkha ecosystems. II: West and Central Asia. *Tasks Veg Sci* 42
13 Khan MA, Weber DJ (eds) (2005) Ecophysiology of high salinity tolerant plants. *Tasks Veg Sci* 40
14 Hodges C (2005) The seawater foundation Inc. (work with *Salicornia*). Available at: http://www.seawater.foundation.org
15 Manzanar Mangrove Initiative; The manzanar project Sato G. Available at: http://www.tamu.edu/ccbn/dewitt/manzanar/default.htm
16 Lieth H (2000) Available at: http://www.usf.uni-osnabrück.de/~h.lieth/projects/expo2000
17 Yensen NP, Available at: http://www. NYPA. Com (work mainly with *Salicornia* and *Distichlis*)
18 Loughland R, Available at: http://www.cerehc.org.ae/CER.htm
19 ICBA, Available at: http://www.biosaline.org
20 Khan MA, Available at: http://www.halophyte.org/halophyte_files/htmls/drkhan.ht

# Index

abiotic stress 201, 287, 297-302
abiotic stress related gene 300
Akçakale 145
*Alhagi pseudoalhagi* 258-263
alkalinity 144, 255
alluvial soil 142, 143, 145
amino acid accumulation 122
*Anabasis brachiata* 226, 227
*Anabasis salsa* 225-230
Anatolia 142-146
Ankara 143, 144
antioxidant activity 44
antioxidant capacity 43, 44
antioxidant defense 62, 68
antioxidant enzyme 21
antiporter 305, 306, 309-314
*Arabidopsis thaliana* 99-105, 276, 278, 279, 281, 283, 288, 317-322, 326, 327, 351
Aral Sea basin 248
Aras 144, 146
arid region 233-239, 241-245
arid soil 143, 144 , 202
aridity stress 178
*Artemisia diffusa* 257-260
*Artemisia gurganica* 229
*Artemisia pauciflora* 225
*Arthrophytum lehmannianum* 228
AsA/DHA ratio 46
AsA/GSH cycle 47
ascorbate 45, 46, 49
ascorbic acid peroxidase (APX) 25
*Asteraceae* 223, 235, 236
*Athrocnemum indicum* 241-245
*Atriplex halimus* 107-113
*Atriplex stocksii* community 166, 172, 173
*AtRNR2,3 bis* 330-332
*AtRNR2,5* gene 330-332

barren sand 239
basaltic soil 143
*Batis maritima* 73-79
berry defense system 46
*Beta vulgaris* 115-127
bicarbonate 54, 55
biodiversity 234, 238, 248, 254
biomass production 64, 249
biphasic activation 330, 332
Birjand 178

$C_3/C_4$ plant 249
$C_4$ photosynthesis pathway 188
cadmium (Cd) 326-332
calcium (Ca) 61, 62, 66, 67
$Ca^{2+}$concentration 121
$Ca^{2+}/Na^+$ ratio 67
calcium chloride 183
camel 237
carbohydrage storage 121
carbon assimilation 126
carbon isotope discrimination 200
carbon stable isotope 200
cardinal temperature 181
carotenoid 54
cash crop 159
cash crop halophyte 116, 345
*Castanea sativa* 202
catalase (CAT) 23
catalase activity 329, 331, 332
cation transport 306, 311
cationic charge 65
cellular compartment 32
cellular compartmentation 127
cellular redox state 45, 47, 49
Central Anatolia 142, 144
chamaephytes 146, 150, 151
Charvak reservoir 266
chemical extraction 359
*Chenopodiaceae* 213, 223, 235, 236
chestnut soil 142
chloride (Cl) 61, 63, 65, 66
$Cl/NO_3^-$ 65
chlorophyll amount 54
chlorosis 65, 67
*Chrithmum maritimum* 73-79
citrus plant 37-42
citrus rootstock 40
$CO_2$, sub-stomatal 29
$CO_2$ assimilation rate 29

$CO_2$ fixation rate 21
$CO_2$ sequestration 357, 358
coastal plant community 164, 207, 208
colluvial soil 142
common bean 201
copper (Cu) 67, 68, 326-332
cotton 201
*Cressa cretica* community 166, 170, 171
crop productivity 33
cryptophyte 146, 150, 151
*Cydonia oblonga* 54
*Cynara cardunculus* 335-342
*Cyperus arenarius* community 165-168

decision-support system (DSS) 266-268, 271
de-epoxidation index 57
deoxyribonucleoside triphosphates (dNTPs) 326, 332
desert greenification 353
desertification 144
differential gene espression 301
DNA damage 325, 332
*DREB*-related gene 288-294
dwarf semishrub 225, 226, 229, 230

East Anatolia 142, 144
ecological restoration 238, 239
Egypt 207-217
electrolytes leakage 187
endemic halophyte 142, 146, 151-153, 234-236
environmental stress condition 50
Euphrates 144, 146
Euro-Siberian floral division 142
evaporation rate 252
exogenous application 21

floral division, Euro-Siberian 142
floral division, Irano-Turanian 146-150
floral division, Mediterranean 142
fluorescein diacetate (FDA) 133, 136-138, 178
food security 22
forage crop 107-113, 159-161, 215, 236-238, 259, 260, 357
forage dry matter percentage 191
forage yield 192
free radicals 32, 331
free radical formation 331
fructose 122
fruit tree cultivation 53, 54

gas exchange 124
genome sequence 353
genomic analysis 351-353
geographical information system (GIS) 266-271
ArcView GIS 3.2 269
geothermal water 251
germination 82-87, 108, 109, 179-185
germination rate 181, 183, 185
germination temperature 86, 179, 180
global net primary production 358, 359
glucose 122
β-glucosidase 133, 137, 138
glutamine 123, 124
glutathione 47
glycinebetaine (GB) 21, 31
GB-induced enhancement 31
glycophyte 115-117
grapevine 297-302, 305-314
growth stimulation, by salt 107-112

halophyte characterization 3
halophyte community 221-230, 252
halophyte diversity 234, 251
halophyte domestication 178
halophyte ecology 253
halophyte usage 215
halophyte, cash crop 116
halophyte, dicotyledonous 116
halophyte, endangered 146, 153
halophyte, endemic 142, 146, 151-153, 234-236
halophyte, facultative 73
halophyte, mineral value 257
halophyte, obligatory 73
halophytic desert 225-230
halophytic rangeland 248, 260
harvest index 202
heavy metal toxicity 326, 332
hemicryptophytes 146, 150, 151
humid soil 143
hydraulic engineering 270
hydrochemical databank 270
hydroecological monitoring 268, 269
hydrogen peroxide 32
hydromorphic soil 144, 145
hydroponic culture 54, 57, 58
hydroxyl radicals 26
hygrohalophyte 146, 150, 151
hyperhalophytic desert221-225

*in vitro* culture 37-42, 54-57
intracellular solute 119
ion homeostasis 121
ion toxicity 9, 82-86, 118, 121
ion uptake 41
Irano-Turanian floral division 146-150
iron (Fe) 67
Fe chlorosis 54
Fe-chelate reductase 54
iron deficiency 53, 55
irrigation, salt accumulation by 144

*Juncus* 213, 214

K concentration 66, 67, 101-105
K value 188
$K^+/Na^+$ ratio 67, 121
$K^+/Na^+$ selectivity 121
$K_2SO_4$ 100-104
Karachi 165, 166
karakul sheep 260
*Kochia scoparia* 177-194, 261

leaf area index (LAI) 28, 186, 187
leaf chlorophyll 30
leaf stomatal conductance 124, 125
leaf succulence 124
leaf transpiration 124
*Leptochloa fusca* 352
lipid peroxidation product 329
lipid peroxide 331
lipid signaling pathway 317-319, 322
livestock feeding 248

macronutrients 63-68
$Mg^{2+}$concentration 67, 121
manganese (Mn) 67, 68
marginal resource 247
Mashad 179
matric potential 10
Mediterranean floral division 142
membrane integrity 121
Mesopotamia 144
metabolic process 127
metallo-halophyte 257
microbial activity variation 132
microbial biomass carbon (MBC) 133, 134
microbial biomass nitrogen (MBN) 133, 135, 136
micronutrients 63-68

Na concentration 65, 68, 101-105
$Na^+$, osmotic role 121
NADPH 64
$NADP^+$/NADPH ratio 47, 49, 64
*Nanophyton erinaceum* 227
*Nicotiana glauca* 131-138
nitrogen (N) 63, 65
$NO3^-$ uptake 65
nitrogen fixation 91-97
nitrogen-fixing capacity 92-94
nodule metabolism 93-95
non-stomatal effects 199, 201
Northeast Anatolia 142
*Nothofagus* 202
Nurek reservoir 266
nutrient imbalance 118
nutrient uptake 61-68, 99, 100, 118
nutrition, human 46, 49
nutritional effect of salinity 73-79, 108-112

*Olea europaea* 131-138
oligonutrient 330
olive 131-138
optimal growth treatment 116, 117
organic soil 143
ornamental plant 360
osmolyte synthesis 10
osmoprotectants 22
osmotic adjustment 9, 65, 102, 119
osmotic effect 83-86
osmotic potential 7, 118-122
osmotic stress 82, 122
osmotically active solute 119
oxidative damage, salt-induced 32
oxidative stress 41, 50, 95-97, 325, 331
oxygen form, activated 64, 65

*Panicum turgidum* 160, 161
pasture rotation 237
pear 54
peroxidase (POD) 23
peroxynitrite 49
phanerophyte 146, 150, 151
phenolic content 335-342
phosphoinositide-specific phospholipases C (PLCs) 318, 319, 322
phospholipase D (PLD) 318-322
phosphorus (P) 61, 63, 66
photosynthesis, non-stomatal limitaton 199

$C_4$ photosynthesis pathway 188
photosynthesis rate 126
photosynthetic apparatus 57
photosynthetic capacity 21
photosynthetic pigment 30
*Phragmites* 213, 214
phytoremediation 247
*Pinus pinaster* 202
*Poaceae* 223, 235, 236, 257-260
*Populus* 201, 202, 281-284
potassium, *see* K
potassium chloride (KCl) 100-104
proline 123, 124
*Prosopis strombulifera* 81-88
psammohalophyte 146, 150, 151
pumping station 267
*Pyrus communis* 54

quick check system (QCS) 5, 6
quince 54

radiation absorption 189, 190
radiation use efficiency (RUE) 190
reactive oxygen species (ROS) 21, 32, 47, 273, 274, 326, 331, 332
ROS-induced DNA 332
real-time RT-PCR (qPCR) 288, 289
reclamation, of salt-affected soil 116, 241-245, 254
regosol soil 143
reproductive growth 187
reproductive trait 254
ribulose-1,5-biphosphate carboxylase (Rubisco) 199
rice 201, 280, 281
risk assessment 265
RNR gene 325, 326, 331
root to shoot ratio 118
ruderal species 150, 151

sabkha greenification 353, 354
salinity, conversion table 347
salinity control 249
salinity tolerance, physiological indicator 200-202
salinity tolerance type 345-349
salinization, loss of land 208
*Salsola arbusculiformis* 228
*Salsola gemmascens* 229
salt accumulation in plants 76
salt compartmentation 116
salt desert 164, 221-230
salt exclusion mechanism 9
salt excretion 263
salt marsh 208, 209, 212-217, 354
salt stress 37-42, 178, 198
salt stress, adaptation mechanism 119
salt stress, influence on plant growth 64, 67, 73-78, 99-101, 107-112
salt stress, physiology 273-275
salt stress effects 47
salt stress response 273, 275, 278, 282
salt stress responsive gene 275, 298-302
salt tolerance, biochemical differences in rootstocks 38
salt tolerance, complexity 3
salt tolerance, genomics 274-284
salt tolerance, inducing in crops 22
salt toxicity 257
salt treatment 336-342
salt-accumulator plant 256
salt-excluder plant 256
Sarez lake 266
sea beet 115-127
seawater irrigation 43, 50, 61-68
secondary salinization 157, 252
secondary water stress 65
seed bank 163-173
seed germination 163, 179-186
seed yield 193
*Sesuvium portulacastrum* 241-245
shelf life, of fruits 46
sodic soil 241
sodium, *see* Na
sodium chloride (NaCl) 99-102, 183
sodium compartmentalization, intracellular 11
sodium sulfate ($Na_2SO_4$) 100-104
soil, arid 143, 144, 202
soil, basaltic 143
soil, chestnut 142
soil, colluvial 142
soil, humid 143
soil, hydromorphic 144, 145
soil, organic 143
soil, sodic 241
soil desalination 241-245
soil deterioration 157
solonchak 221-225, 255

solonetz 225, 255
solute potential 9
Southeast Anatolia 141, 142, 144, 146
split root system 74-78
steppe vegetation 141
stomatal conductance 21, 29
stomatal effect 199, 201
stomatal resistance 8
stress tolerance 318
*Suaeda fruticosa* 241-245
sucrose 122
sugar beet 115-127
sugarcane 201
sunflower plant 61-68
superoxide anion 47, 68
superoxide anion elimination 68
superoxide dismutase (SOD) 23, 68
superoxide scavenging activity 335-342
suppressive subtractive hybridization (SSH) 298, 300
sustainable agriculture 2, 178, 345
sustainable utilization 178, 345
symbiosis 92-95

*Tamarix* 202, 215, 216, 360
tap root 118, 121
*TdDRF1* 287-294
*Thellungiella halophila* 276, 278, 317, 319-322
thermal unit 187
therophyte 146, 150, 151
threshold level, of salinity tolerance 116, 117
Tigris 144, 146
tissue culture 349, 350
tocopherol 45, 48-50
Toktogul reservoir 266
tomato 43-50, 326, 328, 330
trace element 250, 330
transcriptome regulation 275, 276
transition metal 331
transpiration rate 21
*Triticum durum* 288-290
Turan, desert of 221-230
Tuz gölü 146
*Typha* 213, 214

Uzbekistan 247-263, 265-271

vacuole 11, 305, 306, 308, 310, 312, 313
Van 143, 144, 146
vitamin E 48
VvNHX1gene 306-314

wasteland, salt-effected 260
water deficit 118
water deficit, as secondary effect of salinity 6
water loss 124
water potential 118
water reservoir 266, 267
water stress 288
water-stress signaling 319
waterlogging 254
water-use efficiency 8, 126, 199
wheat 201
windbreak 215, 216, 263

xantophyll 54
xerohalophyte 146, 150, 151
xerophyte 146, 150, 151

yeast heterologous system 305, 306

zinc (Zn) 67, 68